43

Advances in
Polymer Science
Fortschritte der Hochpolymeren-Forschung

Editors: H.-J. Cantow, Freiburg i. Br. · G. Dall'Asta, Colleferro · K. Dušek,
Prague · J. D. Ferry, Madison · H. Fujita, Osaka · M. Gordon, Colchester
J. P. Kennedy, Akron · W. Kern, Mainz · S. Okamura, Kyoto
C. G. Overberger, Ann Arbor · T. Saegusa, Kyoto · G. V. Schulz, Mainz
W. P. Slichter, Murray Hill · J. K. Stille, Fort Collins

Polymerizations and Polymer Properties

With Contributions by
V. S. C. Chang, G. K. Elyashevich, A. Fradet
A. Guyot, E. Heidemann, J. P. Kennedy
E. Maréchal, W. Roth

With 94 Figures

Springer-Verlag
Berlin Heidelberg New York 1982

Editors

Prof. Hans-Joachim Cantow, Institut für Makromolekulare Chemie der Universität, Stefan-Meier-Str. 31, 7800 Freiburg i. Br., BRD
Prof. Gino Dall'Asta, SNIA VISCOSA – Centro Studi Chimico, Colleferro (Roma), Italia
Prof. Karel Dušek, Institute of Macromolecular Chemistry, Czechoslovak Academy of Sciences, 162 06 Prague 616, ČSSR
Prof. John D. Ferry, Department of Chemistry, The University of Wisconsin, Madison, Wisconsin 53706, U.S.A.
Prof. Hiroshi Fujita, Department of Polymer Science, Osaka University, Toyonaka, Osaka, Japan
Prof. Manfred Gordon, Department of Chemistry, University of Essex, Wivenhoe Park, Colchester C04 3SQ, England
Prof. Joseph P. Kennedy, Institute of Polymer Science, The University of Akron, Akron, Ohio 44325, U.S.A.
Prof. Werner Kern, Institut für Organische Chemie der Universität, 6500 Mainz, BRD
Prof. Seizo Okamura, No. 24, Minami-Goshomachi, Okazaki, Sakyo-Ku, Kyoto 606, Japan
Prof. Charles G. Overberger, Department of Chemistry, The University of Michigan, Ann Arbor, Michigan 48 104, U.S.A.
Prof. Takeo Saegusa, Department of Synthetic Chemistry, Faculty of Engineering, Kyoto University, Kyoto, Japan
Prof. Günter Victor Schulz, Institut für Physikalische Chemie der Universität, 6500 Mainz, BRD
Dr. William P. Slichter, Chemical Physics Research Department, Bell Telephone Laboratories, Murray Hill, New Jersey 07 971, U.S.A.
Prof. John K. Stille, Department of Chemistry, Colorado State University, Fort Collins, Colorado 805 23, U.S.A.

ISBN-3-540-11048-8 Springer-Verlag Berlin Heidelberg New York
ISBN-0-387-11048-8 Springer-Verlag New York Heidelberg Berlin

Library of Congress Catalog Card Number 61-642

This work is subject to copyright. All rights are reserved, whether the whole or part of the material is concerned, specifically those of translation, reprinting, re-use of illustrations, broadcasting, reproduction by photocopying machine or similar means, and storage in data banks. Under § 54 of the German Copyright Law where copies are made for other than private use, a fee is payable to the publisher, the amount to "Verwertungsgesellschaft Wort", Munich.

© Springer-Verlag Berlin Heidelberg 1982
Printed in Germany

The use of general descriptive names, trademarks, etc. in this publication, even if the former are not especially identified, is not to be taken as a sign that such names, as understood by the Trade Marks and Merchandise Marks Act, may accordingly be used freely by anyone.
Typesetting and printing: Schwetzinger Verlagsdruckerei. Bookbinding: Brühlsche Universitätsdruckerei, Gießen.
2152/3140 – 543210

Table of Contents

Carbocationic Synthesis and Characterization of Polyolefins with Si–H and Si–Cl Head Groups
   J. P. Kennedy, V. S. C. Chang and A. Guyot 1

Kinetics and Mechanisms of Polyesterifications. I. Reactions of Diols with Diacids
   A. Fradet and E. Maréchal 51

Synthesis and Investigation of Collagen Model Peptides
   E. Heidemann and W. Roth 143

Thermodynamics and Kinetics of Orientational Crystallization of Flexible-Chain Polymers
   G. K. Elyashevich 205

Author Index Volumes 1–43 247

Carbocationic Synthesis and Characterization of Polyolefins with Si–H and Si–Cl Head Groups

Joseph P. Kennedy, V. S. C. Chang and A. Guyot*

Institute of Polymer Science, The University of Akron, Akron, Ohio 44325, USA

The objective of this research was to review and examine the preparation of macroreagents consisting of polyolefins carrying Si–Cl or Si–H termini. These polymers were thought to combine the desirable physical-mechanical properties offered by these polyhydrocarbons with the chemical versatility of Si–Cl and Si–H bonds. A general synthetic method has been developed for the preparation of polyolefin macroreagents with silane head groups. The key to the synthesis was the finding that initiators containing Si–Cl or Si–H groups and a cationogenic moiety are able to induce transferless cationic polymerizations under conditions such that the silicon-functional groups survive. Extensive model experiments provided guidance for subsequent polymerizations. Experimental conditions have been worked out under which none of the electrophilic species (i.e., initiating and propagating carbenium ion) that arise during polymerization would destroy the Si–H function. Isobutylene and α-methylstyrene polymerizations were carried out under transferless conditions and the effect of solvent polarity, temperature, and monomer and initiator concentrations on polymerization details and product characteristics have been determined. Quantitative H^1 NMR and IR studies demonstrate the survival of the Si–H function during polymerization and the syntheses of polyisobutylene and poly(α-methylstyrene) carrying 1 Si–H terminus per chain. These macroreagents may be of great interest for the addition of polyolefin sequences to small molecules for purposes of modification or to polymers for the preparation of block and graft copolymers.

I.	Introduction	3
II.	Experimental	4
	A. Materials	4
	B. Experimental Techniques	4
	1. Polymerization Studies	4
	2. Characterization Methods	5
	C. Syntheses of Initiators	5
	D. Model Syntheses and Derivatizations	10
	E. Block Copolymers by Coupling Technique	13
	F. Model Experiments on the Stability of Si–H Bonds in Carbocationic Polymerization	14

* Visiting Scientist; Permanent address: Centre National De La Recherche Scientifique, Laboratoire Des Materiaux Organiques, B.P. 24, 69390 Vernaison, France

III.	**Results and Discussion**	15
	A. Cationic Polymerization with Si–Cl Containing Initiator/Et$_2$AlCl Initiating Systems	15
	1. Introduction	15
	2. Syntheses of Si–Cl Containing Initiators	16
	3. α-Methylstyrene Polymerization and Characterization	17
	4. Conclusions	20
	B. Cationic Polymerization with Si–H Containing Initiator/Me$_3$Al Initiating Systems	21
	1. Introduction	21
	2. The Stability of Si–H Bonds under Carbenium Ion Polymerization Conditions	21
	3. α-Methylstyrene Polymerization with Si–H Containing Initiator/Me$_3$Al Initiating System	31
	a) Introduction	31
	b) Effect of Reaction Conditions on the Polymerization of α-Methylstyrene	31
	c) Head-group Characterization	38
	d) Conclusions	39
	4. Isobutylene Polymerization with Si–H Containing Initiator/Me$_3$Al Initiating System	39
	a) Introduction	39
	b) Effect of Reaction Conditions on the Polymerization of Isobutylene	40
	c) Head-group Characterization	46
	d) Model Derivatization Experiments	47
	e) Conclusions	48
IV.	**References**	48

I. Introduction

Polymers with reactive endgroups are most desirable intermediates for polymer derivatizations[1-10] and block and/or graft copolymer syntheses[8-18], and are therefore of great current scientific and technological interest. Polymers carrying one or two reactive terminal functions may be regarded as polymeric reagents, macroreagents, capable of modifying small molecules or adding to other polymers and thus to yield new block or graft copolymers. Polymers with two reactive termini, α,ω-difunctional polymers, are termed telechelic polymers[19] and are used in a great variety of applications, e.g. polyurethanes[20, 21], networks[22].

The aim of this research was the preparation of unique silicon-functional macroreagents, particularly linear polyolefins carrying one or two Si–Cl or Si–H termini and thus to combine the excellent physical properties offered by these polyhydrocarbons with the versatility and chemical reactivity of the Si–Cl and Si–H bonds.

A thorough search through the scientific and patent literature failed to provide information concerning the synthesis of –SiCl and –SiH terminated polymers by free radical and cationic techniques. In contrast, anionic and condensation polymerization methods have been employed for the preparation of these macroreagents. Thus, polystyrene, polyisoprene, polybutadiene and poly(methyl methacrylate) bearing one or two terminal –SiH or –SiCl functional groups have been obtained by living anionic polymerization followed by capping the living polymers with $RSi(CH_3)_2Cl$ (R = H or Cl). These polymers have subsequently been used for the preparation of graft and block copolymers[8-11]. Linear polysiloxanes containing one or two –SiH or –SiCl termini have been obtained by various condensation techniques also. Plumb and Atherton[18] have thoroughly reviewed this subject. Terminal functional polysiloxanes have been used for synthesis of various block copolymers through Si–O–C or Si–C linkages, e.g., poly(siloxane-b-oxyalkylene)[23]. These materials are outstanding foam stabilizers in the one-shot flexible polyurethane foam process.

Since a thorough review of the scientific and patent literature failed to produce any information relative to the preparation of silicon-functional polyolefins by cationic methods, this untapped field appeared particularly attractive for systematic research. Added justification for this objective was provided by the possible exploitation of the vast synthetic possibilities offered by a combination of the chemistries of cationic polymerization and that of organosilanes[24].

This paper concerns the direct cationic synthesis of new aliphatic and aromatic macroreagents, specifically polyisobutylenes and poly(α-methylstyrenes), carrying Si–Cl and Si–H head groups. To this end new silicon-functional initiators containing a Si–Cl or Si–H group and a cationogenic moiety, $-C_6H_4CH_2Cl$, have been synthesized. A representative example of such an initiator is $H(CH_3)_2SiCH_2CH_2C_6H_4CH_2Cl$. The cationogenic group in conjunction with suitable Lewis acids, e.g., diethylaluminum chloride or trimethylaluminum, is able to initiate efficient polymerization of vinyl monomers, e.g., isobutylene[26-29] or α-methylstyrene[30], without chain transfer to monomer, and thus leads to silicon-functional head groups in the polymer. Conditions have been worked out under which the Si–Cl and Si–H head groups survive the cationic synthesis step.

II. Experimental

A. Materials

Isobutylene (Linde Division, Union Carbide): was dried by passing the gas through a glass column packed with Molecular Sieves (3 Å, powder). *α-Methylstyrene (Aldrich Chemical Co.):* was washed with 10% NaOH solution, distilled water, dried with anhydrous calcium chloride, then distilled over calcium hydride and stored in a freezer. It was freshly redistilled before use. *Trimethyl Aluminum (Ethyl Co.):* was distilled under dry nitrogen atmosphere and stored in a freezer. *Vinyl Benzyl Chloride (Dow Chemical Co.):* was used as received or freshly distilled before use. According to the manufacturer the material is a mixture of 60% meta and 40% para isomers. The separation of these isomers is very difficult. *Dimethylchlorosilane, Methyldichlorosilane, Trichlorosilane (Petrarch Systems, Inc.), Methanol, Chloroplatinic Acid (Fisher Scientific Co.), Calcium Hydride, and Lithium Aluminum Hydride (Alfa Division, Ventron Corp.):* were used as received without further purification. *1-Chloroethyl Benzene (Pfaltz & Bauer, Inc.):* was refluxed over calcium hydride for 3 h under nitrogen, distilled, and stored in a freezer. *n-Hexane and n-Heptane (Fisher Scientific Co.):* were refluxed with fuming sulfuric acid, washed with distilled water, dried with anhydrous calcium chloride, and freshly distilled over calcium hydride under dry nitrogen before use. *Methylene Chloride (Fisher Scientific Co.):* was refluxed over calcium hydride overnight and freshly distilled before use. *Tetrahydrofuran (THF) (Fisher Scientific Co.):* was refluxed over lithium aluminum hydride overnight and freshly distilled before use. *1,2-Dimethoxyethane (Aldrich Chemical Co.): was dried over* Molecular Sieves (4 Å). *2,4,4-Trimethyl-1-pentene (Aldrich Chemical Co.):* was refluxed over calcium hydride for 3 h under nitrogen then distilled and stored in a dessicator.

B. Experimental Techniques

1. Polymerization Studies

Typically, reactions were carried out by the use of culture tubes with Teflon lined caps in a stainless steel dry box equipped with a cooling bath and inlets for gaseous reagents. Cooling was achieved by passing liquid nitrogen through copper coils immersed in *n*-heptane heat exchanger. The moisture level in the box was kept below ~50 ppm by continuously flushing with nitrogen dried by passing the gas through glass columns packed with barium oxide and molecular sieves.

Glassware was cleaned with aqueous carbitol, rinsed with dilute HCl and with distilled water. The culture tubes were dried in an oven at 180 °C overnight and cooled to room temperature in the dry box under nitrogen.

Polymerizations were generally carried out as follows: Gaseous reagents, i.e., isobutylene, methyl chloride, were passed through drying columns and condensed in the bath and collected in culture tubes.

Precooled coinitiator solution was rapidly added to precooled solution of monomer and initiator while shaking the culture tube. Polymerizations were terminated generally after 5 to 10 min of shaking by adding 5 ml prechilled methanol. The reaction mixture was shaken with 0.5 N cold HCl to remove coinitiator, washed with distilled water, and transferred to a preweighed aluminum dish. Most of the solvent was removed by evaporating at room temperature under a hood and finally the polymer was dried in vacuo at 50 °C overnight.

In experiments in which the effect of monomer concentration was studied the polarity of the medium was maintained by replacing aliquots of the monomers by n-hexane cosolvent, so that the total volume of n-hexane and monomer remained constant. This technique was also used in model studies.

2. Characterization Methods

Molecular weights were determined by gel permeation chromatography GPC using a Waters Associates' instrument equipped with a Waters Model M-6000A pump, five μ-Styragel columns of pore sizes of 10^6, 10^5, 10^4, 10^3, and 500 Å, a Model 440 UV detector, and a Model R 401 differential refractometer. Calibration curves were prepared by using (a) fractionated polyisobutylenes with narrow molecular weight dispersities ($M_w/M_n \leq$ 1.5), and (b) commercial polystyrene standards. The molecular weights of poly(α-methylstyrene) were calculated by using the polystyrene calibration curve. The pressure was 1000 psi, and flow rate was 2 ml/min. The concentration of the polymer samples was 0.2 g/dl THF.

Proton nuclear magnetic resonance spectra of 15–20% solutions of polymers in CCl_4 were obtained with Varian T-60 or HR-300 spectrometers. Chemical shifts are reported relative to tetramethylsilane standard.

Infrared spectra were recorded on a Perkin-Elmer Model 521 spectrophotometer and 10–15% (w/v) polymer solutions were used. A pair of Perkin-Elmer solution cells with sodium chloride windows were used to hold the sample solution and reference solvent.

Analytical gas chromatography was carried out using a Hewlett-Packard Model 5750 instrument and a SE-30 column.

C. Syntheses of Initiators

Table 1. Novel silicon-functional initiators synthesized

1	$ClSi(CH_3)_2CH_2CH_2\varphi CH_2Cl$	$ClSi(CH_3)_2CH(CH_3)\varphi CH_2Cl$
2	$Cl_2Si(CH_3)CH_2CH_2\varphi CH_2Cl$	$Cl_2Si(CH_3)CH(CH_3)\varphi CH_2Cl$
3	$Cl_3SiCH_2CH_2\varphi CH_2Cl$	$Cl_3SiCH(CH_3)\varphi CH_2Cl$
4	$HSi(CH_3)_2CH_2CH_2\varphi CH_2Cl$	$HSi(CH_3)_2CH(CH_3)\varphi CH_2Cl$
5	$H_2Si(CH_3)CH_2CH_2\varphi CH_2Cl$	$H_2Si(CH_3)CH(CH_3)\varphi CH_2Cl$
6	$H_3SiCH_2CH_2\varphi CH_2Cl$	$H_3SiCH(CH_3)\varphi CH_2Cl$

Table 1 shows the structures of the novel silicon-functional initiators synthesized. All the initiators are mixtures of 80–85% main product (left column) plus 15–20% isomer (right column). The subsequent sections give synthesis details.

1. [2-(p-Chloromethylphenyl)ethyl]dimethylchlorosilane (1)

$$Cl(CH_3)_2SiH + CH_2=CH\varphi CH_2Cl \xrightarrow{H_2PtCl_6} Cl(CH_3)_2SiCH_2CH_2\varphi CH_2Cl$$

To a 300 ml three-neck flask equipped with condenser, thermometer, stirrer, dropping funnel, and dry nitrogen inlet tube was charged 50.8 ml (0.36 mol) vinyl benzyl chloride. After the charge was warmed to 45 °C, 0.1 ml chloroplatinic acid solution, prepared from 1 g chloroplatinic acid in 9 ml 1,2-dimethoxyethane and 1 ml absolute alcohol[31], was added while dry nitrogen was bled through the flask. Subsequently 50 ml (0.45 mol) chlorodimethylsilane was added dropwise. In the course of addition the temperature was kept at 45–50 °C by cooling with an ice bath and by controlling the addition rate of chlorodimethylsilane. The temperature was kept at 45–50 °C for 3 h to complete hydrosilation. The color of the charge turned from light yellow to deep brown. The mixture was allowed to cool to room temperature. The condenser and dropping funnel were disconnected and a vacuum distilling-head was attached to the flask. Vacuum distillation at 104 °C, 0.09 mm, yielded 85 g (95%) of 1. The product was stored in a freezer. The structure of 1 was characterized by H^1 NMR spectroscopy. Figure 1 shows the spectrum and assignments. According to analysis of the H^1 NMR spectrum ~19% of the product is the $Cl(CH_3)_2SiCH(CH_3)\varphi CH_2Cl$ isomer.

2. [2-(p-Chloromethylphenyl)ethyl]methyldichlorosilane (2)

$$Cl_2(CH_3)SiH + CH_2=CH\varphi CH_2Cl \xrightarrow{H_2PtCl_6} Cl_2(CH_3)SiCH_2CH_2\varphi CH_2Cl$$

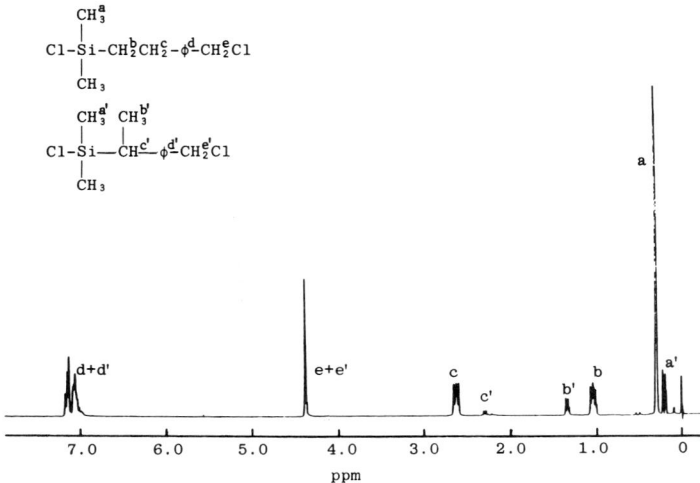

Fig. 1. H^1 NMR spectrum of [2-(p-chloromethylphenyl)ethyl]dimethylchlorosilane

Carbocationic Synthesis and Characterization of Polyolefins with Si–H and Si–Cl Head Groups 7

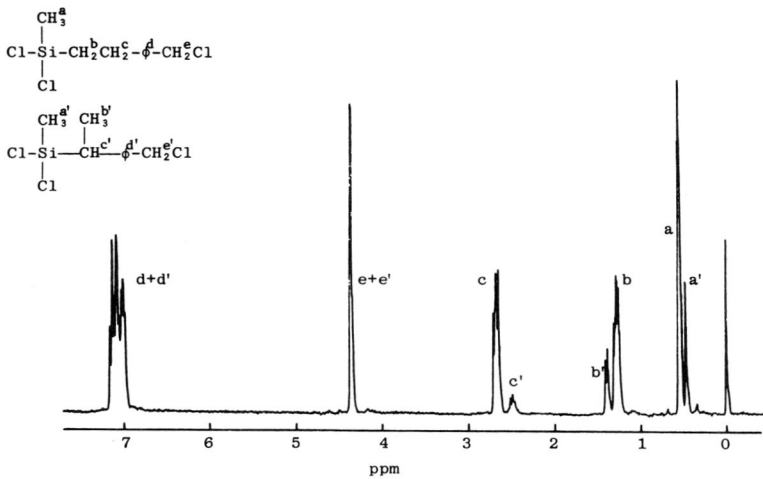

Fig. 2. H^1 NMR spectrum of [2-(p-chloromethylphenyl)ethyl]methyldichlorosilane

The synthesis of 2 was carried out by the same procedure as that used for the synthesis of 1, except 50 ml (0.48 mol) methyldichlorosilane and 55 ml (0.4 mol) vinyl benzyl chloride were employed. Yield was 100 g (93%). Bp. 134 °C, 0,4 mm. The H^1 NMR spectrum of 2 together with assignments is shown in Fig. 2. Approximately 20% of the product is the Cl$_2$(CH$_3$)SiCH(CH$_2$)φCH$_2$Cl isomer.

3. [2-(p-Chloromethylphenyl)ethyl]trichlorosilane (3)

Cl$_3$SiH + CH$_2$=CHφCH$_2$Cl $\xrightarrow{\text{H}_2\text{PtCl}_6}$ Cl$_3$SiCH$_2$CH$_2\varphi$CH$_2$Cl

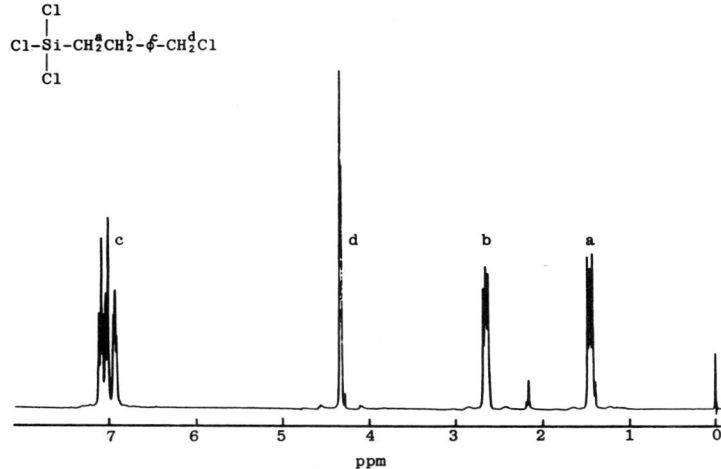

Fig. 3. H^1 NMR spectrum of [2-(p-chloromethylphenyl)ethyl]trichlorosilane

Similarly, *3* was synthesized by the same procedure as that used for the synthesis of *1* except in this case 50 ml (0.52 mol) trichlorosilane and 63.5 ml (0.45 mol) vinyl benzyl chloride was charged. The yield was 121.8 g (94%). Bp. 113–5 °C, 0.1 mm. The H^1 NMR spectrum is shown in Fig. 3. Again, ~ 18% of the product is $Cl_3SiCH(CH_3)\varphi CH_2Cl$.

4. [2-(p-Chloromethylphenyl)ethyl]dimethylsilane (4)

$$Cl(CH_3)_2SiCH_2CH_2\varphi CH_2Cl \xrightarrow[\text{ether, 0 °C}]{\text{LAH}} H(CH_3)_2SiCH_2CH_2\varphi CH_2Cl$$

A 300 ml three-neck flask equipped with condenser, stirrer, dropping funnel, dry nitrogen inlet tube, and containing 5.5 g (0.145 mol) lithium aluminum hydride LAH suspension in 100 ml anhydrous diethyl ether, was placed in an ice bath. Over a period of 25 min 30 ml (0.123 mol) *1* was added dropwise into the stirred suspension. The mixture was stirred for an additional hour at 0 °C, then poured over a mixture of 50 ml ether, 100 g crushed ice, and 50 ml ice water with stirring. When necessary more crushed ice was added to cool the mixture. The layers were separated, and the organic layer was concentrated first by distillation over calcium hydride, then by vacuum distillation over calcium hydride. The yield of *4* was 22.5 g (86%). Bp. 78–9 °C, 0.4 mm. The product was stored in a freezer. The structure of *4* was confirmed by its H^1 NMR spectrum as shown in Fig. 4.

5. [2-(p-Chloromethylphenyl)ethyl]methylsilane (5)

$$Cl_2(CH_3)SiCH_2CH_2\varphi CH_2Cl \xrightarrow[\text{ether, 0 °C}]{\text{LAH}} H_2(CH_3)SiCH_2CH\varphi CH_2Cl$$

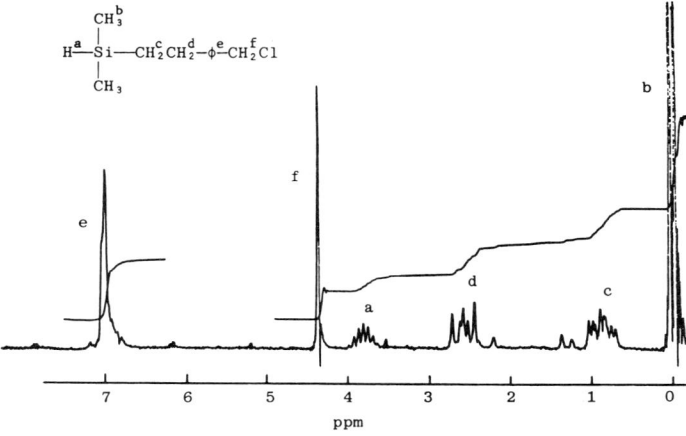

Fig. 4. H^1 NMR spectrum of [2-(p-chloromethylphenyl)ethyl]dimethylsilane

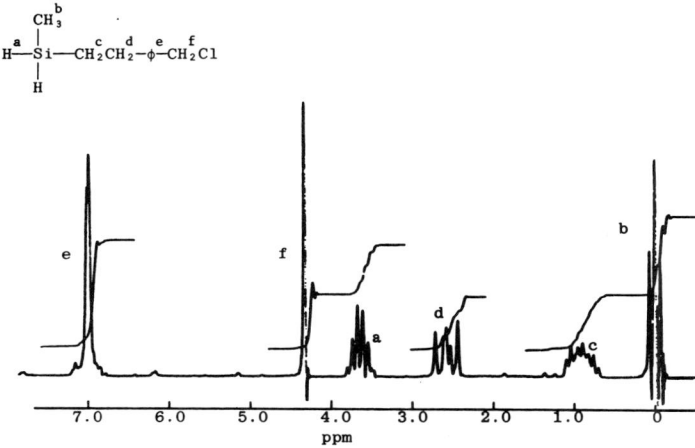

Fig. 5. H¹ NMR spectrum of [2-(p-chloromethylphenyl)ethyl]methylsilane

The synthesis of 5 was carried out by the same procedure as that used for the synthesis of 4, except in this case 30 ml (0.12 mol) 2 and 9.1 g (0.24 mol) LAH were charged. Yield was 21 g (88%). Bp. 72–3 °C, 0.4 mm. The product was stored in a freezer. Its H¹ NMR spectrum is shown in Fig. 5.

6. [2-(p-Chloromethylphenyl)ethyl]silane (6)

$$Cl_3SiCH_2CH_2\varphi CH_2Cl \xrightarrow[\text{ether, 0 °C}]{\text{LAH}} H_3SiCH_2CH_2\varphi CH_2Cl$$

The synthesis of 6 was undertaken by the same method used for the synthesis of 4, except in this case 30 ml (0.135 mol) 3 and 15.4 g (0.4 mol) LAH were charged. The yield was 20 g (80%). Bp. 68.5 °C, 0.5 mm. The structure of 6 was identified by its H¹ NMR spectrum shown in Fig. 6.

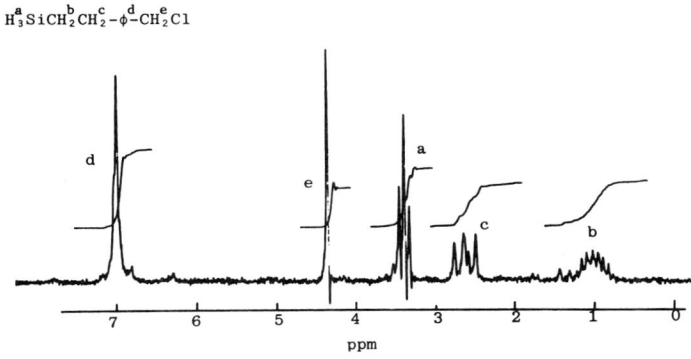

Fig. 6. H¹ NMR spectrum of [2-(p-chloromethylphenyl)ethyl]silane

D. Model Syntheses and Derivatizations

1. [2-(p-Methylphenyl)ethyl]dimethylsilane (7)

$$Cl(CH_3)_2SiCH_2CH_2\varphi CH_2Cl \xrightarrow[\text{THF, reflux}]{\text{LAH}} H(CH_3)_2SiCH_2CH_2\varphi CH_3$$

A 250 ml three-neck flask equipped with a condenser, magnetic stirrer, dropping funnel, dry nitrogen inlet tube, was charged with 3.8 g (0.1 mol) LAH suspension in 100 ml anhydrous THF. Slowly, 10 ml (0.04 mol) *1* was added dropwise with stirring. The addition rate was controlled to have the reaction mixture boil gently. The mixture was then poured over a stirred mixture of 50 ml ice water. When necessary more crushed ice was added to cool the system. The product was extracted twice with 50 ml ether, the organic solutions were combined and concentrated first by distillation over calcium hydride then by vacuum distillation over calcium hydride. The yield was 5.2 g (72%). Bp. 62 °C, 0.2 mm. Figure 7 shows the H^1 NMR spectrum of 7.

2. [2-(p-Methylphenyl)ethyl]dimethylchlorosilane (8)

$$CH_2=CH\varphi CH_2Cl \xrightarrow[\text{THF, reflux}]{\text{LAH}} CH_2=CH\varphi CH_3$$

$$CH_2=CH\varphi CH_3 + Cl(CH_3)SiH \xrightarrow{H_2PtCl_6} Cl(CH_3)_2SiCH_2CH_2\varphi CH_3$$

p-Methylstyrene was prepared by using the procedure employed for the synthesis 7, except in this case 30 ml (0.21 mol) vinyl benzyl chloride and 9.5 g (0.25 mol) lithium aluminum hydride were charged. The yield was 18.8 g (76%). Bp. 47 °C, 2.2 mm. Subsequently, *8* was synthesized by employing the same method as described before for the synthesis of *1*, except, in this instance, 18 g (0.15 mol) p-methylstyrene and 22.3 ml

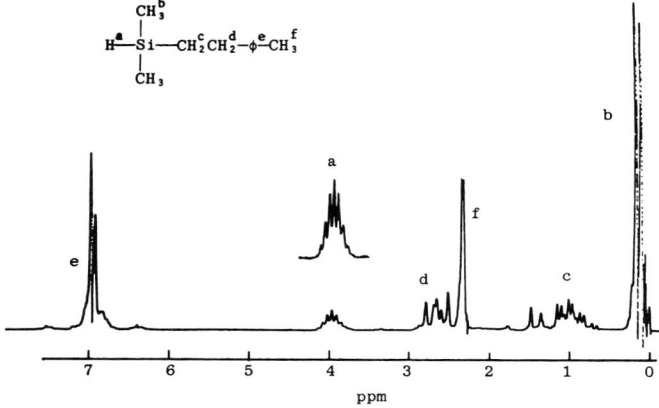

Fig. 7. H^1 NMR spectrum of [2-(p-methylphenyl)ethyl]dimethylsilane

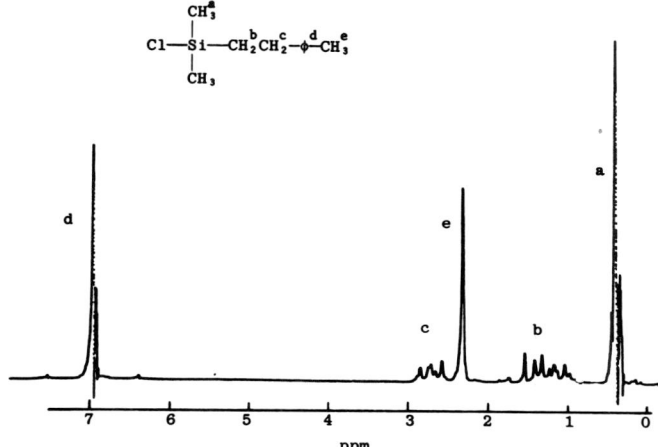

Fig. 8. H^1 NMR spectrum [2-(p-methylphenyl)ethyl]dimethylchlorosilane

(0.2 mol) dimethylchlorosilane were charged. The yield was 28.7 g (90%). Bp. 50 °C, 0.2 mm. Figure 8 shows the H^1 NMR spectrum of 8.

3. [2-(p-Methylphenyl)ethyl]silane (9)

$$Cl_3SiCH_2CH_2\varphi CH_2Cl \xrightarrow[\text{THF, reflux}]{\text{LAH}} H_3SiCH_2CH_2\varphi CH_3$$

9 was prepared by following the procedure described before for the synthesis of 7, except in this case, 15 ml (0.067 mol) 3 and 10 g (0.26 mol) LAH were charged. The yield was 7.5 g (74%). Figure 9 shows the H^1 NMR spectrum of 9.

4. 2,4,4-Trimethylpentyl[2-(p-methylphenyl)ethyl]dimethylsilane (10)

$$\underset{\underset{CH_3}{|}}{\overset{\overset{CH_3}{|}}{CH_3-C-CH_2}}-\overset{\overset{CH_3}{|}}{C=CH_2} + H(CH_3)_2SiCH_2CH_2\phi CH_3 \xrightarrow{H_2PtCl_6}$$

$$\underset{\underset{CH_3}{|}}{\overset{\overset{CH_3}{|}}{CH_3-C-CH_2}}-\overset{\overset{CH_3}{|}}{CH}-CH_2-\overset{\overset{CH_3}{|}}{\underset{\underset{CH_3}{|}}{Si}}CH_2CH_2\phi CH_3$$

n-Heptane solvent was used to simulate conditions of polymer end-group coupling. A 500 ml three-neck flask equipped with condenser, magnetic stirrer, dropping funnel, and gas inlet was charged with 3.8 ml (0.024 mol) 2,4,4-trimethl-1-pentene and 100 ml n-heptane (freshly distilled over calcium hydride). The flask was maintained under dry nitrogen throughout the reaction. After heating the solution to reflux, 0.02 ml of chloroplatinic acid solution, prepared as in the synthesis of 1, was added. Subsequently 5 ml

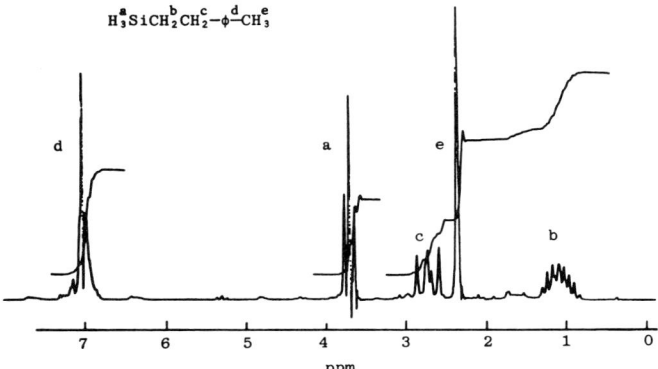

Fig. 9. H¹ NMR spectrum of [2-(p-methylphenyl)ethyl]silane

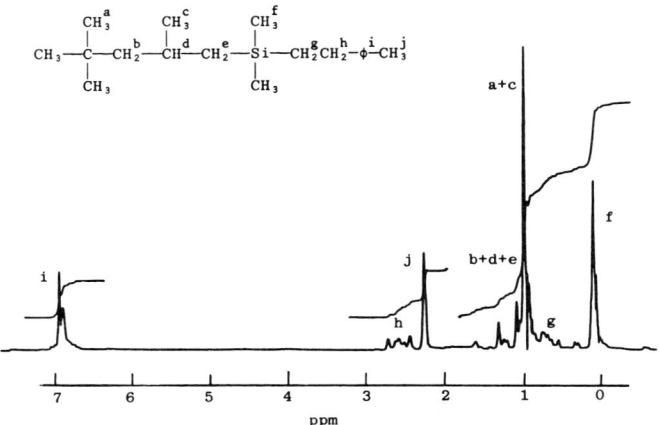

Fig. 10. H¹ NMR spectrum of 2,4,4-trimethylpentyl[2-(p-methylphenyl)ethyl]dimethylsilane

(0.024 mol) 7 was added dropwise over a period of 15 min and the reaction mixture was refluxed for 20 h. The crude product was fractionally distilled. However, the yield was less than ~10%. Benzene and tetrahydrofuran have also been tried as solvents, but yields remained poor. Figure 10 shows the H¹ NMR spectrum of *10*.

5. 3-Glycidoxypropyl[2-(p-methylphenyl)ethyl]dimethylsilane (11)

$$CH_2-CH-CH_2-O-CH_2CH=CH_2 + H(CH_3)_2SiCH_2CH_2\phi CH_3 \xrightarrow{H_2PtCl_6}$$
$$\underset{O}{\vee}$$

$$CH_2-CH-CH_2-O-CH_2CH_2CH_2(CH_3)_2SiCH_2CH_2\phi CH_3$$
$$\underset{O}{\vee}$$

The synthesis of *11* followed a procedure similar to that described in the preceding experiment for the synthesis of *10*, except in this case, 10 ml (0.045 mol) 7 and 7.2 ml

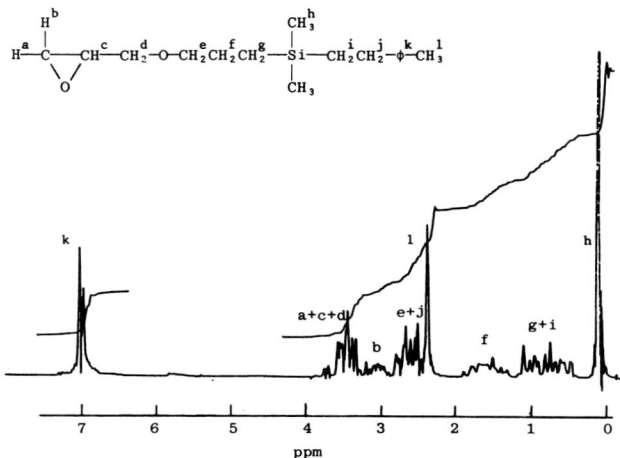

Fig. 11. H¹ NMR spectrum of 3-glycidoxypropyl[2-(p-methylphenyl)ethyl]dimethylsilane

(0.06 mol) allyl glycidyl ether in 100 ml freshly distilled *n*-hexane were used. After 20 h of refluxing, an aliquot of the solution was analyzed by glpc. The conversion to *11* was quantitative. Vacuum distillation yielded 10.3 g (78%) of *11*. Bp. 148 °C, 0.05 mm. Figure *11* shows the H¹ NMR spectrum of 11.

E. Block Copolymers by Coupling Technique

1. Attempted Coupling with Living Polyisoprenyllithium[1]

Poly(α-methylstyrene) bearing a Si–Cl head-group was purified by precipitation in dry *n*-heptane in a dry box. After filtration, the polymer was transferred into a flask equipped with a break-seal, a graduated side tube and a vacuum stopcock. The polymer was degassed on vacuum line and benzene was distilled into the flask to dissolve the polymer. The system was freeze-dried to remove the last trace of CH_2Cl_2 co-solvent used in the polymerization of α-methylstyrene. The polymer was again dissolved in benzene and the flask was sealed. An aliquot was poured into the graduated side tube and sealed for effective functionality determination (Sect. III.A.3.b.). The arm carrying the break-seal was then connected to a large graduated cylinder and four smaller side cylinders. Each side cylinder was equipped with a break-seal and could be separated later. After purging the system under high vacuum the break-seal was broken to allow the solution be poured into the large cylinder. The solution was divided into the smaller side tubes (measured volumes) and sealed for coupling reactions. One of the cylinders was connected by a break-seal to a flask with a similar side cylinder which contained a known amount of living polyisoprenyllithium in benzene.

After the flask was purged under high vacuum overnight, dry benzene was distilled into the flask. Then the coupling reaction was performed by breaking both break-seals

---

1 Dr. J. M. Heuchen's help in carrying out this experiment is gratefully acknowledged

and allowing the reaction mixture to stir overnight at 50 °C. Finally, one of the arms of the flask was broken in a beaker containing methanol and the polymer was precipitated and dried.

The crude product was extracted sequentially with *n*-pentane and acetone, and analyzed by GPC and H^1 NMR.

2. Attempted Coupling with α,ω-Disodium Polyisobutylene Glycolate

α,ω-Disodium polyisobutylene glycolate was prepared by reacting α,ω-dihydroxypolyisobutylene (obtained by ozonolysis of a copolymer of isobutylene and isoprene followed by reduction with LAH[5]) with sodium dispersion in gently refluxing toluene under nitrogen. The clear toluene solution of α,ω-disodium polyisobutylene glycolate was separated from excess sodium and mixed under dry nitrogen with a toluene solution of poly(α-methylstyrene) with Si–Cl head-group. After reaction excess *n*-butyllithium was added to destroy any unreacted Si–Cl head-group. Finally, the polymer was precipitated into methanol and dried. It was then analyzed by GPC.

F. Model Experiments on the Stability of Si–H Bonds in Carbocationic Polymerization

1. The $HSi(CH_3)_2CH_2CH_2\varphi CH_3/Me_3Al/CH_2Cl_2$ Model

In a 70 ml culture tube equipped with a Teflon lined cap, 10 ml 0.2 M *7* was mixed with 10 ml 1 M Me$_3$Al in CH$_2$Cl$_2$ solvent at 22 °C in a dry box under nitrogen atmosphere. After one hour the Me$_3$Al was destroyed by slowly adding methanol to the mixture. The tube was withdrawn from the dry box, and the reaction mixture was washed with ice-cold 0.5 N HCl solution and water, separated, dried over anhydrous MgSO$_4$, solvent was evaporated under vacuum, and the products were analyzed by H^1 NMR spectroscopy.

2. The $HSi(CH_3)_2CH_2CH_2\varphi CH_2Cl/Me_3Al/CH_2Cl_2$ Model

Using the same technique employed in the previous model study, a series of experiments have been carried out by mixing 15 ml 0.1 M *4* and 15 ml 1.0 M Me$_3$Al in CH$_2$Cl$_2$ at 22° for 15 min; at −20° for 10 min; and at −50° for 5, 10, 20, and 60 min. The reactions were terminated by slowly adding prechilled methanol. The products were washed with ice-cold 0.5 N HCl and distilled water, separated, dried over MgSO$_4$, solvent removed under vacuum, and analyzed by H^1 NMR and IR spectroscopies.

3. The $H_3SiCH_2CH_2\varphi CH_2Cl/Me_3Al/CH_2Cl_2$ Model

Following the same procedure as in the previous model study, a series of experiments have been carried out by mixing 15 ml 0.05 M *6* and 15 ml 0.5 M Me$_3$Al in CH$_2$Cl$_2$ at 22° for 5 min; at −50° for 15 and 60 min; at −65° for 15 and 60 min; and at −80° for 10 and

60 min. The reactions were terminated and products purified as in the previous model. The products were analyzed by H^1 NMR and IR spectroscopies.

4. The $HSi(CH_3)_2CH_2CH_2\varphi CH_3/(CH_3)_3CCl$ or $CH_3CH(\varphi)Cl/Me_3Al/CH_2Cl_2$ and/or n-C_6H_{14} Model

Following the same procedure as in the previous model study, a series of experiments have been carried out by mixing 1.7 mmol 7, 1.7 mmol $(CH_3)_3CCl$, and 8.5 mmol Me_3Al in solvent combinations of CH_2Cl_2/n-C_6H_{14} = 30 ml/0 ml, 15 ml/15 ml, and 0 ml/30 ml at $-50°$ for 10 min. The reactions were terminated and the products purified as in the previous model. The products were analyzed by H^1 NMR spectroscopy. The experiments were repeated by using $CH_3CH(\varphi)Cl$ instead of $(CH_3)_3CCl$.

5. The $H_3SiCH_2CH_2\varphi CH_3/(CH_3)_3CCl$ or $CH_3CH(\varphi)Cl/Me_3Al/CH_2Cl_2$ and/or n-C_6H_{14} Model

Following the same procedure as in the previous model study, a series of experiments have been carried out by mixing 2.0 mmol 9, 2.0 mmol $(CH_3)_3CCl$ (or $CH_3CH(\varphi)Cl$), and 5.0 mmol Me_3Al in solvent combinations of CH_2Cl_2/n-C_6H_{14} = 30 ml/0 ml, 15 ml/15 ml, and 0 ml/30 ml at $-50°$ for 10 min. The reactions were terminated and the products were purified as in the previous model. The products were analyzed by H^1 NMR spectroscopy.

III. Results and Discussion

A. Cationic Polymerization with Si–Cl Containing Initiator/Et$_2$AlCl Initiating Systems

1. Introduction

Block copolymers can be readily produced by anionic polymerization using "blocking from", i.e., sequential polymerization of a second monomer initiated by a living polymer, or "coupling" of two living polymers with a suitable coupling agent or of a living polymer with another polymer carrying a suitable functional end-group for coupling. Block copolymers have been synthesized by carbenium ion polymerization only in recent years by Kennedy and co-workers[32, 33], who developed a "blocking from" technique making use of polymers with suitable cationically initiating end-groups.

It was theorized that cationic initiators containing Si–Cl functions in conjunction with alkylaluminum compounds would lead to polymers with Si–Cl head-groups which subsequently could be useful for the preparation of block copolymers by coupling. The following equations help to visualize this proposition:

ClSi(CH₃)₂CH₂CH₂φCH₂Cl $\xrightarrow[\text{monomer A}]{\text{Et}_2\text{AlCl}}$

ClSi(CH₃)₂CH₂CH₂φ CH₂—[Polymer A] $\xrightarrow{\begin{array}{c}\text{Polymer B}\!-\!\text{OH}\\\hline -\text{HCl}\end{array}}$

[Polymer B]—O—Si(CH₃)₂CH₂CH₂φ CH₂—[Polymer A]

In order to carry out these syntheses a key requirement was that the Si–Cl bond of the Si–Cl functional initiators should not react with the Et₂AlCl coinitiator and the propagating carbenium ion, but be able to effect condensations. Synthetic possibilities obtainable with polymers having Si–Cl termini are shown in Scheme 1.

Scheme 1. Derivatization and synthesis possibilities offered by ClCH₂φCH₂CH₂Si(CH₃)₃₋ₙClₙ and polymer-CH₂φCH₂CH₂Si(CH₃)₃₋ₙClₙ

$$R-Si(CH_3)_{3-n}Cl_n \begin{cases} + Li-R' \xrightarrow[-n\,LiCl]{Cond.} R-Si(CH_3)_{3-n}R'_n \\ + HO-R' \xrightarrow[-n\,HCl]{Cond.} R-Si(CH_3)_{3-n}(OR')_n \\ + HO-\!\text{(Si)} \xrightarrow[-n\,HCl]{Cond.} R-Si(CH_3)_{3-n}(O-\!\text{(Si)})_n \\ \qquad\qquad\qquad\qquad\qquad\text{(Porcupine Polymer)} \end{cases}$$

Where HO—(Si) silica filler, inorganic solids carrying surface OH groups,
R = –CH₂CH₂φCH₂Cl or –CH₂CH₂φCH₂-Polymer,
R' = small or large (polymer) group, n = 1, 2 or 3.

2. Syntheses of Si–Cl Containing Initiators

After a thorough literature research and analysis[25–29, 31, 34, 35] of the objective, it was concluded that among the most promising Si–Cl containing functional initiators for carbenium ion polymerization would be [2-(p-chloromethylphenyl)ethyl]dimethylchlorosilane *(1)*, [2-(p-chloromethylphenyl)ethyl]methyldichlorosilane *(2)*, and [2-(p-chloromethylphenyl)ethyl]trichlorosilane *(3)*, whose formulas are shown in Table 1. These new initiators combine the highly reactive Si–Cl functional group (reported to be inert to alkylaluminum compounds at ambient temperature[34]) and the benzyl chloride group which is able to initiate carbenium ion polymerization in conjunction with suitable coinitiators[25, 36].

The syntheses of these initiators is described in Sect. II.C. According to detailed H¹ NMR analysis hydrosilylation yielded 15–20% isomers along with the major products *1*, *2*, and *3* as shown in Figs. 1–3. The presence of the isomers, listed in Table 1, should not affect initiating efficiency by the benzyl chloride group, in view of the structure and virtually identical H¹ NMR chemical shifts of the chloromethyl groups.

3. α-Methylstyrene Polymerization and Characterization

α-Methylstyrene polymerization was carried out with these new Si–Cl functional initiators in hope to obtain poly(α-methylstyrene) with Si–Cl head-groups (ClSi-PαMe-St), which subsequently could be coupled with rubbery polymers, i.e., α,ω-disodium polyisobutylene glycolate or α,ω-dilithium polyisoprene.

a) The Stability of Si–Cl Bonds toward Et$_2$AlCl and Propagating Polyisobutylene Carbenium Ion

The key requirements for using Si–Cl functional initiators to produce polymers carrying Si–Cl termini by carbenium ion polymerization are: i) Si–Cl should be inert toward alkylaluminum coinitiators, ii) Si–Cl should not react with propagating carbenium ions, iii) chain transfer to monomer should be negligible so as to end up with one Si–Cl head-group per polymer chain.

The stability of Si–Cl toward Et$_2$AlCl coinitiator and propagating isobutylene carbenium ion has been investigated. Table 2 shows the results obtained in experiments in which the polymerization of isobutylene was induced by the (CH$_3$)$_3$CCl/Et$_2$AlCl initiating system in the presence of φSiCl$_3$. According to these data, the presence of φSiCl$_3$ did not reduce polymer yield, which is construed as evidence that Si–Cl bonds are inert toward Et$_2$AlCl and the propagating polyisobutylene carbocation.

b) Fundamental Aspects of α-Methylstyrene Polymerization Using Si–Cl Containing Initiator/Et$_2$AlCl Initiating System

A series of orienting experiments have been carried out to find suitable conditions for the synthesis of ClSi-PαMeSt with desired molecular weight and terminal functionality. Conditions and results are shown in Table 3.

Table 2. Polymerization of isobutylene using the (CH$_3$)$_3$CCl/Et$_2$AlCl initiating system in the presence of φSiCl$_3$

φSiCl$_3$ mol × 10^6	Conversion %
–	11
3.6	7
7.1	22
10.0	23
25.4	29

i-C$_4$H$_8$ = 5 ml, (CH$_3$)$_3$CCl = 1.0 × 10^{-5} mol, Et$_2$AlCl = 1.45 × 10^{-4} mol, EtCl = 15 ml, −45°, 10 min

Table 3. Polymerization of α-methylstyrene by the $ClSi(CH_3)_2CH_2CH_2\varphi CH_2Cl/Et_2AlCl$ initiating system

αMeSt g	Initiator I mole $\times 10^5$	Et_2AlCl mole $\times 10^4$	Solvent ml φCl	CH_2Cl_2	$n\text{-}C_7H_{16}$	Temp. °C	Time min	Conv. %	$\overline{M}_n{}^a \times 10^{-3}$	$\overline{M}_w{}^a \times 10^{-3}$	$\dfrac{\overline{M}_w}{\overline{M}_n}{}^a$
1.83	15.6	4.5	9.2	3.8	–	–25	2	100	–	–	–
0.91	1.5	4.0	–	1.7	8.0	–33	20	46	32.6	76.8	2.4
0.91	1.1	4.0	–	3.5	8.0	–50	20	61	109	224	2.1
2.73	1.1	4.0	–	3.5	8.0	–50	20	16.7	100	258	2.6
2.73	0.43	4.0	–	3.5	8.0	–50	20	6.5	39.2	208	5.3
2.73	1.5	4.0	8.0	1.7	–	–50	5	89	44.6	109	2.5
2.73	1.5	4.0	8.0	1.7	–	–50	20	100	38.1	106	2.8
0.91	0.64	4.0	–	1.7	6.5	–70	10	51	98.9	201	2.0
0.91	1.5	4.0	–	1.7	6.5	–70	10	82	81.8	200	2.4
0.68	2.2	1.9	–	4.0	20.0	–70	20	58	150	–	–
0.68	1.1	0.95	–	4.0	20.0	–70	20	17.3	135	–	–
0.68	2.2	1.9	–	4.0	20.0	–70	20	11.2	125	–	–

[a] By GPC

Results of these orienting experiments compiled in Table 3 in regard to the effect of temperature, medium polarity, initiator concentration, monomer concentration, and coinitiator concentration are similar to those reported by others[36–39] for cationic polymerization of α-methylstyrene. For example, decreasing temperature, the molecular weight increases; and increasing medium polarity, the yield increases.

Head-group functionality of ClSi-PαMeSt prepared by this Si–Cl containing functional initiator *1* in conjunction with Et$_2$AlCl was investigated. The amount of Si–Cl head-group was determined by conductometry of the HCl generated by hydrolysis of ClSi-PαMeSt, and the theoretical amount of Si–Cl was calculated from the sample weight and $\overline{M}_n$ of ClSi-PαMeSt. The results are shown in Table 4, where the "Effective Functionality" E. F. is[2]

$$\text{E. F.} = \frac{\text{Amount of Si–Cl Found}}{\text{Theoretical Amount of Si–Cl}}$$

According to these results the yields and E. F. decrease with the increasing number of Si–Cl bonds in the functional initiators. It is possible that Si–Cl bonds may slowly react with the diethylaluminum chloride coinitiator and thus reduce the expected E. F. as follows[40]:

$$\equiv\text{SiCl} + \text{R}_2\text{AlCl} \longrightarrow \equiv\text{Si} \underset{\text{Cl}}{\overset{\text{R}}{\diagdown}} \text{Al} \underset{\text{R}}{\overset{\text{Cl}}{\diagdown}} \longrightarrow \equiv\text{SiR} + \text{RAlCl}_2$$

It is also possible that Si–Cl bonds may be consumed slowly by chain transfer process as follows[41, 42]:

$$\equiv\text{SiCl} + \sim\!\sim\!\text{CH}_2\text{–}\overset{\text{CH}_3}{\underset{\phi}{\text{C}}}{}^{\oplus}\ (\text{R}_2\text{AlCl}_2^{\ominus})$$

$$\longrightarrow \equiv\text{Si}^{\oplus}\ (\text{R}_2\text{AlCl}_2^{\ominus}) + \sim\!\sim\!\text{CH}_2\text{–}\overset{\text{CH}_3}{\underset{\phi}{\text{C}}}\text{–Cl}$$

$$\longrightarrow \equiv\text{SiR} + \text{RAlCl}_2$$

However, at this point it is not clear which process caused the reduction of effective functionality yet.

Table 4. Effective functionality of ClSi-PαMeSt obtained by the use of Cl$_n$Si(CH$_3$)$_{3-n}$CH$_2$CH$_2$φCH$_2$Cl/Et$_2$AlCl initiating system

αMeSt g	Initiator n	mol × 10^5	Et$_2$AlCl mol × 10^4	CH$_2$Cl$_2$ ml	n-C$_7$H$_{14}$ ml	Temp. °C	Time min	Conv. %	E.F.[a]
0.91	3	0.55	1.9	1.7	6.5	−70	7	29	0.19
0.91	2	0.59	1.9	1.7	6.5	−70	5	50	0.33
0.91	1	0.64	1.9	1.7	6.5	−70	3	72	0.82

[a] Effective Functionality = $\dfrac{\text{Amount of Si–Cl Found}}{\text{Theoretical Amount of Si–Cl}}$ ± 5%

[2] Dr. T. Nagy's help in performing conductometric measurement is gratefully acknowledged

c) Block Copolymer Synthesis

Although the effective functionality of ClSi-PαMeSt was less than one, attempts were made to obtain block copolymers by coupling with living polyisoprenyllithium and α,ω-disodium polyisobutylene glycolate. These particular coupling reactions were selected because it is known that Si–Cl bonds readily react with organolithium compounds and sodium alkoxides[43–45].

i. *Coupling with Living Polyisoprenyllithium*[43, 44]. Coupling ClSi-PαMeSt with polyisoprenyllithium has been described in Sect. II.E.1. The product was analyzed by H^1 NMR and GPC. However, the results indicated only negligible coupling with low molecular weight ClSi-PαMeSt and experimentations along these lines were discontinued.

ii. *Coupling with α,ω-Disodium Polyisobutylene*[45]. Coupling ClSi-PαMeSt with α,ω-disodium polyisobutylene has been described in Sect. II.E.2. The product and the starting ClSi-PαMeSt exhibited virtually identical UV and RI GPC traces, which indicated the absence of significant coupling reaction. In view of these negative results, further coupling experiments with Si–Cl bearing polymers were discontinued.

4. Conclusions

It has been postulated and subsequently proven that polymers carrying Si–Cl groups can be synthesized by the use of initiators combining in one molecule Si–Cl and benzyl chloride groups. The latter moiety was used to initiate, in conjunction with Et_2AlCl coinitiator, the polymerization of isobutylene and α-methylstyrene. According to a few orienting experiments Si–Cl bonds may survive isobutylene polymerization but quantitative information is absent. Experiments showed that Si–Cl bonds may not quantitatively survive α-methylstyrene polymerization. Although ClSi-PαMeSt has been obtained, subsequent coupling experiments with living polyisoprenyllithium and α,ω-disodium polyisobutylene failed. Failure of coupling may be due to a variety of reasons, e.g., loss of Si–Cl bonds during purification (Si–Cl groups are extremely moisture sensitive and they could have been destroyed by residual moisture in *n*-heptane, solvent used during precipitation of the polymer), or to the fact that coupling conditions were unfavorable by the use of the nonpolar solvent toluene, and extremely low molar concentration of Si–Cl bonds (the $\overline{M}_n$ of PαMeSt was ~90 000).

While these exploratory investigations demonstrate that the original objective, i.e., the synthesis of polyolefins carrying Si–Cl head-groups, is attainable, this objective has not been pursued further because concurrent studies aimed at the synthesis of Si–H termini bearing polymers promised to yield more valuable intermediates with less experimental difficulties.

B. Cationic Polymerization with Si–H Containing Initiator/Me₃Al Initiating Systems

1. Introduction

In view of the discouraging results obtained with Si–Cl containing functional cationic initiators, attention was turned toward exploiting the possibilities offered by the related Si–H group. Scheme 2 summarizes possible avenues toward derivatization and syntheses of novel sequential polymers by the use of $-Si(CH_3)_2H$, $-Si(CH_3)H_2$, or $-SiH_3$ groups attached to $-CH_2CH_2\varphi CH_2Cl$ or $-CH_2CH_2\varphi CH_2$-polymer. Indeed, by combination of various hydrosilylations and condensations numerous new polymer structures including derivatized polymers, sequential copolymers, stars, etc., can be prepared. For example, poly(methyl methacrylate-b-isoprene)[10], poly(styrene-b-dimethylsiloxane-b-styrene)[10], and various grafted copolymers of polystyrene and poly(methyl methacrylate)[10].

Si–H containing functional initiators have been obtained by reduction with LAH of the corresponding Si–Cl containing compounds. Conditions have been developed under which the benzylic chloride remains unchanged while the Si–Cl bond is converted to Si–H (for experimental details see Sect. II.C.). The initiators obtained are shown in Table 1.

Scheme 2. Derivatization and synthesis possibilities offered by $ClCH_2\varphi CH_2CH_2Si(CH_3)_{3-n}H_n$ and polymer-$CH_2\varphi CH_2CH_2Si(CH_3)_{3-n}H_n$

$$R-Si(CH_3)_{3-n}H_n \begin{cases} + n\ CH_2=CH-R' \xrightarrow[H_2PtCl_6]{\text{Hydrosilylation}} R-Si(CH_3)_{3-n}(CH_2CH_2-R')_n \\ + n\ LiR' \xrightarrow[-n\ LiH]{\text{Condensation}} R-Si(CH_3)_{3-n}R'_n \end{cases}$$

Where $R = -CH_2CH_2\varphi CH_2Cl$ or $-CH_2CH_2\varphi CH_2$-Polymer, R' = small or large (polymer) group, $n = 1, 2$ or 3.

Prior to polymerization studies extensive model experiments have been carried out for guidance in selecting suitable preparative conditions. The next section concerns model experiments; it is followed by two sections concerning polymerization of α-methylstyrene and isobutylene, respectively.

2. The Stability of Si–H Bonds under Carbenium Ion Polymerization Conditions

To synthesize polymers carrying Si–H end-groups by using the $HSi(CH_3)_2CH_2CH_2\varphi CH_2Cl/Me_3Al$ initiating system, conditions must be found under which the Si–H bonds are stable in the presence of carbenium ions and coinitiators, i.e., carbenium ion polymerization conditions must be found under which the Si–H bonds will not react with the coinitiator, the initial benzyl carbenium ion, and the propagating carbenium ion. Experiments have been carried out to guide the selection of suitable polymerization conditions.

a) The Stability of Si–H in the Presence of Me₃Al Coinitiator: The HSi(CH₃)₂CH₂CH₂φCH₃/Me₃Al/CH₂Cl₂/−50° to 22° System

[2-(p-Methylphenyl)ethyl]dimethylsilane *(7)* is a model for the Si–H containing functional initiator [2-(p-chloromethylphenyl)ethyl]dimethylsilane *(4)* in regard of the Si–H functional group.

The Me₃Al coinitiator was chosen in these model studies because according to Kennedy et al.[26]. Me₃Al is an efficient methylating agent and termination of carbocationic polymerization occurs only by methylation in the presence of Me₃AlCl$^\ominus$. Coinitiators such as Et₃Al or Et₂AlCl give rise to more complex terminations and products rendering the identification of the products of these model studies difficult.

The model compound *9* was treated with Me₃Al under conditions simulating a polymerization experiment (for details see Sect. II.F.1.) and the product was examined by H¹ NMR spectroscopy. According to the H¹ NMR analysis of spectra obtained before (Fig. 7) and after (Fig. 12) Me₃Al treatment at −50° and 22°, *7* did not change after contact with Me₃Al.

Evidently the Si–H bond is stable toward Me₃Al under conditions similar to those prevailing during a carbocationic polymerization.

b) The Stability of Si–H in the Presence of Benzyl Carbenium Ion:

i. *The HSi(CH₃)₂CH₂CH₂φCH₂Cl/Me₃Al/CH₂Cl₂/−50° to 22° System*. The stability of Si–H bond in the presence of initial benzyl cation was investigated by model experiments in which the [2-(p-chloromethylphenyl)ethyl]dimethylsilane *(4)* initiator was mixed with Me₃Al coinitiator. Scheme 3 summarizes possible reaction pathways in this model exper-

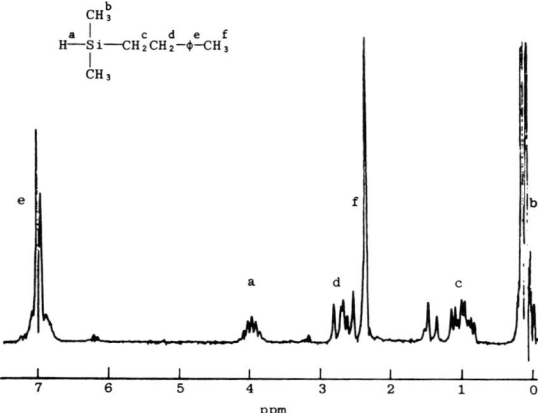

Fig. 12. H¹ NMR spectrum of [2-(p-methylphenyl)ethyl]dimethylsilane recovered from experiment III.G.1

iment. The initially formed benzyl cation may be methylated by $Me_3AlCl^{\ominus}$ to give *12* plus Me_2AlCl, or hydridated by Si–H bonds which would lead to *7* plus the transitory siliconium ion which in turn would instantaneously be methylated by $Me_3AlCl^{\ominus}$ to yield *13*.

Scheme 3. Possible pathways and products with the $HSi(CH_3)_2CH_2CH_2\varphi CH_2Cl/Me_3Al$ model system

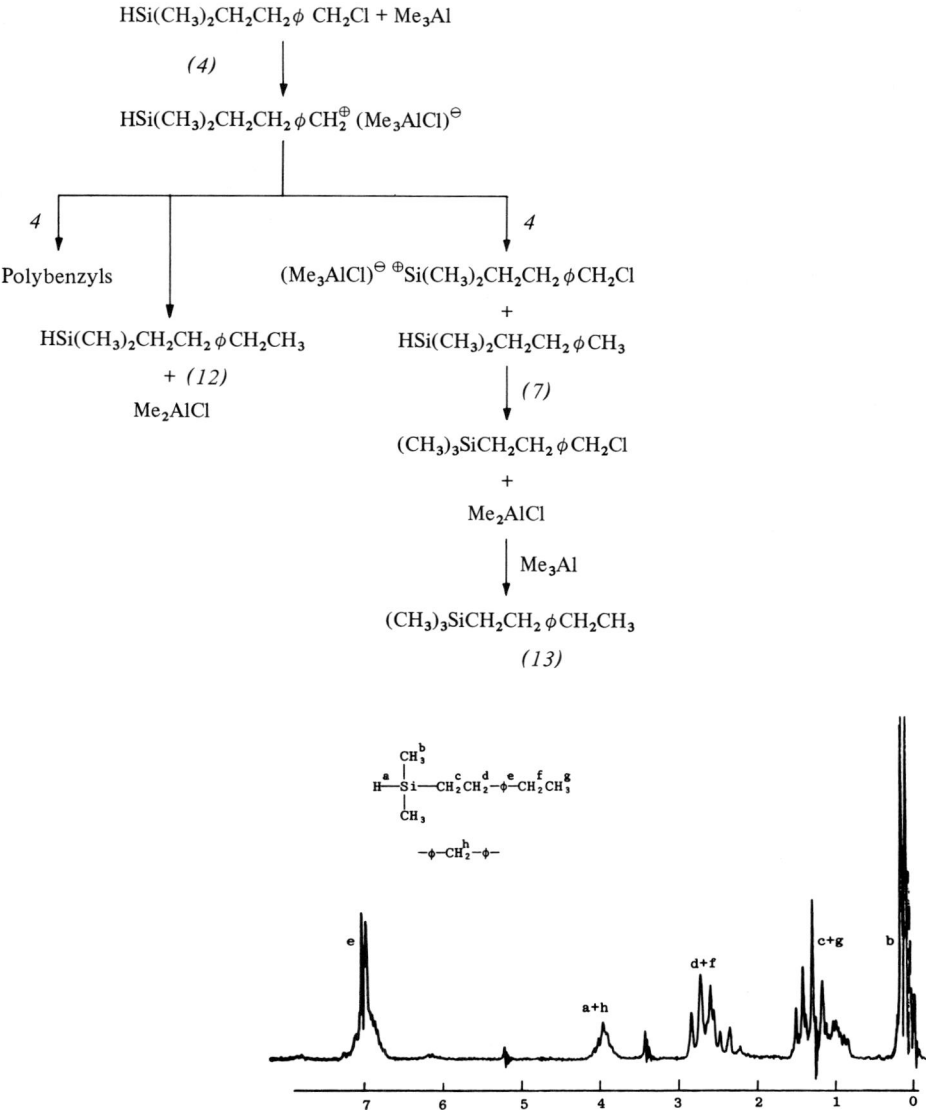

Fig. 13. H^1 NMR spectrum of products formed from $HSi(CH_3)_2CH_2CH_2\varphi CH_2Cl/Me_3Al/CH_2Cl_2/22°$ 15 min model experiment

Experimental details and conditions are given in Sect. II.F.2. The products were analyzed by H^1 NMR and quantitative IR spectroscopy. H^1 NMR analysis was inconclusive mainly because resonances associated with methylene protons between two phenyl rings, $-\varphi-CH_2-\varphi-$ (due to ring benzylation and polyphenylation), overlapped with those of silyl protons. A representative spectrum together with major assignments of the products of experiment carried out at 22° for 15 min is shown in Fig. 13. New peaks appearing between 1.0–2.9 ppm were assigned to the ethyl group in 12. Even if there were some 7 and 13 formed, the amount, as judged from the H^1 NMR spectra of the products, should be negligible, since there were no new peaks which would be the characteristic of $-\varphi-CH_3$ and $-Si(CH_3)_3$. The H^1 NMR spectra of the products obtained in the other runs were similar, except those prepared at –20° and –50° in which the methylene peak of the benzyl chloride group disappeared much slower than at 22° because of the carbocationation rate of benzyl chloride is slower at lower temperature.

Conclusive evidence for the stability of Si–H bond toward the benzyl carbenium ion was obtained by quantitative IR measurements.

The Si–H bond is a strong IR absorber and this property has frequently been used in quantitative studies[46]. Figure 14 shows the curve constructed by the use of model compounds 7 and 9. A satisfactory linear relation exists between the absorbance of the Si–H

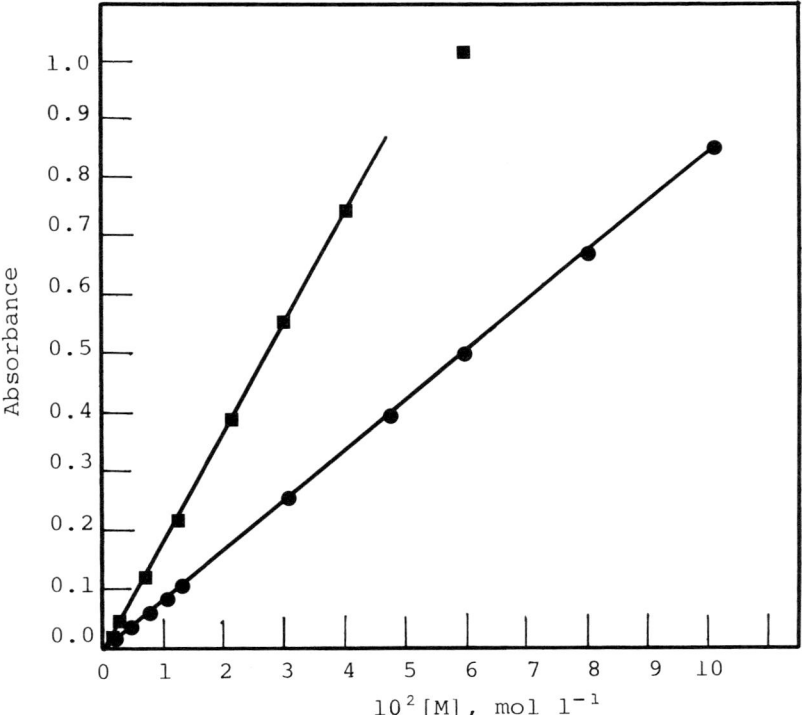

Fig. 14. IR calibration curve of $HSi(CH_3)_2CH_2CH_2\varphi CH_3$ (○) and $H_3SiCH_2CH_2\varphi CH_3$ (●) at 2115 cm^{-1} in CCl$_4$

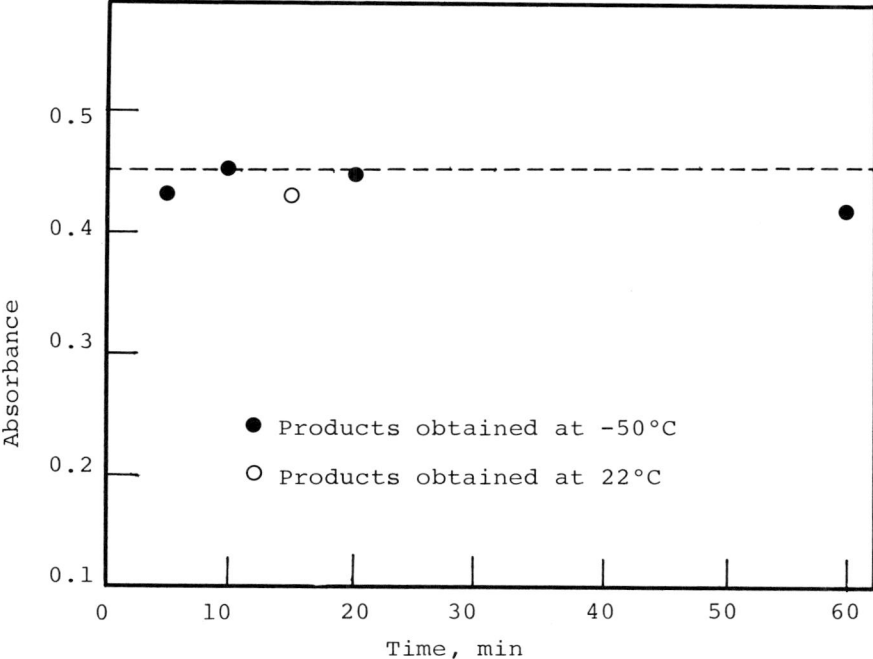

Fig. 15. IR absorbances of products from the $HSi(CH_3)_2CH_2CH_2\varphi CH_2Cl/Me_3Al/CH_2Cl_2$ model measured at 2115 cm^{-1} with concentration of products 5.0×10^{-2} M. The *dotted line* indicates the corresponding absorbance obtained from calibration curve of $HSi(CH_3)_2CH_2CH_2\varphi CH_3$ in CCl_4 (Fig. 36)

stretching band at 2115 cm^{-1} and concentration in CCl_4. Model experiments have been carried out in which *4* was mixed with Me_3Al in CH_2Cl_2 at −50° and 22° for various times (see Experimental Sect. II.F.2). The products of these experiments have been isolated and 5.0×10^{-2} M solutions were prepared (assuming the density and molecular weight remain unchanged), and subjected to IR analysis. Figure 15 shows the results. Evidently absorbances due to Si–H bonds in *4* did not change upon treatment with Me Al within experimental variation. The dotted line in Fig. 15 indicates the theoretical absorbance of 5.0×10^{-2} M Si–H based on the calibration curve shown in Fig. 14. The experimental points are consistently somewhat below the theoretical line and this may be due to a systematic error, e.g., slight change in density of the products from *4*, or to a systematic loss of Si–H bond during purification. According to this evidence Si–H bonds remain essentially unaffected in the presence of benzyl cations under conditions similar to those prevailing during carbocationic polymerizations.

ii. *The $H_3SiCH_2CH_2\varphi CH_2Cl/Me_3Al/CH_2Cl_2/-80°$ to 22° System.* A further series of experiments have been carried out with model *6*, which carries the –SiH group. It was theorized that the –SiH group may be less protected sterically than *4* and consequently would be more reactive than the Si–H bond in –Si(CH$_3$)$_2$H (in model *4*). Experimental details are given in Sect. II.F.3. Scheme 4 shows possible pathways and products that

Scheme 4. Possible pathways and products with the $H_3SiCH_2CH_2\varphi CH_2Cl/Me_3Al$ model system

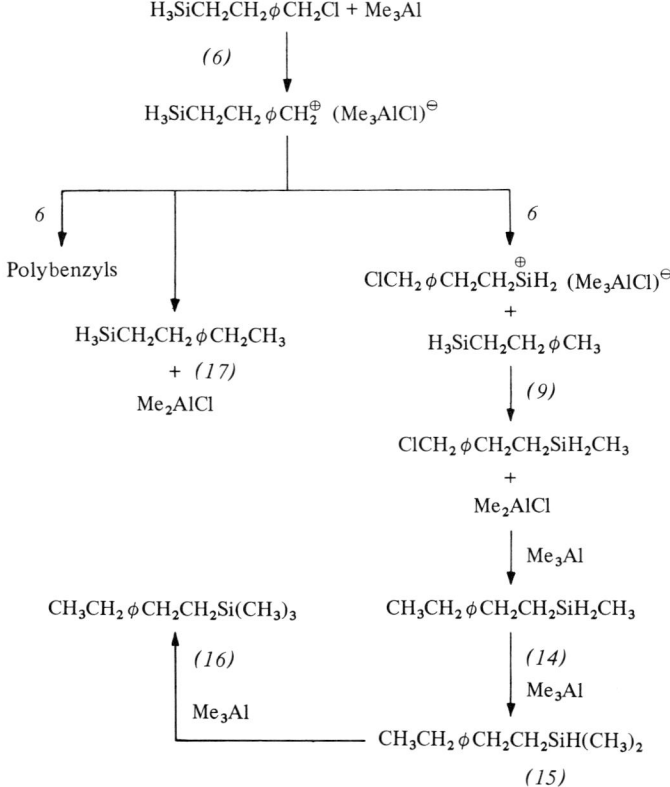

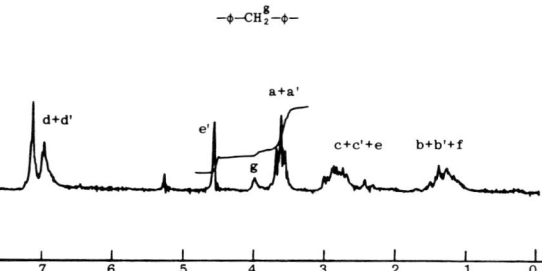

Fig. 16. H^1 NMR spectrum of products formed from $H_3SiCH_2CH_2\varphi CH_2Cl/Me_3Al/CH_2Cl_2/-50°/$ 60 min model experiment

may form in this model experiment. The formation of products *9, 14, 15,* and *16* would be expected in case of reaction between Si–H and benzyl carbenium ion occurred.

A representative H^1 NMR spectrum of the products formed at $-50°$ after 60 min is shown in Fig. 16. The resonance at 3.9 ppm, associated with methylene protons between two phenyl rings ($-\varphi$–C$\underline{H}_2$–φ–), is indicative of the formation of polybenzyls. Resonances at 1.0–1.5 ppm and 2.5–3.0 ppm are assigned to the ethyl group of *17*. According to integration of the resonances at 4.5 ppm and 3.5 ppm, assigned to the methylene protons of the benzyl chloride group are –Si$\underline{H}_3$ proton peaks, respectively, ~48% of the original benzyl group has reacted with Me$_3$Al. Absence of resonances in the 0.0–0.5 ppm region indicates the absence of Si–CH$_3$ group, i.e., the absence of produce *14, 15,* and *16* in Scheme 4.

Subsequently the products of these experiments have been analyzed by quantitative IR spectroscopy. The calibration curve showing the stretching absorbance of the SiH$_3$ group in *9* versus concentration is shown in Fig. 14. Model experiments were carried out by mixing *6* with Me$_3$Al at $-65°$, $-50°$, and $-22°$ under the conditions described in Sect. II.F.3. of Experimental and 2.2×10^{-2} M solutions of the products of these experiments were analyzed by IR. Results are shown in Fig. 17. Assuming that the products of model experiments and that of *6* have the same density and formula weight, IR absorbances show that benzyl cations are essentially inert toward the –SiH$_3$ group.

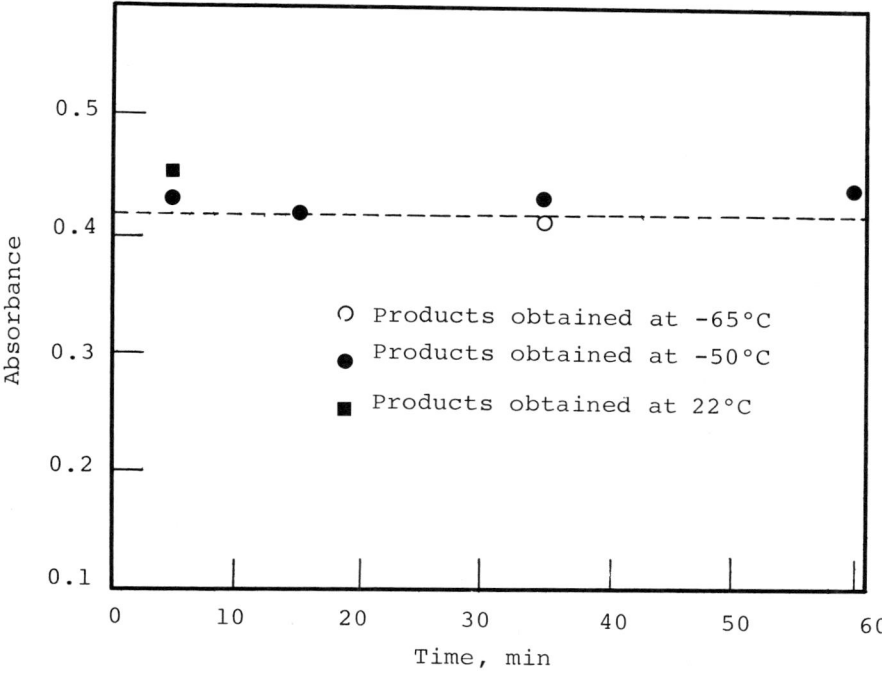

Fig. 17. IR absorbance of products from H$_3$SiCH$_2$CH$_2$$\varphi$CH$_2Cl/Me_3Al/CH_2Cl_2$, measured at 2115 cm^{-1} with concentration of products 2.2×10^{-2} M. The *dotted line* indicates the corresponding absorbance obtained from the calibration curve of H$_3$SiCH$_2$CH$_2$$\varphi$CH$_3$ in CCl$_4$ (Fig. 36)

c) The Stability of Si–H in the Presence of Propagating Carbenium Ion

Having demonstrated that the Si–H bond is quite stable in the presence of Me$_3$Al coinitiator and benzyl carbenium ion, further model experiments have been carried out to explore the stability of the Si–H bond in the presence of propagating carbenium ions.

i. *The HSi(CH$_3$)$_2$CH$_2$CH$_2\varphi$CH$_3$/Me$_3$Al/(CH$_3$)$_3$CCl or CH$_3$(φ)CHCl/n-C$_6$H$_{14}$ and/or CH$_2$Cl$_2$/$-50°$ System*. In these model experiments propagating carbenium ions of isobutylene and styrene were simulated by carbenium ions generated from *t*-butyl chloride and 1-chloroethyl benzene in conjunction with Me$_3$Al. Reaction conditions are described in Sect. II.F.4.

The IR calibration line shown in Fig. 36 may not be valid with the products formed in these series of experiments since in this case the densities and formula weight of the products may not be assumed the same as that of 7, which was used to construct the calibration line. Thus only H^1 NMR spectroscopy has been employed to explore the stability of Si–H bonds in the presence of carbenium ions simulating propagating carbenium ions of isobutylene and styrene.

H^1 NMR spectra of products obtained in runs with *t*-butyl chloride carried out under simulated isobutylene polymerization conditions did not show loss of Si–H bonds. Figure 18 shows a representative H^1 NMR spectrum together with pertinent assignments of the run with 1:1 CH$_2$Cl$_2$/n-C$_6$H$_{24}$. The ratio of the peaks associated with the Si–H proton (at 4.0 ppm) and Si–CH$_3$ protons (at 0.1 ppm) remained approximately 1:6. Similarly H^1 NMR spectra of products obtained in runs with 1-chloroethyl benzene indicated that most Si–H bonds survived in the presence of styryl cations. A representative H^1 NMR spectrum of the products obtained in the presence of 1:1 CH$_2$Cl$_2$/n-C$_6$H$_{14}$ is shown in Fig. 19. An exception was noted: In the case the experiment was carried out in pure CH$_2$Cl$_2$ some Si–H bonds seemed to be consumed. It is conceivable that in pure CH$_2$Cl$_2$ siliconium ions[42, 47, 48] are formed by hydride transfer from Si–H to carbenium ions; the latters are formed from the organic chloride plus Me$_3$Al.

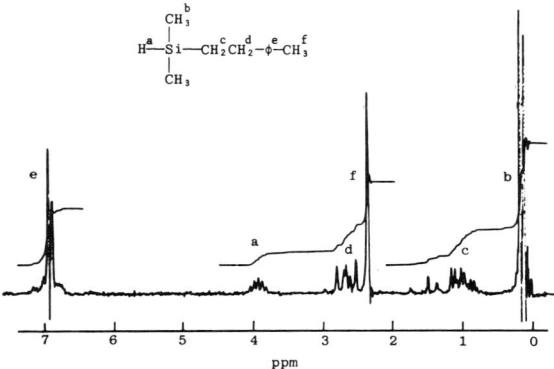

Fig. 18. H^1 NMR spectrum of products formed from HSi(CH$_3$)$_2$CH$_2$CH$_2\varphi$CH$_3$/Me$_3$Al/(CH$_3$)$_3$CCl/ CH$_2$Cl:n-C$_6$H$_{14}$ = 1:1/$-50°$/30 min model experiment

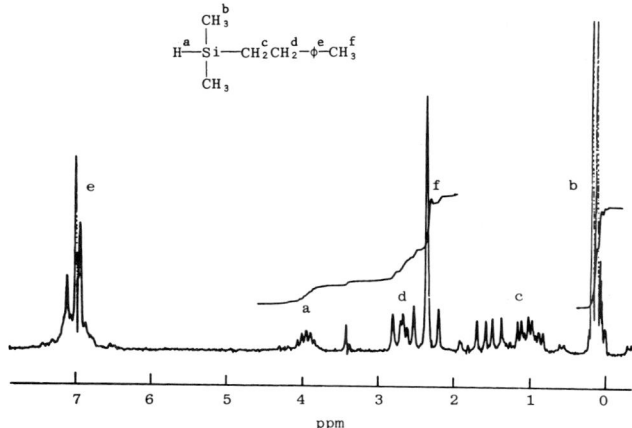

Fig. 19. H^1 NMR spectrum of products formed from HSi(CH$_3$)$_2$CH$_2$CH$_2$φCH$_3$/Me$_3$Al/ CH$_3$(φ)CHCl/n-C$_6$H$_{14}$ = 1 : 1/−50°/30 min model experiment

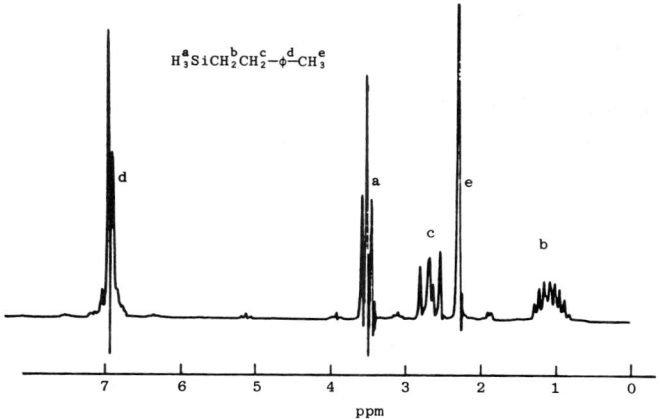

Fig. 20. H^1 NMR spectrum of products formed from H$_3$SiCH$_2$CH$_2$φCH$_3$/Me$_3$Al/(CH$_3$)$_3$CCl/ CH$_2$Cl$_2$: n-C$_6$H$_{14}$ = 1 : 1/−50°/30 min model experiment

ii. *The H$_3$SiCH$_2$CH$_2$φCH$_3$/Me$_3$Al/(CH$_3$)$_3$CCl or CH$_3$(φ)CHCl/n-C$_6$H$_{14}$ and/or CH$_2$Cl$_2$/ −50° System.* The stability of Si–H bond in [2-(p-methylphenyl)ethyl]silane *(11)* has been investigated in the presence of simulated propagating carbenium ions of isobutylene and styrene. Reaction conditions are described in Sect. II.F.5. Generally the results were similar to those observed with the HSi(CH$_3$)$_2$CH$_2$CH$_2$φCH$_3$/Me$_3$Al/(CH$_3$)$_3$CCl or CH$_3$(φ)CHCl/n-C$_6$H$_{14}$ and/or CH$_2$Cl$_2$/−50° system. Based on H^1 NMR analyses Si–H bonds are not consumed except in runs with 100% CH$_2$Cl$_2$ in which ~3.7% Si–H was consumed. H^1 NMR spectra together with assignments of the products obtained in runs in which 1 : 1 CH$_2$Cl$_2$/n-C$_6$H$_{14}$ have been used are shown in Fig. 20 and 21. According to

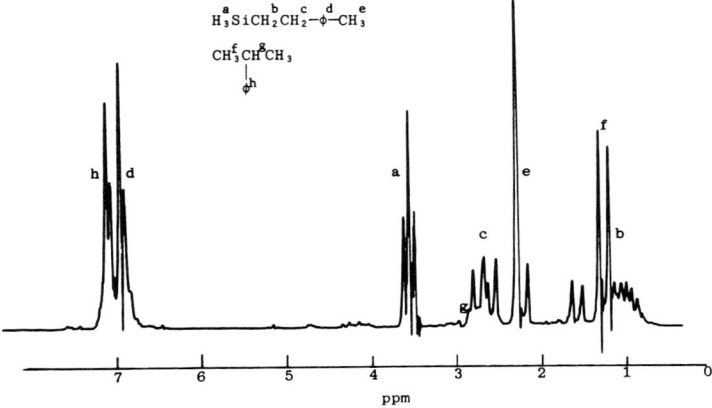

Fig. 21. H^1 NMR spectrum of products formed from H$_3$SiCH$_2$CH$_2$φCH$_3$/Me$_3$Al/CH$_3$(φ)CHCl/ CH$_2$Cl$_2$: n-C$_6$H$_{14}$ = 1 : 1/−50°/30 min model experiment

these spectra Si–H bonds are quite stable under simulated isobutylene and styrene polymerization conditions.

It should be emphasized that Si–H containing compounds should be carefully handled during purification so as to avoid hydrolysis of Si–H bonds. An effective method to suppress hydrolysis of Si–H bonds is to reduce the polarity of the medium by the addition of a large amount of n-hexane before the aluminum compound is removed by washing with dilute cold HCl solution.

d) Conclusions

As a prelude to polymerization studies, extensive model experiments have been carried out to define conditions under which the Si–H bonds in H$_n$Si(CH$_3$)$_{3-n}$CH$_2$CH$_2$φCH$_2$Cl,

Table 5. The effect of polarity on the polymerization of αMeSt using the HSi(CH$_3$)$_2$CH$_2$CH$_2$φCH$_2$Cl/Me$_3$Al initiating system

CH$_2$Cl$_2$ %	n-C$_7$H$_{16}$ %	Yield g	Conv. %	$\overline{M}_n \times 10^{-4}$	$\overline{M}_w \times 10^{-4}$	MWD[a]	$\overline{P}$[b] $\times 10^6$
100	0	1.51	42.65	4.23	7.13	1.69	35.7
80	20	0.82	23.13	3.09	5.46	1.77	26.5
60	40	0.17	4.72	2.31	4.30	1.86	7.5
40	60	0.12	3.45	2.05	4.01	1.96	5.9
20	80	0.04	0.90	1.43	3.00	2.09	2.4
0	100	0.00	0.00	–	–	–	–

[αMeSt] = 1.0 M, [HSi(CH$_3$)$_2$CH$_2$CH$_2$φCH$_2$Cl] = 2.5 × 10^{-3} M, [Me$_3$Al] = 5.0 × 10^{-2} M, Total Volume = 30 ml, −50 °C, 15 min
[a] By GPC
[b] $\overline{P}$ = moles of polymer chains = yield/$\overline{M}_n$

n = 1–3, would survive carbocationic polymerization conditions. Three key requirements have been examined: survival of Si–H bonds in the presence of i) Me$_3$Al coinitiator, ii) benzyl carbenium ions, and iii) propagating carbenium ions of isobutylene and styrene (simulated by *tert.*-butyl and styryl cations). According to the results of H^1 NMR and quantitative IR analyses, convenient conditions exist (from 22° to −60 °C, mixed CH$_2$Cl$_2$/n-C$_6$H$_{14}$ media, broad concentration ranges) under which Si–H bonds quantitatively survive cationic polymerizations. The use of pure CH$_2$Cl$_2$ should be avoided. Also great care should be exercised during the purification of Si–H containing polymers on account of their sensitivity toward hydrolysis.

3. α-Methylstyrene Polymerization with Si–H Containing Initiator/Me$_3$Al Initiating System

a) Introduction

An objective in this research was the synthesis of polymers with Si–H head-groups. Thus guided by the results of model experiments (Sect. II.F.) the synthesis of poly(α-methylstyrene)(PαMeSt) with Si–H head-group (HSi-PαMeSt) was undertaken.

Effects of solvent polarity, counter-anion nucleophilicity, temperature, and monomer concentration on the carbenium ion polymerization chemistry have been extensively studied[29, 36–38, 49]. Based on previous knowledge[26–29] Me$_3$Al was chosen because with this coinitiator undesired chain transfer to monomer processes are absent. Preliminary experiments showed that Et$_3$Al coinitiator did not yield PαMeSt, possibly because the nucleophilicity of the counter-anion Et$_3$AlCl$^\ominus$ is too high and thus termination by hydridation is faster than propagation[36].

Research described in this section concerns effects of solvent polarity, temperature, monomer and initiator concentration on the polymerization of α-methylstyrene with Si–H containing initiator/Me$_3$Al system for the synthesis of HSi-PαMeSt and of desirable molecular weight.

b) Effect of Reaction Conditions on the Polymerization of α-Methylstyrene

i. *Effect of Solvent Polarity.* The effect of solvent polarity on the polymerization of αMeSt induced by the HSi(CH$_3$)$_2$CH$_2$CH$_2$φCH$_2$Cl/Me$_3$Al system has been investigated at −50° by the use of various CH$_2$Cl$_2$/n-C$_6$H$_{14}$ solvent compositions. Table 5 shows the series of solvents investigated together with other pertinent experimental details and results. The effect of solvent composition on $\overline{M}_n$ and percent conversion is further illustrated in Fig. 22.

Evidently both HSi-PαMeSt molecular weights and yields are greatly affected by medium polarity. Highest $\overline{M}_n$ and yields are obtained in pure CH$_2$Cl$_2$ and the addition of n-C$_6$H$_{14}$ strongly reduced $\overline{M}_n$ and yields.

These effects, however, are not unexpected. Previous authors have also found a reduction in $\overline{M}_n$ and conversion with decreasing medium polarity. Conceivably, reduced $\overline{M}_n$'s at reduced polarity are due to relatively faster termination, i.e., the nucleophilicity of the Me$_3$AlCl$^\ominus$ counter-anion increases in relatively nonpolar media which results in

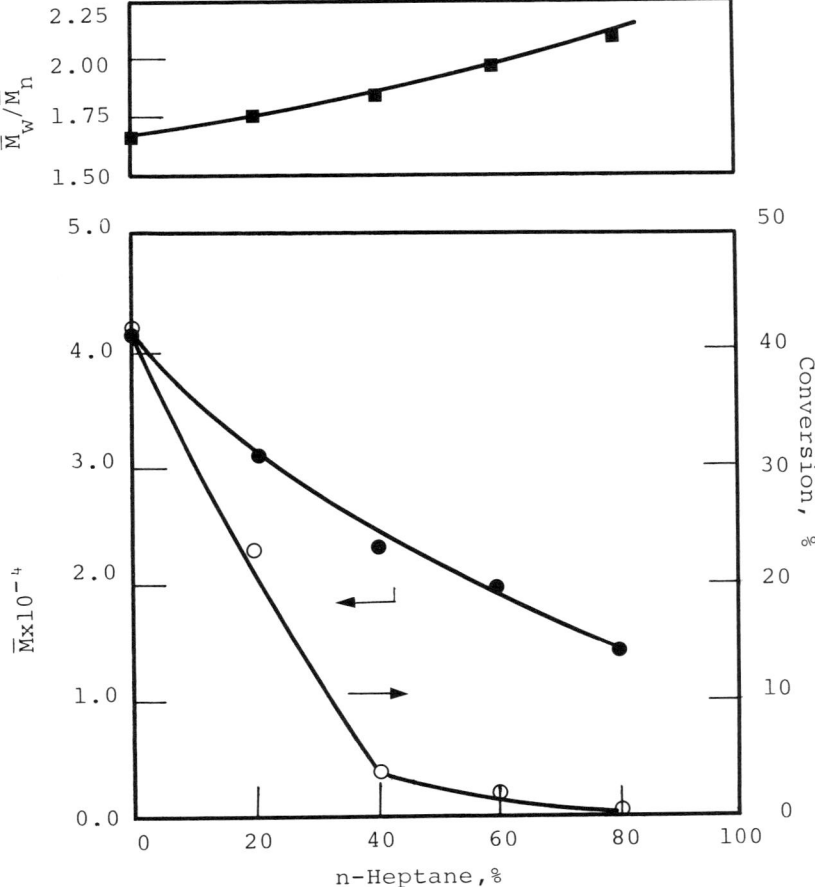

Fig. 22. The effect of solvent composition on the molecular weight (●), conversion (○), and molecular weight dispersity (■) of PαMeSt prepared using the HSi(CH$_3$)$_2$CH$_2$CH$_2$φCH$_2$Cl/Me$_3$Al initiating system (See Table 5 for reaction conditions)

acceleration of methylation[26–29, 36, 49]. The reduced yields at reduced polarity may be attributed to both the $\overline{M}_n$ and the moles polymer chains ($\overline{P}$) decrease when the medium polarity decreases, in which the lowering of $\overline{P}$ may in turn be due to slower carbocationation and faster termination of the initiator in relatively nonpolar media[25–27, 29].

The effect of changing solvent composition (polarity) on molecular weight dispersity is noteworthy. $\overline{M}_w/\overline{M}_n$ is quite low (1.69) in the experiment carried out by the use of 100% CH$_2$Cl$_2$ and it increases monotonically with increasing n-C$_6$H$_{14}$ content. It is very difficult to interpret these data at this time.

ii. *Effect of Temperature.* The effect of temperature on αMeSt polymerization initiated by the HSi(CH$_3$)$_2$CH$_2$CH$_2$φCH$_2$Cl/Me$_3$Al system has been studied in the range from −20° to −80 °C at two polarity levels, i.e., with CH$_2$Cl$_2$/n-C$_7$H$_{16}$ = 70/30 and 90/10 ratios.

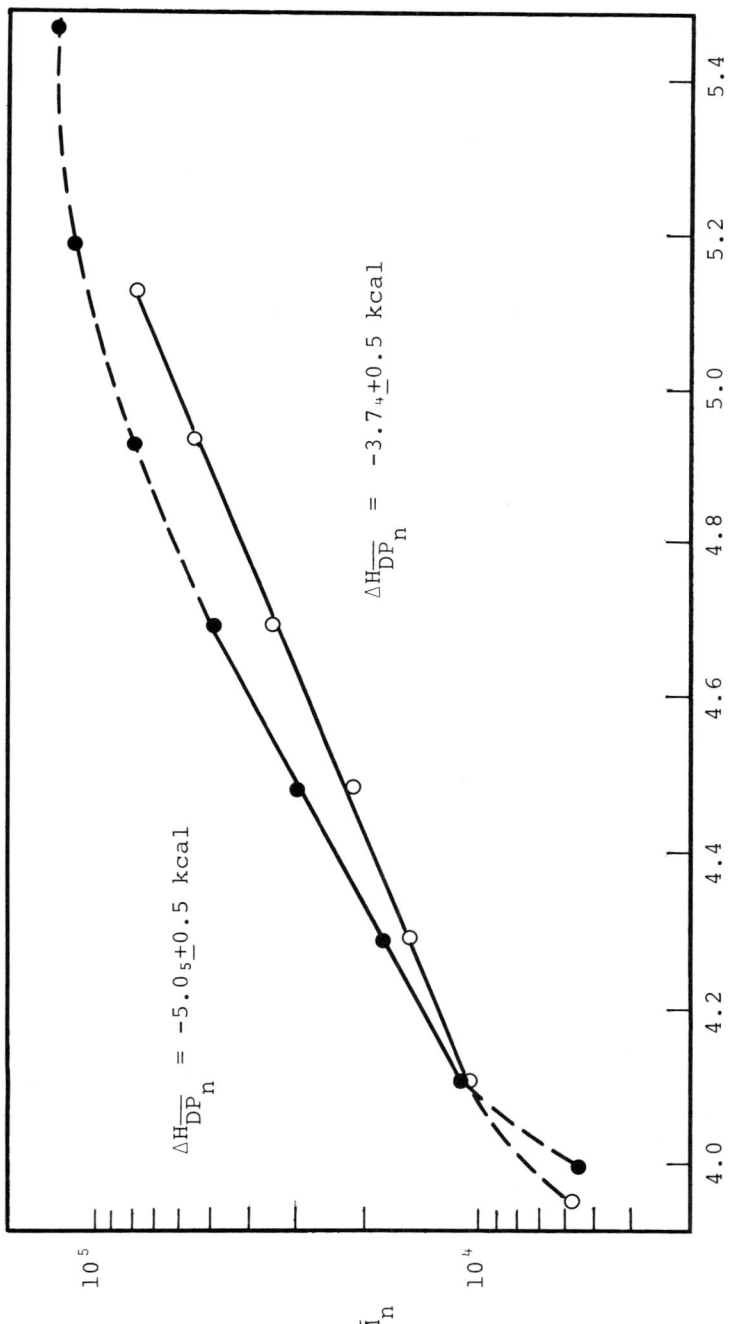

Fig. 23. The effect of temperature on the molecular weight of PαMeSt prepared using the HSi(CH$_3$)$_2$CH$_2$CH$_2$φCH$_2$Cl/Me$_3$Al initiating system at solvent composition: CH$_2$Cl$_2$/n-C$_7$H$_{16}$ = 90/10 (●) and 70/30 (○) (See Table 6 for reaction conditions)

Table 6. The effect of temperature on the polymerization of αMeSt using the $HSi(CH_3)_2CH_2CH_2\varphi CH_2Cl/Me_3Al$ initiating system

Temp. °C	Yield g	Conv. %	$\overline{M}_n^a \times 10^{-4}$	$\overline{M}_w^a \times 10^{-4}$	MWD^a	$\overline{P}^b \times 10^6$
Solvent Composition: $CH_2Cl_2/n\text{-}C_7H_{16}$ = 70/30; Time = 15 min						
−20	0.302	8.5	0.61	1.48	2.42	49.4
−30	0.216	6.1	1.10	2.34	2.13	19.7
−40	0.225	6.3	1.53	3.31	2.17	14.7
−50	0.221	6.2	2.13	4.23	1.98	10.4
−60	0.289	8.2	3.55	6.30	1.78	8.2
−70	0.458	13.0	5.54	9.40	1.70	8.3
−80	0.592	16.7	7.52	12.30	1.64	7.9
Solvent Composition: $CH_2Cl_2/n\text{-}C_7H_{16}$ = 90/10; Time = 5 min						
−20	0.281	7.9	0.58	1.15	1.99	48.5
−30	0.228	6.4	1.13	2.06	1.82	20.2
−40	0.282	7.9	1.84	3.23	1.78	15.3
−50	0.393	11.1	3.01	4.93	1.64	13.0
−60	0.603	17.0	4.82	7.86	1.63	12.5
−70	0.966	27.3	7.25	11.54	1.59	13.3
−80	1.491	42.1	10.43	16.42	1.57	14.3
−90	1.908	53.8	11.87	19.21	1.62	16.1

[αMeSt] = 1.0 M, $[HSi(CH_3)_2CH_2CH_2\varphi CH_2Cl]$ = 2.5 × 10^{-3} M, [Me$_3$Al] = 5.0 × 10^{-2} M, Total Volume = 30 ml
^{a} By GPC
^{b} $\overline{P}$ = moles of polymer chains = yield/$\overline{M}_n$

Pertinent experimental details together with results are shown in Table 6 and illustrated in Fig. 23.

Figure 23 shows the effect of temperature on $\overline{M}_n$ in the form of Arrhenius plots[36–8]. The negative slope of Arrhenius lines is in conformity with the usual observation, i.e., molecular weights invariably increase with decreasing temperature. The "overall activation energy differences of molecular weights" $\Delta H_{\overline{DP}_n}$, calculated for the linear (drawn in by solid lines) portions of the Arrhenius plots, are $\Delta H_{\overline{DP}_n} = -3.7_4 \pm 0.5$ and $-5.0_5 \pm 0.5$ kcal/mol for the experiments with $CH_2Cl_2/n\text{-}C_7H_{16}$ = 70/30 and 90/10, respectively. Although these $\Delta H_{\overline{DP}_n}$ values differ only by −1.3 kcal/mol, in view of the good repeatability of these data the small $\Delta H_{\overline{DP}_n}$ difference may be real, i.e., $\Delta H_{\overline{DP}_n}$ is higher in the more polar medium[49].

The effect of temperature on molecular weight dispersity is noteworthy as shown in Fig. 23. Evidently, lowering the temperature narrows molecular weight dispersity and $\overline{M}_w/\overline{M}_n$ falls much below 2.0, characteristic of the most probable distribution, at both polarity levels investigated. The effect is particularly pronounced with $CH_2Cl_2/n\text{-}C_7H_{16}$ = 90/10. It seems as though decreasing the temperature tends to narrow molecular weight distribution. A similar phenomenon has recently been observed by Kennedy and Chou[50, 51], but it is very difficult to explain these observations at the present time.

It is also of interest that $\overline{P}$ decreased gradually with decreasing temperature. Conceivably the initiation rate decelerated with decreasing temperature[26, 29, 52, 53] (see Sect. III.B.2.b.i.).

iii. *Effect of Monomer Concentration.* The effect of monomer concentration on the molecular weight of HSi-PαMeSt has been studied. Experimental conditions and results are shown in Table 7 and are illustrated in Fig. 24. Medium polarity was maintained by the use of *n*-hexane as co-solvent, i.e., keeping the total volume of α-methylstyrene plus *n*-hexane constant (see Experimental Sect. II.B.1).

As shown by the data in Fig. 46 the molecular weight of HSi-PαMeSt increases with increasing monomer concentration. Thus in addition to temperature and solvent polarity, the molecular weight of HSi-PαMeSt can be efficiently controlled by changing the monomer concentration.

In order to examine the kinetic meaning of Fig. 24, however, it is necessary to review some previous works of Mandal and Kennedy[52, 53] and of Plesch[54]:

Assuming that the number average degree of polymerization ($\overline{DP}_n$) is determined by chain transfer to monomer and assuming unimolecular termination relative to propagation (i.e., chain breaking due to solvent, polymer, impurities are absent), the simple Mayo equation[55]

$$\frac{1}{\overline{DP}_n} = \frac{k_{tr,M}}{k_p} + \frac{k_t}{k_p} \cdot \frac{1}{[M]}$$

has been extensively used to elucidate the mechanism of cationic polymerization[52–54, 56, 57]. Thus the Mayo plot of $1/\overline{DP}_n$ versus $1/[M]$ provides relative rate constants $k_{tr,M}/k_p$ and k_t/k_p and yields quantitative insight as to the molecular weight controlling events. The Mayo plot is particularly useful when it traverses the origin. The absence of an intercept is proof positive for the absence of chain transfer to monomer. According to Mandal and Kennedy's[52, 53] analysis, however, the opposite is not true, i.e., the presence of an intercept does not necessarily prove chain transfer to monomer. Thus it is not too surprising that Fig. 46 shows small intercepts for the simple Mayo plots of the polymerizations of αMeSt induced by HSi(CH$_3$)$_2$CH$_2$CH$_2$φCH$_2$Cl/Me$_3$Al system. According to model studies (see Sect. III.B.2.b.i.) carbocationation by the benzyl chloride initiating moiety in conjunction with Me$_3$Al coinitiator in relatively nonpolar medium is inefficient and $\overline{P}$ is generally much less than the number of initiator molecules charged. Since these

Table 7. The effect of monomer concentration on the polymerization of αMeSt using the HSi(CH$_3$)$_2$CH$_2$CH$_2$φCH$_2$Cl/Me$_3$Al initiating system

Temp. °C	[αMeSt] M	Conversion %	$\overline{M}_n^a \times 10^{-4}$	$1/\overline{DP}_n \times 10^3$
−50	1.0	2.22	1.55	7.64
−50	1.5	2.12	2.15	5.50
−50	2.0	2.07	2.65	4.46
−50	2.5	2.03	3.03	3.90
−60	1.0	3.46	2.44	4.84
−60	1.5	3.17	3.41	3.47
−60	2.0	3.26	4.93	2.40
−60	2.5	3.32	4.62	2.56

[HSi(CH$_3$)$_2$CH$_2$CH$_2$φCH$_2$Cl] = 2.5 × 10^{-3} M, [MeAl] = 5 × 10^{-2} M, CH$_2$Cl$_2$ Solvent, *n*-Hexane Cosolvent, Total Volume = 30 ml, 5 min
[a] By GPC

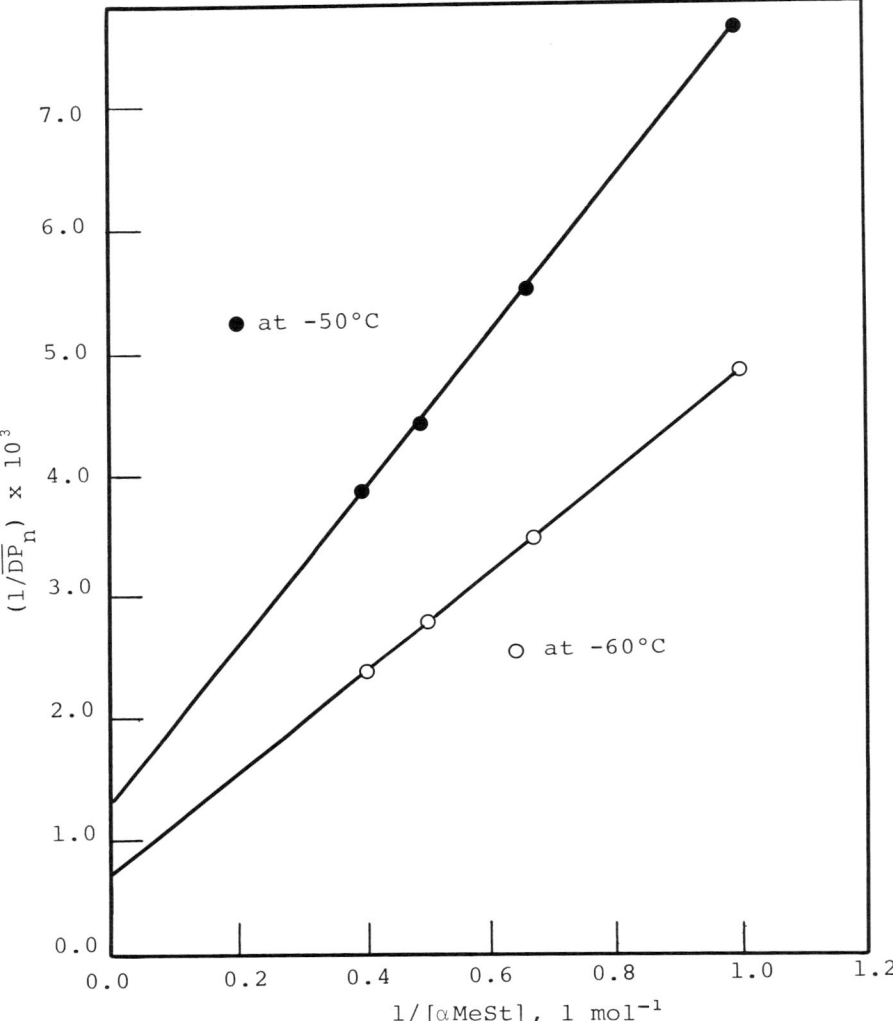

Fig. 24. "Mayo Plots" of αMeSt polymerization using the HSi(CH$_3$)$_2$CH$_2$CH$_2$φCH$_2$Cl/Me$_3$Al initiating system (See Table 7 for reaction conditions)

results and those observed by other authors[25, 26, 52, 53] are in agreement, it is concluded that the presence of small intercepts in Fig. 46 is not due to chain transfer to monomer but rather to slow initiation and to destruction by methylation. Further support for the suggestion that in the system under investigation chain transfer to monomer is absent comes from direct quantitative end-group analysis by IR, to be discussed in Sect. III.B.3.c.

iv. *Effect of Initiator Concentration.* The effect of initiator concentration on the molecular weight of PαMeSt has been investigated. Experimental conditions and results are

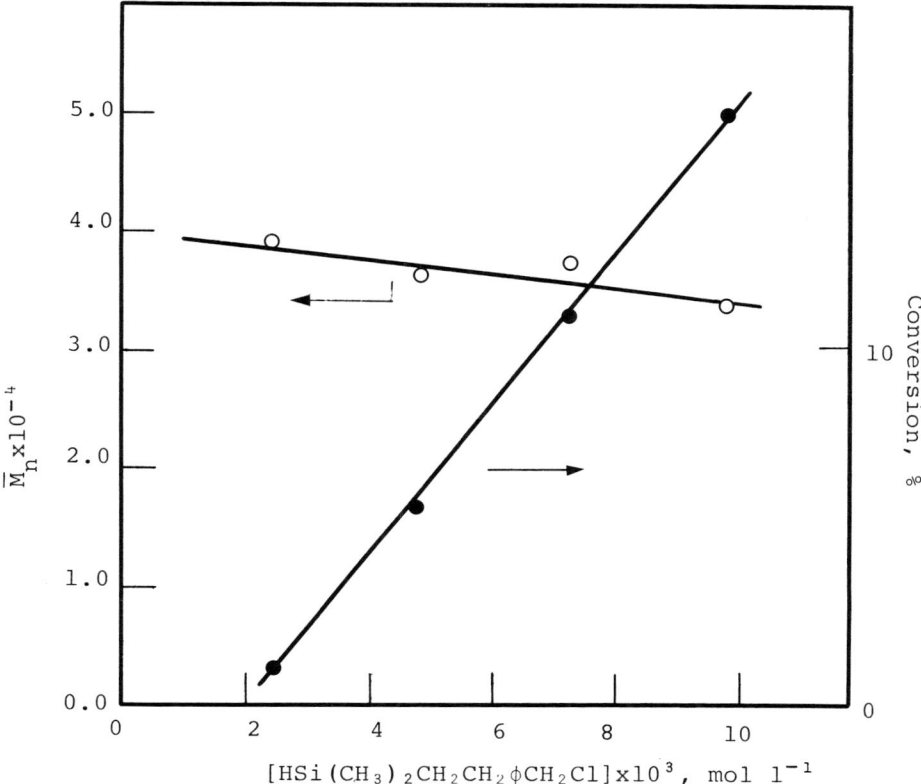

Fig. 25. The effect of initiator concentration on molecular weight and conversion of PαMeSt prepared using the $HSi(CH_3)_2CH_2CH_2\varphi CH_2Cl/Me_3Al$ initiating system (See Table 8 for reaction conditions)

Table 8. The effect of initiator concentration on the polymerization of αMeSt using the $HSi(CH_3)_2CH_2CH_2\varphi CH_2Cl/Me_3Al$ initiating system

$HSi(CH_3)_2CH_2CH_2\varphi CH_2Cl$ $M \times 10^3$	Me_3Al $M \times 10^2$	Conv. %	$\overline{M}_n^a \times 10^{-4}$
2.5	5.0	2.1	3.88
5.0	10.0	5.1	3.59
7.5	15.0	10.7	3.78
10.0	20.0	16.4	3.36

[αMeSt] = 1.0 M, Solvent: CH_2Cl_2/n-C_6H_{14} = 60/40, −50 °C, 10 min
[a] By GPC

compiled in Table 8 and are illustrated in Fig. 25. Evidently, the yields increase with increasing initiator concentration, while molecular weights remain essentially unchanged within what is considered to be experimental variation. These phenomena can be explained by assuming that initiation is favored by increasing initiator concentration (initiation is energetically favorable as it involves the attack of αMeSt on an initial benzyl cation with the simultaneous formation of a relatively more stable tertiary benzyl cation[58]), however, molecular weights remain essentially unchanged because increasing initiator concentration does not change the rate of chain breaking events.

$$HSi(CH_3)_2CH_2CH_2\phi CH_2^\oplus \qquad \sim\sim\sim CH_2\underset{\phi}{\overset{CH_3}{\underset{|}{\overset{|}{C}}}}\!\!^\oplus$$

Initial   Propagating
Carbenium Ion   Carbenium Ion

Chain transfer by the $HSi(CH_3)_2CH_2CH_2\varphi CH_2Cl$ would be energetically unfavorable as it would involve $Cl^\ominus$ transfer from the initiator to the propagating cation, i.e., the creation of a relatively less stable initial benzyl carbocation by sacrificing a more stable propagating tertiary benzyl cation[58].

c) Head-Group Characterization

H^1 NMR spectroscopy was found to be unsuitable for head-group analysis of HSi-PαMeSt. The resonance associated with the Si–H proton at the head-group is broadened by multiple splitting and the resonances of the aromatic protons of the initiator fragment are buried in the aromatic proton resonances of phenyl rings of the αMeSt repeating units.

Quantitative information as to Si–H head-group concentration was obtained by IR spectroscopy. To this end we determined IR absorbances at 2115 cm^{-1} of a well purified HSi-PαMeSt (reprecipitated twice from dry toluene into dry n-heptane) obtained by $HSi(CH_3)_2CH_2CH_2\varphi CH_2Cl/Me_3Al$ system, and the Si–H functionality was calculated by using the calibration curve of the model $HSi(CH_3)_2CH_2CH_2\varphi CH_3$ (see Fig. 14) in conjunction with the $\overline{M}_n$ obtained by GPC. A representative set of analytical data is shown in Table 9. The functionality of 0.92 shown is quite satisfactory and is well within experimental error of 1.0 ± 0.1. Thus PαMeSt carrying one Si–H group per molecule (HSi-PαMeSt) can be prepared by the $HSi(CH_3)_2CH_2CH_2\varphi CH_2Cl/Me_3Al$ initiating system.

Table 9. Functionality analysis of HSi-PαMeSt prepared by the $HSi(CH_3)_2CH_2CH_2\varphi CH_2Cl/Me_3Al$ initiating system

$\overline{M}_n$ (GPC) of HSi-PαMeSt (Fig. 26)	1.74×10^4
HSi-PαMeSt in CCl$_4$ (w/v)	0.0337 g/0.5 ml
[HSi-PαMeSt]^a	3.87×10^{-3} M
IR Absorbance of 3.87×10^{-3} M HSi-P MeSt at 2115 cm^{-1} (Fig. 27)	0.035
IR Absorbance of 3.87×10^{-3} M $HSi(CH_3)_2CH_2CH_2\varphi CH_3$ at 2115 cm^{-1}	0.038^b
Si–H Functionality	0.9 ± 0.1

^{a} Prepared at [αMeSt] = 0,5 M, [$HSi(CH_3)_2CH_2CH_2\varphi CH_2Cl$] = 1.0×10^{-2} M, Me$_3$Al = 1.0×10^{-2} M, CH$_2Cl_2$/n-C$_6H_{14}$ = 60/40, Total Volume = 500 ml, $-60°$, 30 min; Purified by twice reprecipitation from toluene into n-hexane
^{b} See Fig. 14: Calibration curve of $HSi(CH_3)_2CH_2CH_2\varphi CH_3$ at 2115 cm^{-1} in CCl$_4$

d) Conclusions

The effects of reaction variables on the polymerization of αMeSt initiated by the HSi(CH$_3$)$_2$CH$_2$CH$_2$φCH$_2$Cl/Me$_3$Al system have been investigated. By increasing the polarity of the medium, i.e., CH$_2$Cl$_2$ concentration in a mixed CH$_2$Cl$_2$/n-C$_6$H$_{14}$ system, conversions and $\overline{M}_n$ increase. Decreasing the temperature from $-20°$ to $-90°$ increases molecular weights and conversions most likely because of "freezing out" termination by methylation. Molecular weight distribution significantly narrows (depending on the polarity of the system) from >2.0 to ~1.6 by reducing the temperature from $-20°$ to $-80°$. Increasing monomer concentrations increases the molecular weight. However, due to complications arising in the interpretation of linear Mayo plots, $k_{tr,M}/k_p$ and $k_t/k_{tr,M}$ cannot be obtained. Increasing the initiator concentration increases the yield, but leaves the molecular weights practically unchanged. Evidently, HSi(CH$_3$)$_2$CH$_2$CH$_2$φCH$_2$Cl is not a chain transfer agent in αMeSt polymerization. By carefully balancing these variables, HSi-PαMeSt's of desirable molecular weights and molecular weight distributions can be prepared at convenient rates.

Head-group characterization by quantitative IR spectroscopy indicated 1.0 ± 0.1 Si–H bond per polymer. This key data is evidence for the correctness of the proposition that PαMeSt carrying a Si–H head-group can be obtained by the use of HSi(CH$_3$)$_2$CH$_2$CH$_2$φCH$_2$Cl/Me$_3$Al initiating system.

4. Isobutylene Polymerization with Si–H Containing Initiator/Me$_3$Al Initiating System

a) Introduction

This section concerns the synthesis of polyisobutylenes (PIB) bearing a Si–H head-group (HSi-PIB) by the use of Si–H containing functional initiator in conjunction with Me$_3$Al coinitiator. First the effect of reaction conditions on the rate and molecular weight have been investigated and subsequently a H^1 NMR method for the quantitative characterization of Si–H groups in HSi-PIB was developed.

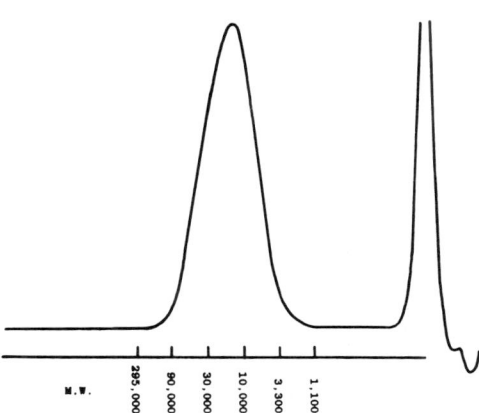

Fig. 26. GPC trace of HSi-PαMeSt (See Table 9 for polymerization condition)

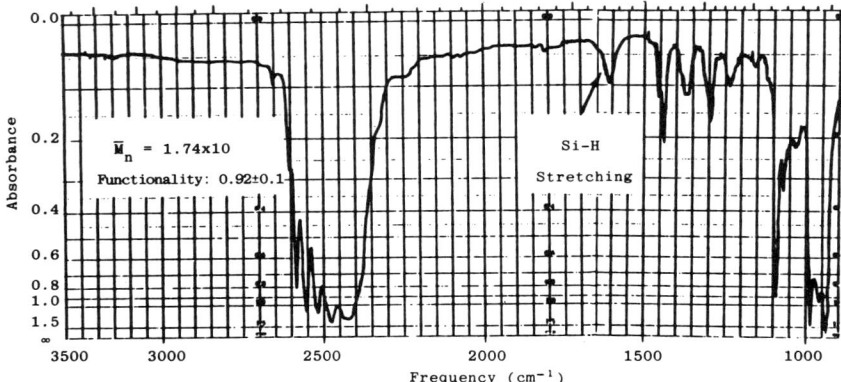

Fig. 27. IR spectrum of HSi-PαMeSt (See Table 9 for polymerization condition)

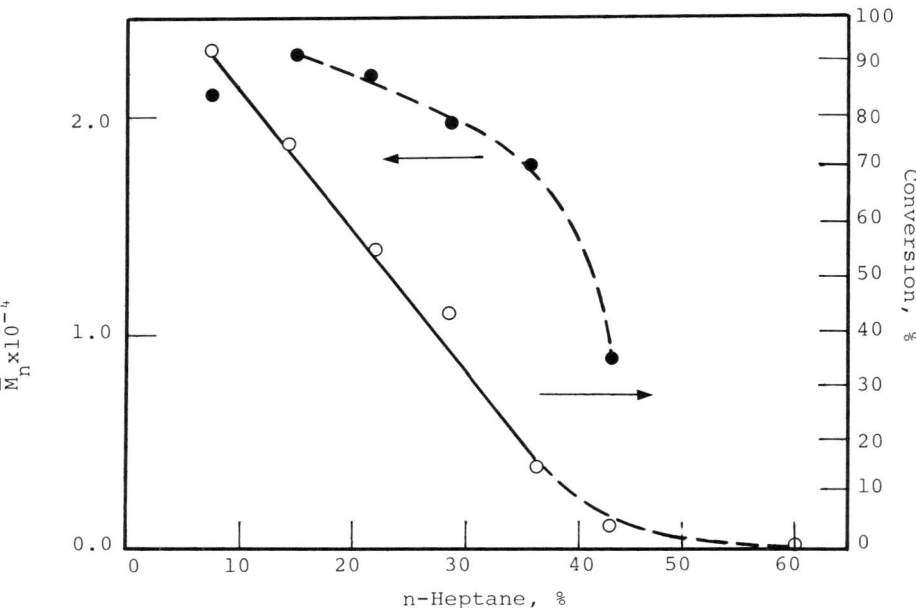

Fig. 28. The effect of solvent composition on molecular weight (●) and conversion (○) of HSi-PIB prepared using $HSi(CH_3)_2CH_2CH_2\varphi CH_2Cl/Me_3Al$ initiating system (See Table 10 for reaction conditions)

b) Effect of Reaction Conditions on the Polymerization of Isobutylene

i. *Effect of Medium Polarity.* The effect of medium polarity on the molecular weight of HSi-PIB prepared by the $HSi(CH_3)_2CH_2CH_2\varphi CH_2Cl/Me_3Al$ initiating system has been investigated by changing the composition of the $CH_2Cl_2/n\text{-}C_6H_{14}$ solvent system. Experimental conditions and results are shown in Table 10 and Fig. 28.

Table 10. The effect of solvent polarity on the polymerization of isobutylene by the HSi(CH$_3$)$_2$CH$_2$CH$_2$φCH$_2$Cl/Me$_3$Al initiating system

CH$_2$Cl$_2$ %	n-C$_7$H$_{16}$ %	Yield g	Conv. %	$\overline{M}_n \times 10^{-4}$	$\overline{M}_w \times 10^{-4}$	MWD[a]	$\overline{P}$[b] × 10^6
92.7	7.3	1.57	93.1	2.13	3.68	1.73	73.6
85.5	14.4	1.28	76.0	2.29	3.65	1.60	55.9
78.2	21.8	0.94	56.0	2.21	3.48	1.57	42.7
70.9	29.1	0.41	24.2	1.99	3.10	1.56	20.5
63.6	36.4	0.12	7.2	1.80	2.79	1.55	6.8
56.4	43.6	0.05	3.4	–	–	–	–
40.0	60.0	0.0	–	–	–	–	–
20.0	80.0	0.0	–	–	–	–	–

[i-C$_4$H$_8$] = 1.0 M, [HSi(CH$_3$)$_2$CH$_2$CH$_2$φCH$_2$Cl] = 2.5 × 10^{-3} M, [Me$_3$Al] = 5.0 × 10^{-2} M, Total Volume = 30 ml, −50 °C, 15 min
[a] By GPC
[b] $\overline{P}$ = Moles of polymer chains = yield $\overline{M}_n$

In general, these results and those obtained in a similar set of experiments with αMeSt are similar, i.e., conversions, molecular weights, and the number of polymer molecules decrease in the presence of increasing amount of n-heptane cosolvent. Conceivably, the rate of termination by methylation increases with decreasing solvent polarity[26–29, 36, 49], i.e., solvation. The effect becomes particularly noticeable at CH$_2$Cl$_2$/n-C$_7$H$_{16}$ ≃ 60/40. The significance of molecular weight data obtained at CH$_2$Cl$_2$/n-C$_7$H$_{16}$ = 92.7/7.3 is questionable in view of the very high (93.1%) conversion obtained.

While in general the effect of medium polarity on isobutylene and α-methylstyrene polymerizations is similar, a closer examination of the data reveals some subtle but important differences. Thus conversion of isobutylene is generally much higher than that of α-methylstyrene under the same conditions. This difference is most likely due to the balance of relative ion stabilities and monomer reactivities involved. According to principles of physical organic chemistry[58] the relative stabilities of propagating αMeSt cation, initiating benzyl cation, and propagating isobutylene carbocation are as follows:

$$\phi-\underset{\underset{CH_3}{|}}{\overset{\overset{CH_3}{|}}{C}}{}^{\oplus} \; > \; \phi CH_2^{\oplus} \; \simeq \; CH_3-\underset{\underset{CH_3}{|}}{\overset{\overset{CH_3}{|}}{C}}{}^{\oplus}$$

Thus with αMeSt, the kinetic chain is relatively short, monomer is consumed mainly by initiation and propagation, and chain transfer by the HSi(CH$_3$)$_2$CH$_2$CH$_2$φCH$_2$Cl initiator is unfavorable (see Sect. III.B.3.b.i.). In contrast, with isobutylene the kinetic chain may live longer because it is sustained by thermodynamically favorable chain transfer by the initiator. Scheme 5 illustrates the mechanism of isobutylene polymerization by the HSi(CH$_3$)$_2$CH$_2$CH$_2$φCH$_2$Cl/Me$_3$Al system. The kinetic chain is sustained by chain transfer loops shown on the left margin of the Scheme.

ii. *Effect of Temperature.* The effect of temperature on the isobutylene polymerization has been investigated at two polarity levels. Results are summarized in Table 11 and illustrated in Fig. 29.

Scheme 5. Mechanism of isobutylene polymerization with the H(CH$_3$)$_2$SiCH$_2$CH$_2$$\varphi$CH$_2Cl/Me_3$Al initiating system

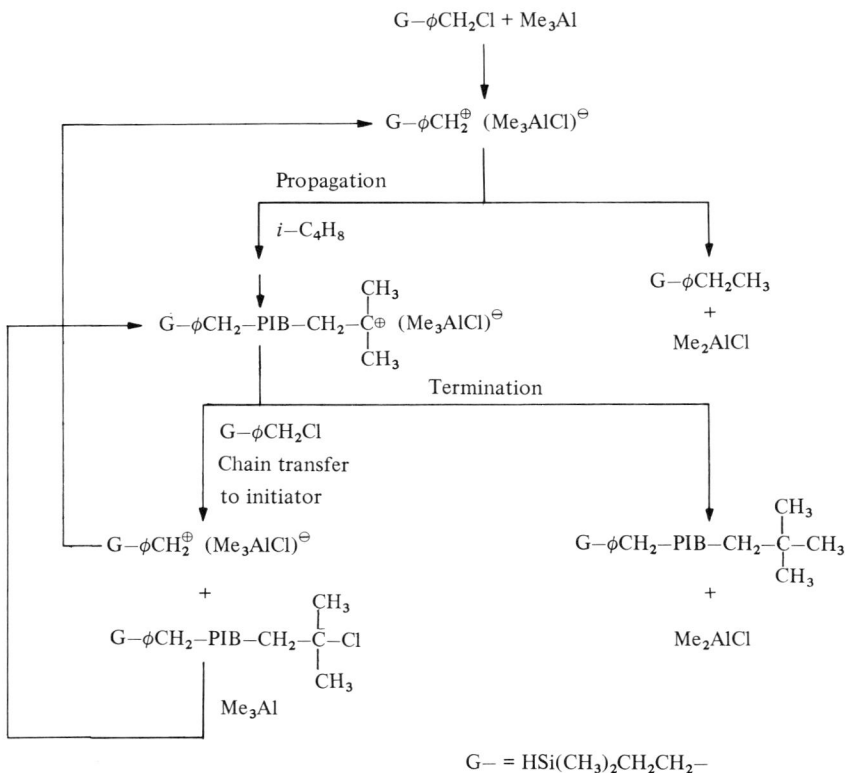

G— = HSi(CH$_3$)$_2$CH$_2$CH$_2$—

The effect of temperature on the molecular weight of PIB exhibits the usual trend[36–38]: decreasing the temperature results in increased molecular weights. Thus $\Delta H_{\overline{DP}_n} = -1.6_7 \pm 0.5$ and $-1.8_4 \pm 0.5$ kcal/mol in CH$_2$Cl$_2$/n-C$_6$H$_{14}$ = 85/15 and 65/35, respectively. However, the difference in $\Delta H_{\overline{DP}_n}$ values obtained at different polarity levels is deemed to be insignificant and is considered to be within experimental error. It is noteworthy that the effect of temperature on $\overline{M}_n$ of PIB is much less than that observed with PαMeSt.

Temperature coefficients of molecular weights, i.e., $\Delta H_{\overline{DP}_n}$ values, are sensitive indicators as to the mechanism of chain breaking in isobutylene polymerization. Thus Kennedy and Trivedi[49] found that $\Delta H_{\overline{DP}_n} = -1.8 \pm 1.0$ kcal/mol indicates a polymerization system in which molecular weight is controlled by termination. Similarly, $\Delta H_{\overline{DP}_n} = -6.6 \pm 1.0$ kcal/mol indicates molecular weight is controlled by chain transfer. The fact that $\Delta H_{\overline{DP}_n} \simeq -1.7 \pm 0.5$ kcal/mol in polyisobutylenes obtained with the HSi(CH$_3$)$_2$CH$_2$CH$_2$$\varphi$CH$_2Cl/Me_3$Al system is strong evidence for molecular weight control by termination, i.e., for polymerization without chain transfer to monomer. The proposition is further substantiated by results of model experiments of Kennedy et al.[26] and H^1 NMR analysis of HSi-PIB to be discussed in Sect. III.B.4.c.

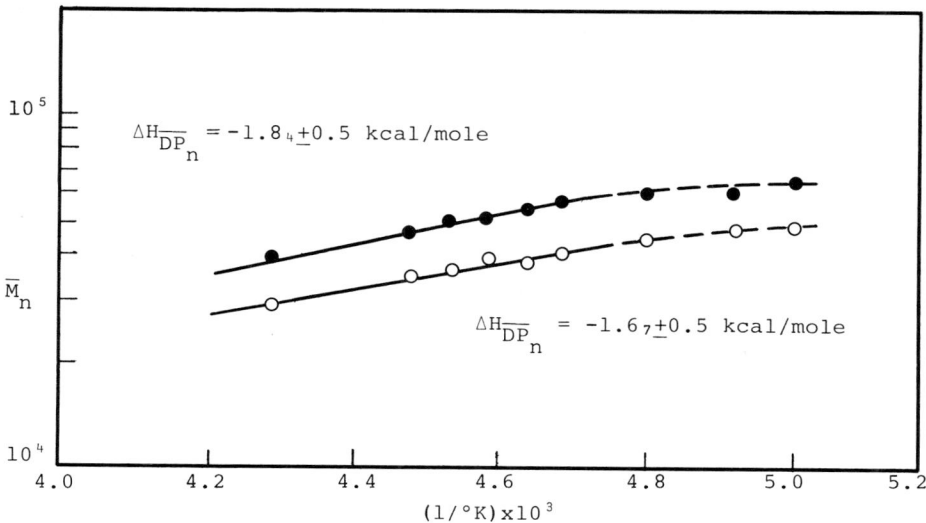

Fig. 29. The effect of temperature on the molecular weight of PIB prepared using the HSi(CH$_3$)$_2$CH$_2$CH$_2$$\varphi$CH$_2Cl/Me_3$Al initiating system with solvent composition: CH$_2$Cl$_2$/n-C$_6$H$_4$ = 85/15 (●) and 65/35 (○) (See Table 11 for reaction conditions)

Table 11. The effect of temperature on the polymerization of isobutylene using the HSi(CH$_3$)$_2$CH$_2$CH$_2$$\varphi$CH$_2Cl/Me_3$Al initiating system

Temp. °C	Yield g	Conv. %	$\overline{M}_n^a \times 10^{-4}$	$\overline{M}_w^a \times 10^{-4}$	MWD^a	$\overline{P}^b$ $\times 10^6$
Solvent composition: CH$_2$Cl$_2$/n-C$_6$H$_{14}$ = 65/35; Time = 10 min						
−40	0.146	8.6	2.88	4.76	1.64	5.1
−50	0.132	7.8	3.43	5.21	1.52	3.9
−55	0.088	5.2	3.97	6.32	1.59	2.2
−60	0.054	3.2	3.70	5.86	1.58	1.4
−65	0.032	1.9	4.43	6.45	1.45	0.7
−75	0.041	2.4	4.47	6.94	1.47	0.8
Solvent composition: CH$_2$Cl$_2$/n-C$_6$H$_{14}$ = 85/15; Time = 10 min						
−40	1.140	67.7	4.01	6.22	1.48	28.4
−50	1.279	76.0	4.54	7.01	1.55	28.2
−55	1.005	59.7	5.08	7.61	1.50	19.8
−65	0.756	44.9	5.78	8.27	1.50	19.8
−70	0.776	46.1	5.80	8.70	1.50	13.4
−75	0.769	45.7	6.22	9.24	1.49	12.4

[HSi(CH$_3$)$_2$CH$_2$CH$_2$$\varphi$CH$_2$Cl] = 2.5 × 10^{-3} M, Me$_3$Al = 5.0 × 10^{-2} M, [i-C$_4$H$_8$] = 1.0 M, Total Volume = 30 ml
^{a} By GPC
^{b} $\overline{P}$ = Moles of polymer chains = yield/$\overline{M}_n$

Table 12. The effect of monomer concentration on the polymerization of isobutylene using the $HSi(CH_3)_2CH_2CH_2\varphi CH_2Cl/Me_3Al$ initiating system

Temp. °C	$i\text{-}C_4H_8$ M	Conversion %	$\overline{M}_n{}^a \times 10^{-4}$	$1/\overline{DP}_n \times 10^3$
−40	0.5	1.28	0.67	8.40
−40	1.0	1.47	1.32	4.42
−40	1.5	2.00	3.44	3.44
−40	2.0	3.78	3.94	2.51
−50	0.5	1.32	1.42	6.76
−50	1.0	2.36	2.03	3.94
−50	1.5	2.91	2.59	2.80
−50	2.0	4.91	8.29	2.16

$[HSi(CH_3)_2CH_2CH_2\varphi CH_2Cl] = 2.5 \times 10^{-3}$ M, $[Me_3Al] = 5 \times 10^{-2}$ M, CH_2Cl_2 Solvent, $n\text{-}C_6H_{14}$ Cosolvent, Total Volume = 30 ml, 5 min
[a] By GPC

Decreasing the temperature below −50° tends to decrease conversions. The effect is particularly noticeable under more polar ($CH_2Cl_2/n\text{-}C_6H_{14} = 85/15$) conditions. It is of interest that this effect is just the opposite to that observed with αMeSt (see Sect. III.B.3.b.i.). Thus although in both cases $\overline{P}$ decreased and $\overline{M}_n$ increased with decreasing temperature, however, with αMeSt a larger temperature effect on $\overline{M}_n$ ($\Delta H_{\overline{DP}_n} \sim -4.0$ kcal/mol) overcompensated the decreasing of $\overline{P}$ and thus increased conversion with decreasing temperature. In contrast, with isobutylene the smaller temperature effect on $\overline{M}_n$ ($\Delta H_{\overline{DP}_n} \sim -1.7$ kcal/mol) was not enough to increase the conversion with decreasing temperature as $\overline{P}$ decreased.

iii. *Effect of Monomer Concentration.* The effect of monomer concentration on isobutylene polymerization with the $HSi(CH_3)_2CH_2CH_2\varphi CH_2Cl/Me_3Al$ initiating system has been investigated at −40° and −50°. n-Hexane was used as cosolvent to maintain medium polarity (see Experimental Sect. II.B.1.). Experimental conditions and results are summarized in Table 12 and Fig. 52.

As discussed previously in Sect. III.B.2.b.i., the carbocationation of the benzyl chloride initiating moiety in conjunction with Me_3Al coinitiator was not efficient[25] and the initiator may be consumed by initiator destruction via methylation and chain transfer[36, 59]; thus the small intercepts in Fig. 30 may not be attributed to chain transfer to monomer but rather to the complication in initiation and propagation[52, 3] that limits the validity of the the conventional Mayo plot. Further evidence (slope of the log $\overline{M}_n$ versus $1/T$ plot, direct quantitative head-group analysis by H^1 NMR) also corooborates this conclusion.

In spite of the failure of Mayo analysis, this work has shown that PIB molecular weights can be efficiently controlled by monomer concentration, in addition to temperature and solvent polarity.

iv. *Effect of Initiator Concentration.* The effect of initiator concentration on the molecular weight of PIB has been investigated. Results are summarized in Table 13 and Fig. 31.

Evidently a linear inverse relationship exists between molecular weights and initiator concentration. This effect is certainly due to efficient chain transfer to the initiator:

∼∼∼C(C)(C)C⊕ + HSi(CH₃)₂CH₂CH₂φCH₂Cl ⟶

∼∼∼C(C)(C)−Cl + HSi(CH₃)₂CH₂CH₂φCH₂⊕

Table 13. The effect of initiator concentration on the polymerization of isobutylene using the HSi(CH₃)₂CH₂CH₂φCH₂Cl/Me₃Al initiating system

[HSi(CH₃)₂CH₂CH₂φCH₂Cl] M × 10³	[Me₃Al] M × 10²	Conv. %	$\overline{M}_n^a$ × 10⁻⁴
2.5	5.0	13.4	2.06
5.0	10.0	15.7	1.09
7.5	15.0	8.0	0.73
10.0	20.0	11.2	0.53

[i-C₄H₈] = 1.0 M, Solvent: CH₂Cl₂/n-C₆H₁₄ = 67/33, −50°, 10 min
[a] By GPC

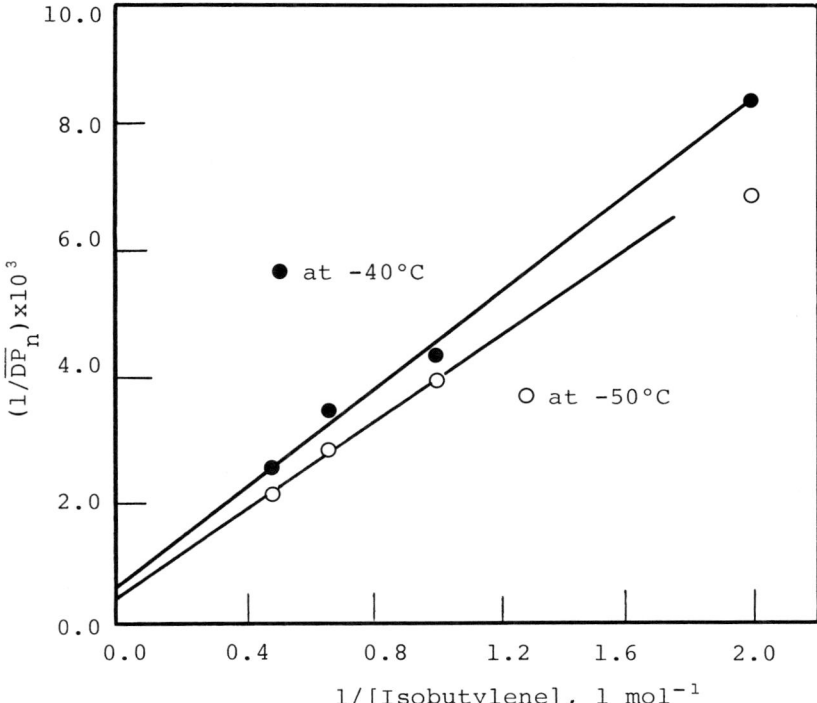

Fig. 30. "Mayo Plots" of PIB prepared using the HSi(CH₃)₂CH₂CH₂φCH₂Cl/Me₃Al initiating system (See Table 12 for reaction conditions)

As shown by the data in Fig. 31, the chain transfer constant of this initiator, $C_I \simeq 1.0$. In this context it is of interest to remember that the effect of initiator concentration on the molecular weight of HSi-PαMeSt was negligible, probably because of unfavorable thermodynamics (Sect. III.B.3.b.iv.). In contrast, with isobutylene chain transfer from the propagating carbenium ion to initiator is thermodynamically favorable (see Sect. III.B.4.b.i.). Thus it is not surprising to find a large C_I. The chain transfer mechanism has been illustrated in Scheme 5.

c) Head-Group Characterization

Conclusive proof as to the survival of Si–H groups in HSi-PIB after cationic polymerization was obtained by H^1 NMR analysis. To this end sufficiently low molecular weight samples were prepared ($\overline{M}_n \simeq 600$) and examined by H^1 NMR spectroscopy. A representative spectrum is shown in Fig. 32.

Resonances associated with the aromatic protons and the Si–$\underline{H}$ proton in the HSi(CH$_3$)$_2$CH$_2$CH$_2$φCH$_2$– head-group, respectively, appear at 6.5–7.0 and 3.5–4.0 ppm.

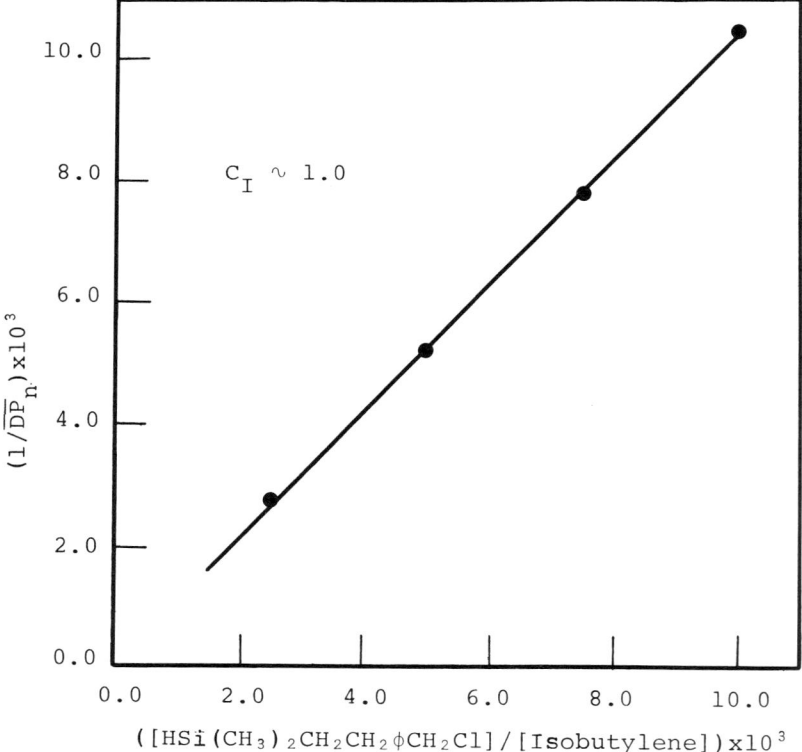

Fig. 31. The effect of initiator concentration on molecular weight of PIB prepared using the HSi(CH$_3$)$_2$CH$_2$CH$_2$φCH$_2$Cl/Me$_3$Cl initiating system (See Table 13 for reaction conditions)

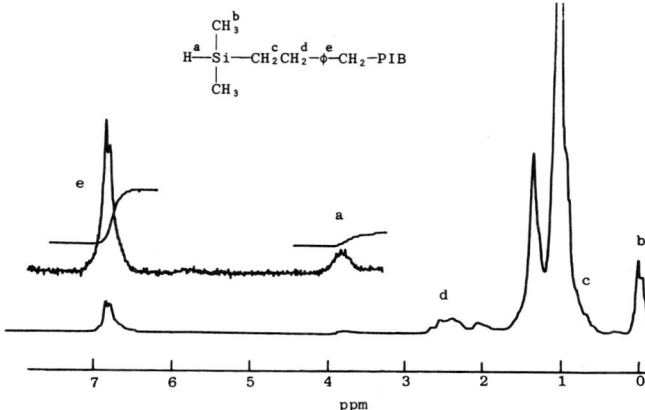

Fig. 32. H^1 NMR spectrum of HSi-PIB prepared at [i-C$_4$H$_8$] = 0.4 M, HSi(CH$_3$)$_2$CH$_2$CH$_2$$\varphi$CH$_2$Cl = 1.0 × 10^{-1} M, [Me$_3$Al] = 1.0 × 10^{-3} M, CH$_2$Cl$_2$/n-C$_6$H$_{14}$ = 64/36, Total Volume = 300 ml, −50°, 30 min

Significantly, integration of aromatic protons/Si–H head-group per polymer and demonstrates survival of this bond during polymerization.

Absence of resonance in the 4.5–5.2 ppm region indicates absence of terminal unsaturation which would arise upon chain transfer to monomer[59–61]. Further, absence of resonances at 1.63 and 1.90 ppm indicates absence of –CH$_2$C(CH$_3$)$_2$Cl terminus[62]. Evidently if there is termination by chlorination in addition to termination by methylation[26, 27], the *tert.*-chloride end-group would have been consumed by methylation by the large excess of Me$_3$Al in the reaction mixture (see Scheme 5).

d) Model Derivatization Experiments

It has been illustrated in Scheme 2 that the Si–H head-group is a potentially useful functional group for block copolymer synthesis or for being derivatized to other useful functional groups. Thus [2-(p-methylphenyl)ethyl]dimethylsilane (7) was employed as a model for the head-group of polymers obtained by HSi(CH$_3$)$_2$CH$_2$CH$_2$$\varphi$CH$_2Cl/Me_3$Al initiating system, and model experiments have been carried out to explore the feasibility of the concept in Scheme II.

i. *Coupling with 2,4,4-Trimethyl-1-pentene.* It was theorized that the desirable triblock copolymer PαMeSt-*b*-PIB-*b*-PαMeSt could be prepared by hydrosilylating α,ω-di(isobutenyl)polyisobutylene[63] *(18)* with 2 moles of HSi-PαMeSt. Thus, a model coupling experiment has been carried out using 2,4,4-trimethyl-1-pentene and 7 as the models for *18* and terminal functional-groups of HSi-PαMeSt, respectively. However, in spite of several attempts (see Experimental Sect. II.D.4.) in which various experimental procedures have been tried (i.e., use of *n*-hexane, THF, *n*-heptane, benzene) only ∼10% of coupling was obtained. Conceivably steric compression reduces the yield.

ii. *Derivatizing with Allyl Glycidyl Ether.* The derivatizability of Si–H head-group to expoxy group by hydrosilylation has been examined. Thus excess allyl glycidyl ether was hydrosilylated with 7 (see Experimental Sect. II.D.5) and the crude product was analyzed by glpc. A quantitative yield was obtained, since the glpc trace of 7 vanished. On the basis of this evidence, it is expected that by the use of the same procedure Si–H head-groups of PαMeSt or PIB (prepared by the $HSi(CH_3)_2CH_2CH_2\varphi CH_2Cl/Me_3Al$ initiating system) could also be derivatized to epoxy head-groups.

e) Conclusions

Isobutylene polymerization with the $HSi(CH_3)_2CH_2CH_2\varphi CH_2Cl/Me_3Al$ initiating system has been investigated under a variety of conditions. Increasing medium polarity (by increasing the relative concentration of CH_2Cl_2 in a $CH_2Cl_2/n\text{-}C_7H_{16}$ solvent system) increases conversions and molecular weights. However, molecular weight distributions of PIB are not as sensitive toward changes in medium polarity as that of PαMeSt. As anticipated, HSi-PIB molecular weights increase with decreasing temperature and $\Delta H_{\overline{DP_n}} \simeq -1.7 \pm 0.5$ kcal/mol irrespective of the solvent compositions used. This data strongly suggests that the system under investigation is chain transfer to monomer absent. In contrast to observations made with αMeSt, isobutylene conversions tend to decrease with decreasing temperature. Increasing isobutylene concentration increases molecular weights, however, conventional Mayo analysis cannot be applied to calculate rate constant ratios. A straight inverse relation exists between initiator concentration and HSi-PIB molecular weights, $C_I \sim 1.0$.

Head-group analysis by H^1 NMR demonstrated the presence of one Si–H group per PIB molecule. These data are self-reinforcing and prove that HSi-PIB can be prepared by the use of the $HSi(CH_3)_2CH_2CH_2\varphi CH_2Cl/Me_3Al$ initiating system.

Model Si–H head-group derivatization demonstrated the potential usefulness of this group. While the Si–H group can be easily hydrosilylated with high yield[24, 35] steric hindrance may reduce the yield.

Acknowledgement. Financial support by the National Science Foundation (Grants DMR-7727618 and INT-7919194) is gratefully acknowledged.

IV. References

1. Madec, P.-J., Marechal, E.: J. Polym. Sci., Polym. Chem. Ed. *16*, 3165 (1978)
2. Wiberg, E., Amberger,E.: "Hydrides of the Elements of Main Groups I-IV", Elsevier Publishing, 1971, p. 513
3. Plueddmann, E. P., Fanger, G.: J. Am. Chem. Soc. *81*, 2632 (1959)
4. Midland Silicones Ltd., Brit. 834, 326 (July 10, 1957)
5. Union Carbide Corp., Brit. 914, 282 (Dec. 12, 1958)
6. General Electric Co., Brit. 935, 534 (Dec. 7, 1959)
7. General Electric Co., Brit. 994, 396 (May 20, 1963)
8. Greber, G., Balciunas, A.: Makromol. Chem. *69*, 193 (1963)
9. Greber, G., Balciunas, A.: Makromol. Chem. *79*, 149 (1964)
10. Greber, G.: Makromol. Chem. *101*, 104 (1967)
11. Greber, G., Reese, E., Balciunas, A.: Farbe u. Lack *4*, 249 (1964)
12. I.C.I. Ltd., Brit. 1, 104, 153 (Dec. 14, 1965)

13. Gobran, R. H.: "Addition Polymers with Reactive Terminals" in "Chemical Reactions of Polymers", Fettes, E. M. (Ed.), Interscience, New York, 1964, p. 295
14. French, D. M.: Rubber Chem. Technol. *42*, 71 (1969)
15. Heitz, W. et al.: "Preparation of Oligomers with Functional Endgroups by Polymerization Reactions" in "Macromolecular Chemistry-8" (Intern. Symp. Macromol., Helsinki, 1972), Saarela, K. (Ed.), Butterworths, London, 1973, p. 65
16. Cameron, G. G., Qureshi, Y.: IUPAC Intern. Symp. on Macromolecules, Florence, Italy, Preprints, Vol. 2, 193 (1980)
17. Vaughn, H. A.: Polym. Lett. *7*, 569 (1969)
18. Plumb, J. B., Atherton, J. H.: "Copolymers Containing Polysiloxane Blocks" in "Block Copolymers", Allport, D. C., Janes, W. H. (Ed.), John Wiley, New York, 1973, p. 305
19. Uraneck, C. A., Hsieh, H. L., Buck, O. G.: J. Polym. Sci. *46*, 535 (1960)
20. Morton, M., Rubio, D. C.: Plastic and Rubber: Materials and Applications *3*, 139 (1978)
21. Zapp, R. L., Serniuk, G. E., Minckler, L. S.: Rubber Chem. Technol. *43*, 1154 (1970)
22. Lin, S. C., Pearce, E. M.: J. Polym. Sci., Polym. Chem. Ed. *17*, 3095 (1979)
23. Union Carbide Corp.: Brit. 892, 136 (Sept. 25, 1958)
24. Noll, W.: Chemistry and Technology of Silicones, Academic Press, New York, 1968
25. Reibel, L. C., Kennedy, J. P., Chung, Y. L.: J. Org. Chem. *42*, 690 (1977)
26. Kennedy, J. P., Desai, N. V., Sivaram, S.: J. Am. Chem. Soc. *95*, 6386 (1973)
27. Sivaram, S.: J. Organometal. Chem. *156*, 55 (1978)
28. Ambrose, R. J., Newell, J. J.: J. Polym. Sci., Polym. Chem. Ed.*17*, 2129 (1979)
29. Kennedy, J. P., Trivedi, P. D.: Adv. Polymer Sci. *28*, 83 (1978)
30. Smith, R. A.: Ph. D. Dissertation, University of Akron, 1981
31. Union Carbide: U.S. Pat. 3, 925, 434 (Dec. 9, 1975)
32. Kennedy, J. P., Feinberg, S. C., Huang, S. Y.: J. Polym. Sci., Polym. Chem. Ed. *17*, 243 (1979)
33. Kennedy, J. P., Smith, R. A.: ibid. *18*, 1539 (1980)
34. Mole, T., Jeffery, E. A.: Organoaluminium Compounds, Elsevier Publishing, New York, 1972, p. 382
35. Lukevics, E. et al.: "Hydrosilylation. Recent Achievements" in J. Organometal. Chem. Library *5*, Seyferth, D. (Ed.), Elsevier Publishing, 1977, p. 1
36. Kennedy, J. P., Gillham, J. K.: Adv. Polymer Sci. *10*, 1 (1972)
37. Pepper, D. C., Reilly, R. J.: J. Polymer Sci. *58*, 639 (1962)
38. Kennedy, J. P., Squires, R. G.: Polymer *6*, 579 (1962)
39. Brown, C. P., Mathieson, A. R.: J. Chem. Soc. 3445 (1958)
40. Borisov, S. N., Voronkov, M. G., Lukevits, E. Ya.: "Organosilicon Heteropolymers and Heterocompounds", Plenum, New York, 1970, p. 314
41. Minoura, Y., Toshima, H.: J. Polym. Sci. A-1 *11*, 1109 (1972)
42. Chojnowski, J., Wilczek, L., Fortuniak, W.: J. Organometal. Chem. *153*, 13 (1977)
43. Morton, M. et al.: J. Polym. Sci. *57*, 471 (1962)
44. Hadjichristidis, H., Guyot, A., Fetters, L. J.: Macromol. *11*, 668 (1978)
45. Van Dyke, C. H.: "Synthesis and Properties of the Silicon-Halogen and Silicon-Halogenoid Bond" in Organometallic Compounds of the Group IV Elements, MacDiarmid, A. G. (Ed.), Vol. 2, Part I, Marcel Dekker, New York, 1972, p. 184
46. Smith, A. L.: Analysis of Silicones, John Wiley, New York, 1974, p. 247
47. Sommer, L. H.: Intra-Science Chem-Rept. *7 (4)* 1 (1973)
48. MacDiarmid, A. G.: ibid. *7 (4)*, 83 (1973)
49. Kennedy, J. P., Trivedi, P. D.: Adv. Polymer Sci. *28*, 113 (1978)
50. Kennedy, J. P., Chou, R. T.: Polymer Prepr. *20 (2)*, 306 (1978)
51. Kennedy, J. P., Chou, R. T.: ibid. *21 (2)*, 149 (1980)
52. Mandal, B. M., Kennedy, J. P.: J. Polym. Sci., Polym. Chem. Ed. *16*, 833 (1978)
53. Mandal, B. M.: Polym. Bull. *2*, 625 (1980)
54. Plesch, P. H.: "Cationic Polymerization" in "Progress in High Polymers", Vol. 2, Robb, J. C. and Peaker, W. H. (Eds.), Heywood Books, London, 1968, p. 137
55. Gregg, R. A., Mayo, F. R.: J. Am. Chem. Soc. *70*, 2373 (1948)
56. Cesca, S. et al.: Makromol., Chem. *178*, 2223 (1977)
57. Mathieson, A. R.: in The Chemistry of Cationic Polymerization, Plesch, P. H. (Ed.) MacMillan, New York, 1963, p. 279

58. Liberles, A.: Introduction to Theoretical Organic Chemistry, MacMillan, New York, 1968, p. 208
59. Kennedy, J. P., Rengachary, S.: Adv. Polymer Sci. *14*, 1 (1974)
60. Manatt, S. L., Ingham, J. D., Miller Jr., J. A.: Org. Mag. Res. *10*, 198 (1977)
61. Puskas, J., Banas, E. M., Nerheim, A. G.: J. Polymer Sci. Symp. No. 56, 191 (1976)
62. Kennedy, J. P., Huang, S. Y., Feinberg, S. C.: J. Polym. Sci. Polym. Chem. Ed., *15*, 2801 (1977)
63. Kennedy, J. P. et al.: Polym. Bull. *1*, 575 (1979)

Kinetics and Mechanisms of Polyesterifications
I. Reactions of Diols with Diacids

Alain Fradet and Ernest Maréchal

Laboratoire de Synthèse Macromoléculaire, Université Pierre et Marie Curie, 12, Rue Cuvier, F-75005 Paris

In this article we critically review most of the literature concerning non-catalyzed, proton-catalyzed and metal-catalyzed polyesterifications. Kinetic data relate both to model esterifications and polyesterifications. Using our own results we analyze the experimental studies, kinetic results and mechanisms which have been reported until now. In the case of $Ti(OBu)_4$-catalyzed reactions we show that most results were obtained under experimental conditions which modify the nature of the catalyst. In fact, the true nature of active sites in the case of metal catalysts remains largely unknown.

1	Introduction	53
2	Main Techniques Used in Kinetic Studies of Polyesterifications	55
	2.1 Procedures and Apparatus	55
	2.1.1 Single Run	55
	2.1.2 Separate Runs	55
	2.1.3 Analytical Methods	56
	2.1.3.1 Discontinuous Measurements	56
	2.1.3.2 Continuous Monitoring	57
	2.2 Studies of Side Reactions	57
	2.2.1 Loss of Volatile Reactants	58
	2.2.2 Degradation Reactions	58
	2.2.3 Reverse Reactions	58
	2.3 Treatment of Experimental Data	58
	2.4 Discrimination Between Possible Values of Reaction Orders	59
	2.5 Determination of Activation Parameters	60
3	Has the Elimination of Water to be Taken into Account in Polyesterification Kinetics?	60
	3.1 The General Kinetic Equation – Case of Stoichiometry	60
	3.2 Flory's Statement	61
	3.3 Corrected Equation by Szabo-Rethy	62
	3.4 Other Approximations	63
	3.5 Erroneous Use of Corrections	64
	3.6 Case of Non-Stoichiometric Balance of the Reactants	64
	3.7 Conclusion	65

4	**Inventory of the Main Catalysts Used in Esterifications and Polyesterifications**	65
5	**The Concept of Equal Reactivity of Functional Groups**	70
6	**Kinetic Relations and Mechanisms**	71
	6.1 Non-Catalyzed and Acid-Catalyzed Reactions	71
	6.1.1 Studies at Low Temperatures Without Removal of Water	71
	6.1.2 Studies at High Temperatures (above 100 °C) Without a Catalyst	74
	6.1.2.1 Esterifications and Polyesterifications in Dilute Solutions	75
	6.1.2.2 Esterifications and Polyesterifications in Concentrated Media	77
	6.1.3 Studies at High Temperatures (above 100 °C) in the Presence of Protonic Catalysts	83
	6.1.4 Activation Parameters of Non-catalyzed and Protonic Acid-catalyzed Esterifications and Polyesterifications	83
	6.1.4.1 Non-catalyzed Reactions	83
	6.1.4.2. Strongly Acid-Catalyzed Esterifications and Polyesterifications	84
	6.2 Titanium Tetralkoxide-Catalyzed Reactions	84
	6.2.1 Main Properties of Alkoxytitaniums	85
	6.2.2 Catalytical Behaviour of Alkoxytitaniums	87
	6.3 Other Catalyzed Reactions	89
7	**Concluding Remarks**	92
8	**References**	92

1 Introduction

Step-growth polymerizations have widely been developed in industrial applications whereas knowledge of their mechanisms and of their kinetics has remained far below that of chain polymerization reactions.

In his classic work on polyesterification Flory[1,2] established many of the basic principles of step-growth polymerization:
- Equal reactivity of functional groups irrespective of the size of the molecule to which the group is attached.
- Prediction of molecular weight distribution
- Calculation of gel points in non-linear system.

However, all the studies have been carried out under the following restrictive conditions:
- Stoichiometric balance of the reactants,
- Consideration of only the later stages of the reaction.

Since Flory's works many other articles have been published on this subject and the two most important assumptions that have been made in interpreting the experimental results on both esterification and polyesterification are:
- Only the later stages of the reaction (say, above 8%) are worth considering.
- The kinetic study can be performed over the whole course of the reaction.

Depending on the method of study, the kinetic results (orders, ...) can be very different as shown in Table 1.

Table 1. Kinetic orders of various non-catalyzed polyesterifications of adipic acid with aliphatic primary diols. The first figure in the 3rd column is the overall order and figures in brackets denote orders with respect to acid and alcohol. 2 + 3 means that kinetics has been treated as resulting from the superposition of two reactions with orders 2 and 3, respectively. The range of conversion which has been studied is given in the 2nd column; for instance, 80–100 means that kinetics has been studied between 80 and 100% conversion

Alcohol	Conversion and experimental conditions	Orders	Ref.
Diethylene glycol	80–100; no solvent	3	1)
1,10-Decanediol	80–100; no solvent	3	1)
Diethylene glycol	0–100; no solvent[a]	2 then 3	4)
1,5-Pentanediol	0–100; solution	2 then 3	5)
1,2-Ethanediol	0–100; no solvent	2.5	7)
Diethylene glycol	0–100; no solvent	2.5	7)
Diethylene glycol	0–100; no solvent[a]	2.5	7)
1,10-Decanediol	0–100; no solvent[a]	2.5	7)
2,2-Dimethyl-1,3-propanediol	0–100; no solvent	2.5	8)
1,6-Hexanediol	0–100; no solvent	2.5	10)
1,10-Decanediol	80–100; no solvent	3	13)
Diethylene glycol	0–100; no solvent[a]	2 + 3	15)
1,2-Ethanediol	0–100; Diol excess	3 (1; 2) or 2 (2; 0)	270, 271)
Diethylene glycol	beginning of the reaction[a]	2.5	18)
1,2-Ethanediol	beginning of the reaction	6	19)

[a] Use of Flory's data[1]

It is not only in the field of kinetic relations that discrepancies exist. When the catalyst is a protonic acid and the reaction is carried out in dilute solution, the mechanisms describing the contribution of the catalyst are relatively well-known. But in most other cases and particularly when the catalyst is a metal derivative (see Chap. 4) none of the proposed mechanisms can be considered as definitive.

According to Solomon[3] the following equations describe the reactions which are usually utilized for the preparation of polyesters:

1. Direct esterification of an alcohol with a carboxylic acid or anhydride:

$$ROH + R_1COOH \rightleftarrows R_1COOR + H_2O$$

2. Ester interchange:
   (a) Alcoholysis:

$$ROH + R_1COOR_2 \rightleftarrows R_1COOR + R_2OH$$

   (b) Acidolysis:

$$RCOOH + R_1COOR_2 \rightleftarrows RCOOR_2 + R_1COOH$$

   (c) Double ester interchange:

$$R_1COOR_2 + R_3COOR_4 \rightleftarrows R_1COOR_4 + R_3COOR_2$$

3. Self-condensation of hydroxy acids:

$$n\,HO-R-\underset{O}{\overset{\|}{C}}-OH \rightleftarrows \left(O-R-\underset{O}{\overset{\|}{C}}\right)_n + n\,H_2O$$

4. Schotten-Baumann reaction:

$$R_1COCl + R_2OH \rightleftarrows R_1COOR_2 + HCl$$

5. Lactone polymerization

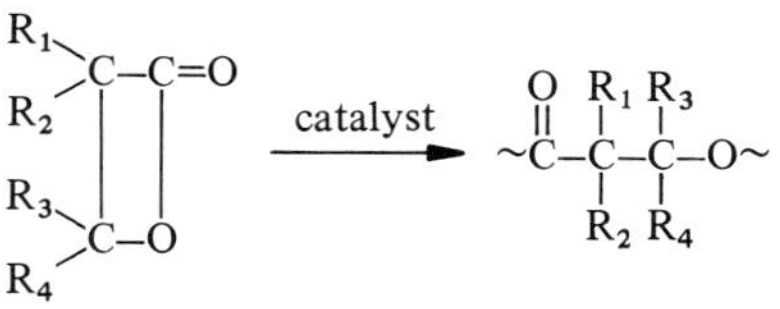

$$\begin{array}{c} R_1 \\ R_2 \end{array}\!\!\!\!\!>\!\!\! \underset{\underset{R_4}{\overset{R_3}{>}C-O}}{C-C=O} \xrightarrow{\text{catalyst}} \sim\!\!\underset{O}{\overset{O}{\overset{\|}{C}}}-\underset{R_2}{\overset{R_1}{C}}-\underset{R_4}{\overset{R_3}{C}}-O\!\sim$$

In addition, the reaction

$$R-\underset{\diagdown O \diagup}{CH-CH_2} + HOOC-R' \longrightarrow R-CHOH-CH_2OOC-R'$$

is being developed.

This review is devoted to the study of the kinetics and mechanisms of the direct esterification of carboxylic acids with alcohols. The results which are analyzed here concern esterifications and linear polyesterifications between small molecules or oligomers in solution or in bulk. We will examine the following points:
- Main techniques used in kinetic studies of polyesterifications, Chap. 2.
- The problem of the elimination of water, Chap. 3.
- Inventory of the main catalysts used in esterifications and polyesterifications, Chap. 4.
- The concept of equal reactivity of functional groups, Chap. 5.
- Kinetic relations and mechanisms, Chap. 6.

2 Main Techniques Used in Kinetic Studies of Polyesterifications

2.1 Procedures and Apparatus

2.1.1 Single Run

This method is the most widely used because it gives a good picture of batch reactions performed in industry. Reactions are carried out in a thermostated flask fitted with constant speed stirrer, inert gaz inlet, sampling device, thermometer, distillation column, and condenser.

At definite times samples are withdrawn and titrated by an appropriate method.

The temperature is controlled either with a constant temperature oil bath or a heating jacket and a P.I.D. regulation with a captor plunged in the reaction medium.

An inert atmosphere is required in all cases and a vacuum line is often used for the removal of water at the end of the polyesterification reaction.

The column must be efficient enough to separate completely the condensation water from the volatile reactants which could be distilled off at the same time. If these conditions are observed the various experimental parameters can be kept constant from the beginning to the end of the reaction.

In the case of a continuous monitoring a special device must generally be used. This apparatus is characteristic of a method which is used and is described further (continuous monitoring).

2.1.2 Separate Runs

Aliquots of the reaction mixture are placed in an inert atmosphere in sealed ampoules or flasks which are plunged in a thermostated oil bath. At different reaction times one ampoule is opened and its content is titrated.

Due to difficulty in carrying out all the experiments under exactly the same conditions, the experimental results show more discrepancies than in the case of a single run.

This technique is mainly used for monoesterifications carried out near room temperature. Moreover, reaction water is not removed; consequently, hydrolysis of ester, which is no longer negligible, must be taken into account.

2.1.3 Analytical Methods

The determination of reaction orders and of activation parameters requires accurate measurements of the concentrations of the different species present in the reaction medium.

If it can be accepted that the stoichiometry is known and that no side reactions take place, the concentration of only one of the products or of the reactants needs to be measured.

Two methods can be used:
discontinuous measurements on samples and continuous following of a physical parameter which, after standardization, can be related to a parameter such as concentration, degree of polymerization, or conversion.

2.1.3.1 Discontinuous Measurements

Determination of carboxy groups. Esterification or polyesterification kinetics is usually followed by this titration which is both easy and accurate. Each sample is dissolved in a solvent or a mixture of solvents ($CHCl_3$, C_6H_6/EtOH or MeOH, toluene/EtOH or MeOH...) and then titrated with alcoholic KOH. The end point is determined either with an indicator (in most cases phenolphthalein) or with a pH-meter. An accuracy of about 0.1–1% can generally be achieved.

Determination of hydroxy groups. Hydroxy groups are rarely titrated. Indeed, this titration is more complex and less accurate than carboxy-group determination. However, it is very helpful to know the hydroxy group content, for example in the case of reactions carried out with a great excess of acid. In this case, the variations of the acid concentration during the esterification are very small and are determined only with poor accuracy.

Various well established procedures are described in the literature[273, 274, 296]. The procedure generally used for polyesters is the phthalylation or acetylation of OH groups with phthalic or acetic anhydrides followed by back titration of excess carboxy groups. Moisture does not interfere but the acidity of the sample must be taken into account. The use of pyromellitic dianhydride is also reported[320].

In addition, some other methods are mentioned. These include:
– Reaction of hydroxy groups with phenyl isocyanate and back titration of the excess isocyanate with dibutylamine. Both moisture and acidity interfere[275].
– Reaction with 3,5-dinitrobenzoyl chloride and back titration with $(C_4H_9)_4N^+$, OH^- [328].
– Direct infrared measurement[276].
– ^{19}F-NMR measurement after acetylation of the hydroxy groups with trifluoroacetic anhydride[277] or after formation of hexafluoroacetone adducts[278].

^{19}F-NMR analysis not only permits an accurate quantitative determination of hydroxy groups but also provides valuable information on their nature and their location. This not the case with ^{1}H-NMR measurements which have been carried out on samples modified by reaction with $(CH_3)_3SiCl$[329] or with naphthalene isocyanate.

Determinations by enthalpimetry[280] and photometry[281] have been reported.

However, none of the methods determining hydroxy groups can be compared to those used for carboxy groups determination both with respect to accuracy and ready execution.

Determination of low molecular weight compounds. The kinetics of monoesterifications or of the first steps of polyesterification can be followed by chromatographic methods. The concentration of each of the species present in the medium can be determined, at least for low molecular weight compounds.

Gas-phase chromatography is mainly used for monoesterifications of low molecular weight compounds[202, 279, 283], high performance liquid chromatography for diesterifications[200], and size exclusion chromatography for more general cases[181, 284, 285].

Monoesters and diesters of terephthalic acid and 1,2-ethanediol can be determined separately by polarography[206, 257].

Determination of $\overline{M}_n$. The end group concentration can be determined from $\overline{M}_n$. However, $\overline{M}_n$ determinations permit only a control of the kinetic plots obtained by other methods. They are not sufficiently accurate to allow unambiguous determinations of reaction orders.

2.1.3.2 Continuous Monitoring

Infrared spectroscopy. Due to experimental difficulties, infrared spectroscopy is used infrequently in these kinetic studies. However, continuous measurements have been carried out by Schumann[286] in the study of the poly(ethylene terepthalate) synthesis.

Determination of water removal rate. This technique is usually applied to the continuous following of esterification kinetics.
- The volume of the water removed in esterifications through azeotropic distillation[287–289] and also in polyesterifications is determined[6, 19, 264, 292]. This method is difficult to carry out because it is necessary to quantitatively collect the water which is released and to determine its volume or its weight. It is important to estimate as accurately as possible the amount of water remaining in the distillation column.
- Another technique has been used by Gordon et al.[258, 267] who measured the pressure increase in the vessel where esterification was carried out under a preestablished vacuum.

Miscellaneous methods. Other techniques, although used much more rarely, can sometimes be helpful. Among these techniques dynamic thermochemical analysis[295], measurements of electric conductivity[297], of ultrasonic absorption[298], and of dielectric constant[299, 300] should be mentioned.

2.2 Studies of Side Reactions

Before studying esterification kinetics, it must be kept in mind that side reactions can interfere with the main reaction. They must either be avoided by changing experimental conditions or taken into account in kinetic calculations. Three types of phenomena can occur:

2.2.1 Loss of Volatile Reactants

Since high temperatures and a nitrogen atmosphere are necessary to obtain measurable rates of polyesterification and to remove the reaction water, a loss of volatile reactants can hardly be avoided, especially in early stages of polyesterification. In the last stages, the decrease of the concentration of the volatile reactants can be of the same order of magnitude as their concentration. Consequently, the ultimate points of the kinetic plot have possibly no significance.

Various precautions must be taken when studying polyesterification kinetics above 85% conversion.
– Saturation of the nitrogen stream with the most volatile reactant[301].
– Discontinuation of a run if volatile reactants are detected in the condenser.
– Use of high molecular weight monomers and of prepolycondensed monomers such as reactive oligomers[13, 228–230, 232].
– Performance of specific runs in order to check if the losses are significant.

An examination of the literature shows that in many studies these precautions have unfortunately not been taken.

2.2.2 Degradation Reactions

In all experiments an inert gas (N_2, Ar) is bubled through the reaction mixture both to avoid oxidation and to eliminate the reaction water. Most studies are carried out in the range 100–250 °C. In such experimental conditions, degradation reactions are low.

2.2.3 Reverse Reactions

During their synthesis esters and polyesters can be modified by the following side reactions: alcoholysis, acidolysis, ester interchange, hydrolysis.

The first three reactions do not interfere with the main reaction since they change neither the number nor the nature of the reactive end groups. On the other hand, it is important to keep in mind that hydrolysis leads to the formation of shorter chains and even to monomers.

At the temperatures at which polyesterifications are performed a stream of dry nitrogen or a vacuum line effectively removes the water.

If polycondensation is carried out at low temperature, removal of the liberated water is impossible. In this case, reverse hydrolysis must be taken into account unless equilibrium is shifted towards esterification by an excess of one of the reactants.

The release of condensation water causes many changes in the reaction medium (volume, weight). This problem has been the subject of so much controversy, both from the theoretical and experimental point of view, that we have discussed it separately (Chap. 3[303]).

2.3 Treatment of Experimental Data

Several textbooks deal with this problem[304, 330]. The main method uses Eq. (2) which is obtained by integration of Eq. (1)

$$-\frac{dC}{dt} = k_d C^m (C + b)^n \tag{1}$$

$$\Rightarrow F_d(C) = -\int_{C_0}^{C} \frac{dC}{C^m (C + b)^n} = k_d(t - t_0) \tag{2}$$

where C and b are respectively [COOH] and the algebraic value of the excess of [OH] with respect to [COOH].

C_0 denotes the initial concentration of the COOH groups (at time $t = 0$); m and n are the orders of carboxy and hydroxy groups respectively; $d = m + n$.

Integral Functions $F_d(C)$ are plotted against time.

The use of a computer is very helpful to carry out a direct processing of the raw experimental data and to calculate the correlation coefficient and the least squares estimate of the rate constant.

In the case of stoichiometric reactions the overall order can be readily estimated from the plot of the fraction α of the reactants, remaining at time t, against $\log_{10} t$ [304, 330]. For a dth order reaction the experimental plot of α vs. $\log_{10} t$ can be superimposed on the dth curve of the following theoretical set of functions α_d:

$$\alpha_d = f(\log_{10} \theta) \tag{3}$$

with:

$$\begin{cases} \theta = \dfrac{1}{1-d}(1 - \alpha^{1-d}) & d \neq 1 \tag{4} \\ \theta = -\ln \alpha & d = 1 \tag{5} \end{cases}$$

For a catalyzed reaction there is a more or less important contribution of the non-catalyzed reaction which must be taken into account unless it has been shown that its contribution is negligible.

If the non-catalyzed reaction is taken into account Eq. (2) becomes:

$$-\frac{dC}{dt} = k_1 C^{m_1}(C + b)^{n_1} + k_2 C^{m_2}(C + b)^{n_2} \tag{6}$$

Indices 1 and 2 denote the catalyzed and non-catalyzed reactions, respectively. If rate constant k_2 and orders m_2 and n_2 have been determined Eq. (3) can be solved by an iterative integration method.

2.4 Discrimination Between Possible Values of Reaction Orders

Unambiguous results can be obtained only if some essential conditions are observed:
— If only narrow ranges of conversion are studied it is impossible to discriminate between close values of reaction orders unless very accurate experimental data have been

obtained. For most common accuracies (0.5–1%) the reaction must be controlled until conversion is 0.55 with at least 10 experimental points.
– If the balance between reactants is stoichiometric only overall orders can be obtained. However, the determination of orders in acid and in alcohol (obtained by non-stoichiometric studies) is highly desirable since it confirms the overall order and provides valuable information on the mechanism.

2.5 Determination of Activation Parameters

The dependence of the rate constant on temperature can be obtained either by the classical Arrhenius equation

$$k = A \exp(-E/RT) \tag{7}$$

where E and A are the activation energy and the preexponential term respectively, or by the transition state theory (see for instance Ref. 304)

$$k = \frac{\varkappa T}{h} \exp(\Delta S^{\neq}/R) \cdot \exp(-\Delta H^{\neq}/RT) \cdot (C^{\neq})^{1-d} \tag{8}$$

where $\Delta H^{\neq}$ and $\Delta S^{\neq}$ are the activation enthalpy and activation entropy respectively, and $C^{\neq}$ is a homogeneity term (equal to 1 unit of concentration) (see Dellacherie[282]).

More and more authors use Eq. (8) which gives $\Delta H^{\neq}$ and $\Delta S^{\neq}$ a more theoretical base than Eq. (7). However, it must be remembered that relation (8) has been established for reactions in the gaseous phase or at the best in highly diluted phase. This is not the case with many of the polyesterification systems studied where concentrations are high and expressed in term of mole per kilogram and not, as normally, in mole per liter. The consequences of these approximations are not clear and it is obvious that a considerable amount of additional work must be carried out on this subject.

3 Has the Elimination of Water to be Taken into Account in Polyesterification Kinetics?

During the course of a polyesterification the volume and the weight of the reaction mixture vary because condensation water is released. In most cases, the progress of the reaction is followed by titration of the acid groups at definite intervals; the carboxy group concentration is expressed in equivalents per kilogram. Consequently, several authors tried to find out if the weight decrease due to the elimination of water must be taken into account.

Flory[1] did not allow for this loss of weight as long as the conversion p was not too low, without explaining this assumption. p is given by the classical relation:

$$p = \frac{N_0 - N}{N_0} \tag{9}$$

where N_0 and N are the number of moles of carboxy groups at times 0 and t respectively.

Most authors used this approximation with no further justification and, moreover, introduced P (apparent extent of reaction) instead of p:

$$P = \frac{C_0 - C}{C_0} \tag{10}$$

C_0 and C are the concentrations of carboxy groups at times 0 and t respectively; they are determined experimentally.

However, Szabo-Rethy[293] showed that Flory's assumption leads to wrong values of rate constants and that the error can be as high as 15 to 35%. This author gave a method of calculation which takes into account the elimination of water. Similar relations have been used recently by Lin and Hsieh[270] but in our opinion erroneously.

In this chapter we summarize the various methods of calculation which are reported in the literature, we show that when the polycondensation is stoichiometric no correction is needed and we propose new relations for cases where the balance between reactive groups is non-stoichiometric.

3.1 The General Kinetic Equation – Case of Stoichiometry

Since [COOH] = [OH] at any time, the general kinetic Eq. (11) which is used by all authors

$$-\frac{d[COOH]}{dt} = k[COOH]^m [OH]^n \tag{11}$$

becomes

$$-\frac{dC}{dt} = k_d C^d \tag{12}$$

where d, m and n are the overall reaction order and the orders relative to acid and alcohol respectively (d = m + n), k_d is the rate constant corresponding to a d overall order, and $C = [COOH]$ in $eq \cdot kg^{-1}$ at time t.

After integration Eq. (12) becomes:

$$\begin{cases} \frac{1}{d-1}\left(\frac{1}{C^{d-1}} - \frac{1}{C_0^{d-1}}\right) = k_d t & d \neq 1 \\ \ln C_0/C = k_1 t & d = 1 \end{cases} \tag{13}$$

where $C_0 = [COOH]$ at initial time.

At any time t, [COOH] = C, and the reaction rate is $k_d C^d$, C being determined by titration. Hence the use of Eq. (13) requires no correction for concentration.

A correction is needed, however, if parameters relating to the number of moles of reactants are introduced (e.g. p or the degree of polymerization $\overline{DP}_n$).

Let m_0 and m be the weight of the reaction mixture at times 0 and t respectively and m_{H_2O} the weight of water released at time t.

Since $N_0 = C_0 \cdot m_0$ (14) and $N = C \cdot m$ (15) the relations

$$m_{H_2O} = 0.018 \, (N_0 - N) \tag{16}$$

$$m = m_0 - m_{H_2O} \tag{17}$$

can be written:

$$m_{H_2O} = 0.018 \, (C_0 m_0 - Cm) \tag{18}$$

$$m = m_0 - 0.018 \, (C_0 m_0 - Cm) \tag{19}$$

thus

$$m = m_0 \frac{1 - 0.018 \, C_0}{1 - 0.018 \, C} \tag{20}$$

From Eqs. (15) and (20) relation (21) can be obtained as

$$N = C \frac{1 - 0.018 \, C_0}{1 - 0.018 \, C} \cdot m_0 \tag{21}$$

and from Eq. (9):

$$p = \frac{C_0 - C \dfrac{1 - 0.018 \, C_0}{1 - 0.018 \, C}}{C_0} \tag{22}$$

Comparison of relations (10) and (22) shows that a "corrected" concentration C^* must be used in place of C with:

$$C^* = C \frac{1 - 0.018 \, C_0}{1 - 0.018 \, C} \tag{23}$$

3.2 Flory's Statement

The fundamental principle of Flory's statement[1] is expressed by relation (12). The data reported in Ref. 1 relate to p for the determination of which the corrections required to take into account the release of water have been carried out implicitly.

$$\overline{DP}_n = \frac{1}{1 - p} = \frac{\overline{M}_n - 18}{\text{mean segment weight}} \tag{24}$$

or

$$\frac{1}{1-p} = \frac{\overline{M_n} - 18}{\frac{M_A + M_B}{2} - 18} \tag{25}$$

where M_A and M_B are the molecular weights of diacid and diol monomers respectively.

Since $\quad \dfrac{1}{C} = \dfrac{\overline{M_n}}{1000} \tag{26}$

the experimental determination of C using Eq. (25) gives $\overline{DP}_n$ and p. This calculation differs slightly from those which have been made in the first part of this chapter but the results are the same:

Since $\quad \dfrac{1}{C_0} = \dfrac{M_A + M_B}{2 \times 1000} \tag{27}$

relation (25) becomes:

$$\frac{1}{1-p} = \frac{\dfrac{1}{C} - 0.018}{\dfrac{1}{C_0} - 0.018} \tag{28}$$

and relation (22) can be obtained from Eq. (28).

Although Flory takes into account the release of water in the determination of p, he does not do so in the following calculations assuming that

$p \approx P \quad (29) \qquad$ or $\quad C \approx C_0 \cdot (1-p) \tag{30}$

which leads to

$$dp/dt = k_d \, C_0^{d-1} \cdot (1-p)^d \tag{31}$$

Rate constants have been obtained by Flory[1] from relation (31).

3.3 Corrected Equation by Szabo-Rethy[293]

If water elimination is taken into consideration in the determination of p it must also be allowed for throughout the calculation. Thus, relation (22) leads to:

$$C_0/C = (1 - 0.018\, C_0) \frac{1}{1-p} + 0.018\, C_0 \tag{32}$$

and from Eq. (10):

$$\frac{1}{1-P} = (1 - 0.018\ C_0)\frac{1}{1-p} + 0.018\ C_0 \tag{33}$$

then

$$P = p\ \frac{1 - 0.018\ C_0}{1 - 0.018\ C_0 \cdot p} \tag{34}$$

From Eq. (12) a relation similar to Eq. (31) can be obtained but it is necessary to replace p by P:

$$\begin{cases} -\dfrac{dC}{dt} = k_d\ C^d \\ \dfrac{1}{1-P} = \dfrac{C_0}{C} \end{cases} \Rightarrow \dfrac{dP}{dt} = k_d\ C_0^{d-1}\ (1-P)^d \tag{35}$$

The corrected relation (35) is the true kinetic expression giving correct k_d values when used with P (apparent conversion) but wrong values when used with p (exact conversion). When p is used instead of P appreciable errors can occur which have been analyzed by Szabo-Rethy[293].

3.4 Other Approximations

In several works published after Flory (e.g. Tang and Yao[7]) elimination of water is not taken into account either in the calculation of the conversion or in the general kinetic relation. However, these two approximations cancel each other and the final relation is correct.

3.5 Erroneous Use of Corrections

Making allowance for the removal of water Lin and Hsieh[270] obtained relation (36) rewritten with our notations:

$$C^* = \frac{C(W_0 - 18) \times 1000}{W_0(1000 - 18\ C)} \tag{36}$$

where W_0 is the total weight of the reaction mixture corresponding to one equivalent of diacid.

Relation (36) and our relation (23) are obviously the same. From Eqs. (10) and (36) Lin and Hsieh[270] obtained values for the conversion p. However, they did not apply the necessary further corrections in Eq. (31).

It is obvious that the use of values of p (exact conversion) in relation (31) is a source of error, in the same way as the use of corrected values of concentration, C^*, in relation (12).

3.6 Case of a Non-Stoichiometric Balance of the Reactants

It seems that this case has never been examined. If only one reactant (e.g. acid) is titrated:

$$[OH] = [COOH] + b_0 \tag{37}$$

where b_0 is the algebraic excess of the hydroxy group concentration at the beginning of the reaction. Thus, relation (11) becomes

$$-\frac{dC}{dt} = k_d C^m (C + b_0)^n \tag{38}$$

When water is released the weight of the reaction medium decreases, and the true excess of the hydroxy group concentration increases since the number of excess hydroxy groups does not change:

$$b_0 \cdot m_0 = b \cdot m \tag{39}$$

where b is the excess of hydroxy group concentration at time t.
The use of Eq. (20) leads to

$$b = b_0 \frac{1 - 0.018\,C}{1 - 0.018\,C_0} \tag{40}$$

which gives the following rate equation:

$$-\frac{dC}{dt} = k_d C^m \left(C + b_0 \frac{1 - 0.018\,C}{1 - 0.018\,C_0}\right)^n \tag{41}$$

It is obvious that if the two reactants are titrated independently Eq. (11) must be used without any correction.

3.7 Conclusion

As long as concentrations are used no correction is needed for stoichiometric reactions. With non-stoichiometric balance, corrections must be carried out.

4 Inventory of the Main Catalysts Used in Esterifications and Polyesterifications

Many compounds have been used as polyesterification catalysts some of which are reported in Table 2. The efficiencies of these catalysts are analyzed in Chap. 6.

Table 2. Catalysts used in esterifications and polyesterifications

Catalyst	Alcohol	Acid	Ref.
	PROTONIC DERIVATIVES		
Protonic acids p–$CH_3C_6H_4SO_3H$, H_2SO_4, $NaHSO_4$, H_3PO_4, H_3PO_2, CF_3SO_3H	They have been used with most systems. Some recent references are given. (For additional references see Table 3.)		20–25, 60, 78, 111–114, 120)
Sulfonated and cation exchange resins	Various alcohols and acids		81–94, 96–102, 107–109, 115, 116, 121)
Oxydized phenolic resins	1-Butanol	Acetic	104, 105)
	NON-PROTONIC DERIVATIVES		
Group IA			
LiOH and various Li compounds	Various alcohols and acids		38, 162, 173)
NaOH and Na_2CO_3	Various alcohols and acids		33, 162)
Na organic derivatives	1,2-Ethanediol	Terephthalic	163, 335, 336)
KOH, K_2O, K_2CO_3	Various alcohols and acids		32, 162)
KCl	1,2-Ethanediol	Succinic	34)
Various alkali metal derivatives	Various dialcohols and diacids		123, 164)
Group IB			
Cu: (organic derivatives)	1,4-Butanediol	Adipic	336)
Group IIA			
Ca: Various organic derivatives	1,2-Ethanediol	Terephthalic	35, 79, 171, 336)
Mg: Various organic derivatives	1,2-Ethanediol	Terephthalic	49, 71, 160)
Be: Various organic derivatives	Various alcohols and acids		69)
Group IIB			
Zn: (Mineral and organic derivatives)	Various alcohols and acids		36, 42, 49, 58, 59, 62, 63, 69, 79, 119, 206, 218, 336)
Leached alloy Al-Ni-Zn	1,2,3-Trihydroxypropane	Fatty acids	118)
Cd: (organic derivatives)	1,2-Ethanediol	Aromatic diacids	165)
Group IIIB			
Al:			
Alumina	C_{2-5} Alcohols	Acetic	126)
Al derivatives on silica gel	1,2-Ethanediol	Terephthalic	36)
$AlCl_3$ on polystyrene	n-Butanol	Acetic	37)
Various Al phosphates	Ethanol	Acetic	125)
Aluminium silicates	Ethanol	Acetic	124)

Table 2 (continued)

Catalyst	Alcohol	Acid	Ref.
Various Al organic derivatives	Amino alcohols	Acrylic, Methacrylic	47, 69, 117)
$Al_2(SO_4)_3$ on aluminosilica carriers	Various alcohols and acids		125, 127, 128)
Leached alloy Al-Ni-Zn	1,2,3-Trihydroxypropane	Fatty acids	118)
Tl:			
TlO_2 and various organic derivatives	1,2-Ethanediol	Terephthalic	184)
Group IVA			
Ti:			
Metal	Various alcohols and diols	Diacids	42, 69)
TiO_2	Ethanol	Benzoic	73)
TiO_2	Various diols	Various diacids	76, 333, 334)
TiO_2 + Silica	1,2-Ethanediol	Methacrylic	51, 70)
TiO_2 + Silica	Various alcohols and acids		72)
TiO_2 + H_2SO_4	Methanol	Acetic	72)
TiO_2 + Sb_2O_3 + H_3PO_4	1,2-Ethanediol	Terephthalic	52)
Ti compounds on silica gel	1,2-Ethanediol	Terephthalic	36)
Ti complex with amino compounds	1,2-Ethanediol	Terephthalic	43, 79)
Ti organic salts	1,2-Ethanediol	Terephthalic	160, 332)
Ti Nitrilotriacetate	1,2-Ethanediol	Terephthalic	46)
$[Ti(RCOO)_3]_2O$	1,2-Ethanediol	Terephthalic	45)
R_2Ti salts	1,2-Ethanediol	Terephthalic	40)
TiR_4	Various diols	Various diacids	42, 65, 69)
$Ti(OR)_4$ or condensed derivatives	Various alcohols and acids		16, 47–50, 53, 54, 71, 74, 75, 77, 117, 187–189, 221–227, 230, 232)
Compounds with phosphinic acid	Various alcohols and acids		331)
Reaction products of a silanol with $Ti(OR)_4$	Diols	Aromatic diacids	41)
Various organic derivatives	1,2-Ethanediol	Terephthalic	44, 66–68)
Zr:			
Various compounds	Various alcohols	Aromatic diacids	68)
Group IVB			
C (associated with various transition metals):	1,2-Ethanediol	Terephthalic	129, 130)
Si:			
SiO_2 (associated with TiO_2)	Various alcohols and acids		51, 70)
Aluminium silicate	Various alcohols and acids		124)
Silica gel as support or alone	Various alcohols and acids		36, 131, 134–137, 175)

Table 2 (continued)

Catalyst	Alcohol	Acid	Ref.
Zeolithes, silicates	Various alcohols and acids		61, 132, 133)
Siloxanes	Alkanediols	Terephthalic	133)
Siloxanes associates with Ti derivatives	Alkanediols	Terephthalic	41)
Ge:			
GeO_2	Various dialcohols and diacids		59, 138–141, 152, 161)
Sn:			
Various organic derivatives	Various dialcohols and diacids		36, 49, 56, 66, 69, 142–147, 149–159, 332
Stannic acid and stannates	Alkylene glycols	Aromatic diacids	148
Pb:			
Various organic derivatives	Diols	Diacids	49, 79, 152, 165, 167
PbO, PbO_2	Diols	Diacids	165
Group VA			
V (various compounds):	Methanol	Aromatic diacids	68
Group VB			
N:			
NR_3	Various alcohols and acids		26, 29
	1,2-ethanediol	Terephthalic	28
NR_3 + $ArSO_2Cl$	Various alcohols and acids		39
HNR_2; $H_2NC_6H_5$; $HN(C_6H_5)_2$	1,2-ethanediol	Fumaric	27
Triethanolamine (in transition metal complexes)	Various diols and diacids		79
Pyridine derivatives	Polyols C_{3-20} (≥ 3 OH)	Polyisobutenyl-succinic	30, 263
	Various alcohols and acids		106
Piperazine derivatives	Alkylene glycols	Various diacids	31
Ammonium salts	Various diols and diacids		80, 103
N-Oxides	Various diols and diacids		110
P:			
Phosphoric acid and its derivatives	Various alcohols and acids		52, 74, 80, 103, 122, 168, 169, 171
Phosphonic acid and its derivatives	Various alcohols and diacids		40, 166

Table 2 (continued)

Catalyst	Alcohol	Acid	Ref.
Phosphinic acid / Ti derivatives	Various alcohols and acids		331
Various phosphorus derivatives	Various alcohols and acids		195, 199
As (various compounds):	Various alcohols and acids		69
Sb:			
Organic derivatives	1,2-Ethanediol	Terephthalic	49, 57, 79, 172
Sb_2O_5 (+ Ti derivatives)	1,2-Ethanediol	Terephthalic	44
Sb_2O_3 alone or associated with organo-metallic derivatives	1,2-Ethanediol	Terephthalic	52, 152, 170
Sb (element)	Methanol	Aromatic diacids	68, 174
Bi:			
Bi on carbon, organic salt	Various alcohols	Unsaturated acids	68, 174, 332
Group VIA			
Mo: various derivatives	Various alcohols and acids		69
W:			
W metal	Various alcohols and acids		176
WO_3-K_2O	Various alcohols and acids	Methacrylic	177
WO_3-SiO_2	Methanol	Methacrylic	178
Various derivatives	Various alcohols and acids		68
Group VIB			
S:			
S(associated with Sn)	Various alcohols and acids		149, 150
$ArSO_2Cl + NR_3$	Various alcohols and acids		39
Se: various compounds	Various alcohols	Unsaturated acids	68, 174
Te: various compounds	Various alcohols and acids	Unsaturated acids	68, 174
Group VIIA			
Mn:			
Aminated complex	Various diols	Various aromatic diacids	79
Organic derivatives	Various diols	Various aromatic diacids	42, 49, 165, 167, 179, 181
MnO_2	Various diols	Various aromatic diacids	165
Various permanganates	Various diols	Various aromatic diacids	180
Group VIII			
Fe:			
Organic derivatives	Methanol	Terephthalic acid	49, 182, 338
$FeCl_3$	Methanol	Various acids	97

Table 2 (continued)

Catalyst	Alcohol	Acid	Ref.
Co:			
Organic derivatives	1,2-Ethanediol	Various diacids	160, 165, 167, 181, 183
Ni:			
Leached Al-Ni-Zn alloy	1,2,3-Trihydroxypropane	Fatty acids	118
Rh (on C):	1,2-Ethanediol	Terephthalic acid	129, 130
Pd (on C):	Various alcohols and acids		129, 130, 174
Pd, organic derivatives	Methanol	Diacid	68, 185
Pt (on C):	1,2-Ethanediol	Terephthalic	129, 130
Lanthanides			
Ce	Methanol	Aromatic diacid	68
Actinides			
U_3O_8 or ThO_2	Methanol	Terephthalic acid	63
Various complexes of transition metals and nitriloacetic acid	Various diols and diacids		186

5 The Concept of Equal Reactivity of Functional Groups

One of the main assumptions which have been made in the study of polyesterifications is the concept of equal reactivity of functional groups. It was first postulated by Flory[1] who, studying various polyesterifications and model esterifications, found the same orders of reaction and almost the same rate constants for the two systems. He concluded that the reaction rate is not reduced by an increase in the molecular weight of the reactants or an increase in the viscosity of the medium. The concept of equal reactivity of functional groups has been fully and carefully analyzed by Solomon[3,13] so that we only discuss here its main characteristics. Flory clearly established the conditions under which the concept of equal reactivity can be applied; these are the following:
– the system must be a true solution of the polymer,
– the neighbourhood of all functional groups (adjacent groups, steric and electronic environment) must be the same,
– the elimination of the side products of the reaction (such as water) must not be hindered by viscosity.

According to Solomon[3], there is no example of a system where, these requirements being fulfilled, the concept of equal reactivity is not observed.

However, there are numerous examples of polyesterifications where the concept of equal reactivity is not observed; some examples have been reported and analyzed by Korshak and Vinogradova[231]. None of them is at variance with Flory's postulates but this shows that great caution is needed in the use of the concept of equal reactivity.

6 Kinetic Relations and Mechanisms

In view of the enormous number of articles and patents concerned with esterifications and polyesterifications, very few kinetic studies on these processes have been reported. In Table 3 (see page 99 to 142) are listed the main kinetic parameters published so far for linear polyesterifications and model esterifications.

The following critical study is mainly devoted to polyesterifications. However, several kinetic analyses relate to monoesterifications whose results are particularly useful for the understanding of polyesterification kinetics and mechanisms. In fact, there is a large number of articles concerned with esterifications and polyesterifications.

6.1 Non-Catalyzed and Acid-Catalyzed Reactions

6.1.1 Studies at Low Temperatures Without Removal of Water

The first studies on esterifications were carried out by Berthelot[233, 234]. Goldschmidt[235–239] studied many proton-catalyzed esterifications in alcohol at relatively low temperatures (below 80 °C) without removal of water. He suggested a pseudo first-order mechanism:

$$R'OH + H^\oplus \rightleftarrows R'OH_2^\oplus$$

$$R'OH_2^\oplus + H_2O \overset{K}{\rightleftarrows} R'OH + H_3O^\oplus$$

$$R'OH_2^\oplus + RCOOH \underset{slow}{\overset{k}{\rightleftarrows}} RCOOR' + H_3O^\oplus$$

where

$$K = \frac{[H_3O^\oplus][R'OH]}{[R'OH_2^\oplus][H_2O]} \quad \text{and} \quad [R'OH] \simeq C^t \tag{42}$$

Let the Goldschmidt parameter r_G be:

$$r_G = \frac{[R'OH_2^\oplus][H_2O]}{[H_3O^+]} \tag{43}$$

Since $[H_3O^\oplus] = [\text{catalyst}] - [R'OH_2^\oplus]$

$$r_G = \frac{[R'OH_2^\oplus][H_2O]}{[\text{catalyst}] - [R'OH_2^\oplus]} \tag{44}$$

And from Eq. (44): $[R'OH_2^\oplus] = r_G[\text{catalyst}]/(r_G + [H_2O])$ \qquad (45)

Now, the rate of acid consumption is given by (46):

$$\text{rate} = -\frac{d[RCOOH]}{dt} = k[RCOOH][R'OH_2^\oplus] \qquad (46)$$

With $[H_2O] = \varkappa$ and initial $[RCOOH] = a$, Eq. (46) can be written:

$$\frac{dx}{dt} = \frac{k(a-x)r_G[\text{catalyst}]}{r_G + \varkappa} \qquad (47)$$

and:

$$kr_G[\text{catalyst}]\, t = (r_G + a) \ln\left(\frac{a}{a-x}\right) - x \qquad (48)$$

r_G can be determined by the trial and error method until linear relationship is obtained. This process is rather restrictive and inaccurate for checking whether the mechanism which is assumed really fits with experimental data.

Goldschmidt's mechanism was widely used by Hinshelwood[240, 241] and Smith[242, 243] for dilute solutions; this corresponds to the experimental conditions for which relation (48) has been established. More recently, Van der Zeeuw[244] applied Goldschmidt mechanism to the reaction of phthalic anhydride with model alcohols in concentrated solutions.

However, in most cases, relation (48) does not account for results obtained under experimental conditions used in industry, i.e. high reactant concentrations. Othmer carried out a detailed study in this field and suggested second-order reactions for the esterifications of n-butanol with acetic acid[245] and monobutyl terephthalate[246] catalyzed by sulfuric acid. Since such relations cannot be established in all cases, no reaction order could be found for the esterification of 2,3-butanediol with acetic acid[247] in the presence of sulfuric acid. Moreover, Othmer's reaction orders were obtained for very concentrated media and in our opinion cannot be connected to a mechanism. In fact, this was not Othmer's objective who established these relations for practical use in industrial esterifications.

In all the preceding studies, the active species was supposed to be $R'OH_2^\oplus$; however, later, many authors, using labelled molecules, considered that it is the protonated acid $RC(OH)_2^\oplus$ instead.

The Ingold[248] classification of esterification and hydrolysis reactions is reported in Table 4. Basic compounds are seldom used as catalysts for esterifications, at least in diluted media. Thus, in Table 4 all arrows are oriented right to left. However, some authors (Naudet[193], Kutepov[27]) carried out base-catalyzed esterifications in concentrated media and proposed mechanisms.

There are four acid-catalyzed processes which are entirely symmetric and reversible. $A_{AL}2$ mechanism has never been observed.

The $A_{AC}1$ mechanism predominates only if the acyl group causes steric hindrance and if the ionizing power of the solvent is high. Participation of the $A_{AC}1$ mechanism in esterifications of acetic acid in CCl_4 has however been reported[249]. The reaction rate is given by

$$v = k[RC(OH)_2^\oplus][H^\oplus] \qquad (49)$$

Table 4. Classification of esterification mechanisms and of ester hydrolysis mechanisms according to Ingold[248]

		Name	Type	
Acid catalysis	Acyl cleavage	$A_{AC}1$	SN_1	$R-\underset{\underset{O}{\parallel}}{C}-OH \xrightleftharpoons{H^\oplus} R-\underset{\underset{OH}{\vert}}{\overset{\oplus}{C}}-OH \xrightarrow[H_2O]{slow} R-\underset{\underset{O}{\parallel}}{\overset{\oplus}{C}} \xrightleftharpoons[slow]{R'OH} R-\underset{\underset{O}{\parallel}}{\overset{\oplus}{C}}-OH\ R' \xrightleftharpoons{} R-\underset{\underset{O}{\parallel}}{C}-OR' \xrightleftharpoons{H^\oplus} R-\underset{\underset{O}{\parallel}}{C}-OR'$
		$A_{AC}2$	Tetrahedral	$R-\underset{\underset{O}{\parallel}}{C}-OH \xrightleftharpoons{H^\oplus} R-\underset{\underset{OH}{\vert}}{\overset{\oplus}{C}}-OH \xrightleftharpoons[R'OH]{slow} R-\underset{\underset{OH}{\vert}}{\overset{\overset{\oplus}{OH_2}}{C}}-OR' \xrightleftharpoons[slow]{H_2O} R-\underset{\underset{OH}{\vert}}{\overset{OH}{C}}-OR' \xrightleftharpoons{} R-\underset{\underset{O}{\parallel}}{C}-OR' \xrightleftharpoons{H^\oplus} R-\underset{\underset{O}{\parallel}}{C}-OR'$
	Alkyl cleavage	$A_{AL}1$	SN_1	$R'OH \xrightleftharpoons{H^\oplus} R'\overset{\oplus}{OH_2} \xrightarrow[H_2O]{slow} R'^\oplus \xrightleftharpoons[slow]{RCOOH} R-\underset{\underset{O}{\parallel}}{C}-OR' \xrightleftharpoons{H^\oplus} R-\underset{\underset{O}{\parallel}}{C}-OR'$
		$A_{AL}2$	SN_2	$R'OH \xrightleftharpoons{H^\oplus} R'\overset{\oplus}{OH_2} \xrightleftharpoons{RCOOH} R-\underset{\underset{O}{\parallel}}{C}-OR' \xrightleftharpoons{H^\oplus} R-\underset{\underset{O}{\parallel}}{C}-OR'$
Basic catalysis	Acyl cleavage	$B_{AC}1$	SN_1	$R-\underset{\underset{O}{\parallel}}{C}-O^\ominus + R'OH \xrightarrow{} R-\underset{\underset{O}{\parallel}}{C}-OH + R'O^\ominus \xrightarrow[OH^\ominus]{slow} R-\overset{\oplus}{\underset{\underset{O}{\parallel}}{C}} + R'O^\ominus \xrightarrow{slow} R-\underset{\underset{O}{\parallel}}{C}-OR'$
		$B_{AC}2$	Tetrahedral	$R-\underset{\underset{O}{\parallel}}{C}-O^\ominus + R'OH \xrightarrow{} R-\underset{\underset{O}{\parallel}}{C}-OH + R'O^\ominus \xrightarrow{} R-\underset{\underset{O^\ominus}{\vert}}{\overset{OH}{C}}-OR' \xrightarrow{slow} R-\underset{\underset{O}{\parallel}}{C}-OR'$
	Alkyl cleavage	$B_{AL}1$	SN_1	$R'OH \xrightarrow{OH^\ominus} R'\overset{\oplus}{OH_2} \xrightarrow{H_2O} R-\underset{\underset{O}{\parallel}}{C}-O^\ominus + R'^\oplus \xrightarrow{slow} R-\underset{\underset{O}{\parallel}}{C}-OR'$
		$B_{AL}2$	SN_2	$R-\underset{\underset{O}{\parallel}}{C}-O^\ominus + R'OH \xrightarrow{OH^\ominus} R-\underset{\underset{O}{\parallel}}{C}-OR'$

The $A_{AL}1$ mechanism is not more often encountered than the $A_{AC}1$ mechanism and is involved in the reaction to a considerable extent only if group R' is allyl, benzyl or tertiary alkyl, in which case

$$v = k[R'OH_2^{\oplus}] \tag{50}$$

The $A_{AC}2$ mechanism is more frequent;

$$v = k[RC(OH)_2^{\oplus}][R'OH] \tag{51}$$

Although the Goldschmidt and Ingold mechanisms take place under the same experimental conditions, they are different:

Ingold ($A_{AC}2$):

$$RC(OH)_2^{\oplus} + R'OH \rightleftarrows RCOOR' + H_2O + H^{\oplus}$$

Goldschmidt:

$$RCOOH + R'OH_2^{\oplus} \rightleftarrows RCOOR' + H_2O + H^{\oplus}$$

A detailed mechanism of Goldschmidt's process has not been given; two reaction paths are possible: either proton transfer to the acid with the formation of $RC(OH)_2^{\oplus}$ (in which case the slow step would be an $A_{AC}2$ Ingold mechanism) or nucleophilic attack of the carbonyl group of the acid on the protonated alcohol. The second mechanism would require an alkyl scission ($A_{AL}1$). In more recent studies[250], it has been shown that scission in most cases is of the acyl type and particularly in the examples studied by Goldschmidt.

It appears that the $A_{AC}2$ mechanism gives a better agreement with experimental results than that of Goldschmidt. This is not in contradiction with the fact that the alcohol is more basic than the acid and that, consequently, the concentration of $R'OH_2^{\oplus}$ is higher than that of $RC(OH)_2^{\oplus}$ since in any case a proton transfer from $R'OH_2^{\oplus}$ to the acid is possible.

When no protonic catalyst is added it is the carboxylic acid which acts as a catalyst; in 1896, Goldschmidt already showed that in this case the reaction order with respect to acid is greater than unity and Hinshelwood[251] suggested the following relation for the esterification rate:

$$v = k_1[RCOOH][R'OH_2^{\oplus}] + k_2[RCOOH]^2 \tag{52}$$

When the temperature increases the second term predominates. However, due to the very low rates of non-catalyzed reactions at low temperatures only very few experiments have been reported under these experimental conditions.

6.1.2 Studies at High Temperatures (above 100 °C) Without a Catalyst

Two different kinds of studies can be distinguished: those which are carried out in dilute solutions and those which are carried out with high concentrations of reactive groups.

6.1.2.1 Esterifications and Polyesterifications in Dilute Solutions

Flory[1, 252-254] studied only the last stage of the reactions, i.e. when the concentration of reactive end groups has been greatly decreased and when the dielectric properties of the medium (ester or polyester) no longer change with conversion. Under these conditions, he showed that the overall reaction order relative to various model esterifications and polyesterifications is 3. As a general rule, it is accepted that the order with respect to acid is two which means that the acid behaves both as reactant and as catalyst. However, the only way to determine experimentally reaction orders with respect to acid and alcohol would be to carry out kinetic studies on non-stoichiometric systems.

The principle of equal reactivity of functional groups originates in the comparison by Flory of the results obtained for monoesterifications and polyesterifications. This means that it is particularly important to check whether Flory hypotheses are correct and whether the study of a polyesterification must be limited to the last stages of the reaction.

Interpretations of Flory's results have given rise to many controversial discussions considering the whole course of the reaction (see below). Therefore, Solomon[13] studied esterifications in dilute media and polyesterifications at high conversions under well-defined experimental conditions. He took particular care to avoid losses of reactants which occur when the reaction starts in a mixture of the acid and alcohol.

Solomon's results[13] on the reaction of 1-dodecanol with dodecanoic acid in dodecyl dodecanoate and on the completion of the polyesterification of an oligo(1,10-decanediyl adipate) prepared under mild experimental conditions confirm Flory's conclusions (reaction order 3) and invalidate other interpretations of Flory's results.

Reaction orders in alcohol and acid were obtained from a study of the first of these reactions under non-stoichiometric conditions. This is not possible for a polyesterification, carried out in an excess of one of the reactants since in this case stoichiometry is required. The orders in acid and alcohol relative to the reaction of 1-dodecanol with dodecanoic acid in dodecyl dodecanoate are 2 and 1 respectively which, according to Solomon, corresponds to an $A_{AC}2$ mechanism. Since the dielectric constant is low, the ions are assumed to be associated as ion pairs:

$$2\,RCOOH \underset{}{\overset{K}{\rightleftarrows}} \underset{A}{RCOO^{\ominus}, RC(OH)_2^{\oplus}} \tag{I}$$

The attack of the alcohol on A is the slow step of the reaction:

since

$$A + R'OH \xrightarrow{k} \underset{\substack{|\\HO\ H}}{\overset{\substack{OH\\|\\\oplus}}{R-C-O-R'}}, RCOO^{\ominus}$$

the reaction rate v is given by:

$$v = k[R'OH][RCOO^{\ominus}, RC(OH)_2^{\oplus}] \tag{53}$$

or

$$v = k\,K[R'OH][RCOOH]^2 \tag{54}$$

Since in such media ionizations are low it can be accepted that [R'OH] and [RCOOH] are the overall alcohol and acid concentration respectively. Due to the higher basicity of alcohol, species B resulting from the protonation equilibrium II

$$\text{RCOOH} + \text{R'OH} \leftrightarrows \underset{B}{\text{RCOO}^{\ominus}, \text{R'(OH)}_2^{\oplus}} \tag{II}$$

has a higher concentration than species A.

However, because ion pairs A and B are not dissociated, the concentration of A does not depend on equilibrium (II).

However, other attempts to treat Flory's results by considering the whole course of the reaction have been carried out and it was recently claimed that these results fit a 2.5th-order rate equation[18] or correspond to the succession of a 2nd- and 3rd-order rate equation[15] (see Sect. 6.1.2.2).

Fradet and Maréchal[228, 229, 232] studied the reactions between functional oligomers in the bulk and between model molecules (1-octadecanol and octadecanoic acid) in various solvents with high boiling points. These reactants made studies possible both in dilute media and under experimental conditions close to those used in industry (high temperatures and some times low pressures). Furthermore, the use of functional oligomers allowed investigations to be performed in macromolecular medium at non-stoichiometric ratios of reactive end groups. Thus, orders with respect to alcohol and acid could be determined for the first time in such a medium. The knowledge of orders relative to each reactant in a polyesterification is of great interest since it confirms the overall order found in polyesterification and the order with respect to each reactant found for the model reactions carried out on small molecules. Moreover, this knowledge can greatly improve the mechanistic interpretation of polyesterification.

Kinetic studies on the bulk polyesterification of α,ω-dicarboxy poly(hexamethylene adipate) with α,ω-dihydroxy polyoxyethylene (both with $\overline{M}_n \approx 1000$)[232], on the bulk esterification of α-methoxy-ω-carboxy polyoxyethylene with α-methoxy-ω-hydroxy polyoxyethylene (both with $\overline{M}_n \approx 1000$)[228] and on the esterification of 1-octadecanol with octadecanoic acid in benzophenone gave orders in acid and alcohol equal to 2 and 1, respectively. This confirms the results found by Flory[1] and Solomon[13] and is in agreement with an autocatalysis by the acid, even in polymeric medium. Solomon's mechanism[13] can be considered as reasonable.

The results on the reaction of 1-octadecanol with octadecanoic acid in octadecyl octadecanoate[229] are quite different from those relative to the same reaction carried out in benzophenone[228] since the order with respect to acid is 1.5 in the first case and 2 in the second. Among the possible explanations of the lowering of the order in acid the most satisfactory is a non-negligible dissociation of the ion pair A and the formation of free $\text{RC(OH)}_2^{\oplus}$. That such a process takes place in a non-polar medium (octadecyl octadecanoate) is rather surprising; however, it can be supposed that all the reactive groups assemble in certain areas where they create a very polar medium and where water tends to be retained. In these areas, the dissociation of ion pairs would be easier and hence the overall order would decrease.

This type of mechanism was postulated by Tang and Yao[7] and many other authors for the whole course of polyesterification (see Sect. 6.1.2.2). The results obtained by Solomon and Fradet and Maréchal for polyesterifications carried out in media analogous to those used by Tang and Yao[7] show that this assumption is not true for the last stages of the reactions where kinetic measurements have a meaning. However, free $RC(OH)_2^\oplus$ may be formed in the first stages of polyesterification where an excess of glycol is present and polarity is very high.

Many other studies have been performed in solution and at high temperatures, the results obtained showing some inconsistencies.

Kemkes[256] assumes that the overall order relative to the esterification of terephthalic acid by 1,2-ethanediol in oligo(1,2-ethanediyl terephthalate) is two; no mechanism has however been suggested. Mares[257] considers that during the esterification of terephthalic acid with 1,2-ethanediol, two parallel kinetic paths take place, one corresponding to a reaction catalyzed by non-dissociated acid and the other to a non-catalyzed process. In fact, Mares[257] is reserved about the existence of protonic catalysis. Some other orders were found for the system terephthalic acid/1,2-ethanediol: 0 (overall)[318]; 2 (acid) and 0 (alcohol)[203]; 1 (acid) and 1 (alcohol)[181]; 1 (acid)[194]. These contradictory results could be partly due to the low solubility of terephthalic acid in 1,2-ethanediol.

Gordon[258] suggested orders 2 with respect to acid and 1 with respect to alcohol for the reactions of adipic acid with pentaerithrytol or with trimethylolpropane in bis(dioxa-3,6-heptyl ether); this reaction is followed by measuring the vapor pressure of the released water.

On the other hand, Davies[5], studying the reaction of adipic acid with 1,5-pentanediol in diphenyl oxide or diethylaniline found an order increasing slowly from two with conversion. From this result he concluded that Flory's[1, 252–254] and Hinshelwood's[240, 241] interpretations are erroneous. Two remarks must be made about the works of Davies[5]; experimental errors relative to titrations are rather high and kinetic laws are established for conversions below 50%. Under such conditions the accuracy of experimental determinations of orders is rather poor.

Manakov and Hasan[307] studied the system pentanoic acid/1-heptanol/n-decane and found an overall reaction order of three.

Several results were reported by Russian authors. They are completely different from those reported above. Sorokin[14] found an overall reaction order of 2 for the system heptanoic acid/1,2-ethanediol/diphenyl oxide. Bolotina[16] studied the reaction of 2-ethylhexyl hydrogenphthalate with 2-ethylhexanol in the corresponding diester and found an order of 1 with respect to acid and of 2 with respect to alcohol.

6.1.2.2 Esterifications and Polyesterifications in Concentrated Media

Several authors have analyzed the kinetics of the whole reaction by using Flory's data. In this case an overall order equal to three is no longer satisfying; however, such a treatment of the problem does not take into account the ideal conditions required by Flory.

Davies[4] found according to Flory's results an order of 2 in the first stages of the polyesterification, then an order of 3.

Tang and Yao[7] using same data in addition to other results found an order equal to 1.5. However, they did not take the overall reaction into consideration since the last four points (extent of reaction above 0.927) do not fit a straight line (see Fig. 1). They suggested the following mechanism:

$$RCOOH \leftrightarrows RCOO^\ominus + H^\oplus$$

$$H^\oplus + RCOOH \leftrightarrows RCOOH_2^\oplus$$

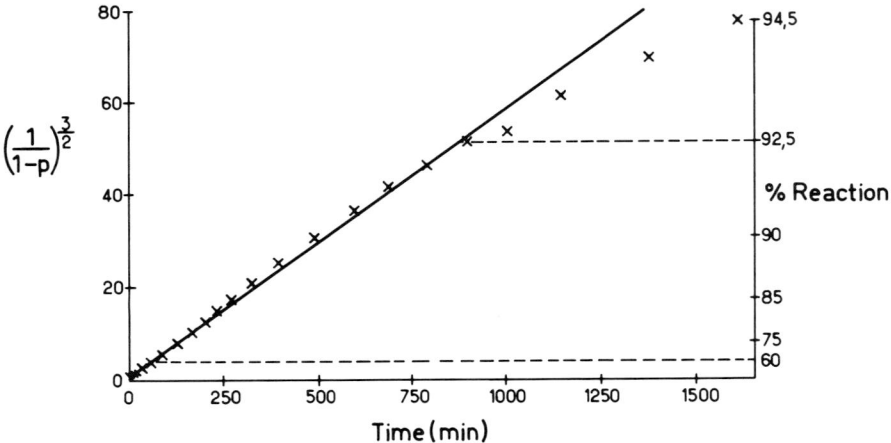

Fig. 1. 2.5-order plot for the reaction of adipic acid with diethylene glycol at 166 °C using Flory's data[1]. Critical study by Solomon[3]

$$H^\oplus + R'OH \rightleftarrows R'OH_2^\oplus$$

$$RCOOH_2^\oplus + RCOOH \rightleftarrows RC(OH)_2^\oplus + RCOOH$$

$$R'OH_2^\oplus + RCOOH \rightleftarrows RC(OH)_2^\oplus + R'OH$$

$$RC(OH)_2^\oplus + R'OH \underset{slow}{\rightleftarrows} R-\overset{\overset{OH}{|}}{\underset{\underset{HO}{|}\;\underset{H}{|}}{C}}-\overset{\oplus}{O}-R' \longrightarrow RCOOR' + H^\oplus + H_2O \nearrow$$

From this pattern they obtained the following relation:

$$\text{Rate} = k[COOH]^{1.5}[OH]\frac{1}{1 + \dfrac{K'[OH]}{[COOH]}} \tag{55}$$

K' is a constant depending on the reaction medium.

In low-polarity solvents which are generally used in polyesterification such ionizations are of low probability at least for the ultimate steps of the reaction. This is why Huang[8] using his own data explained the 2.5th order as being due to interactions between hydrogen-bonded dimers. However, in Huang's mechanism which is very complex a free-charge species is always present in the rate-determining step. The presence of this species, which is similar to the Tang and Yao complex A, considerably decreases the interest in Huang's treatment.

The following remarks can be made about Huang's mechanism as on any mechanism involving hydrogen-bonded dimers such as in [19] where the existence of the equilibrium

$$2\,RCOOH \underset{}{\overset{K}{\rightleftarrows}} R-C\underset{O-H\cdots O}{\overset{O\cdots H-O}{\diagup\diagdown}}C-R$$

is postulated at the beginning of polyesterification.

In all calculations [RCOOH] is a variable parameter and the final rate equation is a function of $[RCOOH]^n$ and of K; n is the overall reaction order when the reaction is carried out with stoichiometric amounts of acid and alcohol. However, it is important to mention that it is the global acidity x of the medium and not [RCOOH] which is measured:

$$x = [RCOOH] + 2[(RCOOH, RCOOH)] \tag{56}$$

or

$$x = [RCOOH] + 2K[RCOOH]^2 \tag{57}$$

which gives

$$[RCOOH] = \frac{-1 + \sqrt{1 + 8Kx}}{4K} \tag{58}$$

When a large amount of dimer is present, K is very high and [RCOOH] is roughly proportional to $x^{0.5}$. RCOOH is proportional to x only when K is very low (no dimer present).

If an order of n with respect to total acid is obtained experimentally, it is out of the question to consider it as an nth order relative to free acid. Unfortunately, this assumption has been made and it can be said that relative studies are "a priori" erroneous.

Carboxylic acids are closely associated even at polyesterification temperatures and in dilute media[260]. The associated species are cyclic dimers or other associated forms[261, 262]. The energy of formation of the intermediary complex being above hydrogen-bonding energy, it seems reasonable not to take the associated forms into consideration in calculations. Our own experimental results[229] are in agreement with this assumption.

Fang et al.[15] suggested an explanation of Flory's results by two successive mechanisms:

A second-order process, which is ionic and takes place mainly at the beginning of the esterification, and a third-order process involving only hydrogen-bonded species and occurring mainly at the end of the esterification.

The second-order process includes the following reactions steps:

$$2\,RCOOH \underset{}{\overset{K_a}{\rightleftarrows}} RC(OH)_2^{\oplus} + RCOO^{\ominus}$$

$$RCOOH \underset{}{\overset{K_e}{\rightleftarrows}} RCOO^{\ominus} + H^{\oplus}$$

$$RC(OH)_2^{\oplus} + R'OH \xrightarrow[\text{slow}]{k_1} R-\underset{\underset{H}{|}}{\overset{\overset{OH}{|}}{C}}-\overset{\oplus}{O}-R' \rightarrow RCOOR' + H^{\oplus} + H_2O$$

The charge equilibrium can be written as follows:

$$[H^\oplus] + [RC(OH)_2^\oplus] = [RCOO^\ominus] \tag{59}$$

which leads to:

$$[H^\oplus] = \frac{K_e[RCOOH]}{\sqrt{K_e[RCOOH] + K_a[RCOOH]^2}} . \tag{60}$$

Since the rate of esterification r_1 is given by

$$r_1 = \frac{k_1 K_a}{K_e}[H^\oplus][RCOOH][R'OH] \tag{61}$$

it becomes

$$r_1 = \frac{k_1 K_a [RCOOH]^2 [R'OH]}{\sqrt{K_e[RCOOH] + K_a[RCOOH]^2}} \tag{62}$$

If it can be assumed that K_e is very small

$$r_1 = k_1 K_a^{1/2}[RCOOH][R'OH] \tag{63}$$

which corresponds to a second-order reaction.

The Tang and Yao[7] mechanism is the same but in fact their method of calculation disregards autoprotonation of the acid in comparison with its dissociation. Since $K_a \ll K_e$, relation (61) becomes:

$$r_1 = k_1 K_a K_e^{-1/2}[RCOOH]^{3/2}[R'OH] \tag{64}$$

Fang's approximations seem more reasonable than those of Tang and Yao.

The 3rd-order mechanism is described by the following set of reactions:

$$RCOOH + R'OH \overset{Kb_1}{\rightleftarrows} \underset{A_1}{RCOOH...HOR'}$$

$$RCOOH + RCOOH \overset{Kb_2}{\rightleftarrows} \underset{A_2}{RCOOH...HOOCR}$$

$$A_1 + RCOOH \xrightarrow{k_4}$$

$$A_2 + R'OH \xrightarrow{k'_4}$$

with a rate of esterification given by:

$$r_2 = k_6(k_4 Kb_1 + k'_4 Kb_2)[RCOOH]^2[R'OH] \tag{65}$$

The global rate of the process is $r = r_1 + r_2$. Of all the authors who studied the whole reaction only Fang et al.[15] took into account the changes in dielectric constant and in viscosity and the contribution of hydrolysis. Flory's results fit very well with the relation obtained by integration of the rate equation. However, this relation contains parameters of which apparently only 3 are determined experimentally independent of the kinetic study. The other parameters are adjusted in order to obtain a straight line. Such a method obviously makes the linearization easier.

Finally, in 1979, Amass[18] reexamined Flory's experimental results and, like Tang and Yao, found that they are in agreement with a 2.5 th-order kinetics.

In contrast to Flory, many authors claimed that it was possible to consider all the results obtained at the beginning until the end of the reaction.

As he did for dilute solutions, Davies[5] suggested a reaction order increasing from 2 at the beginning of the reaction to 3 at the end.

Several authors[192, 264–266, 309, 310, 315, 317, 326] are in favor of bimolecular processes.

Gordon[267], studying the reaction of dodecanoic acid with pentaerythritol in the bulk found an order of 2 relative to acid and of 1 relative to alcohol.

Robins[6] investigated the reaction of 1,2-propanediol with maleic acid and found an overall reaction order of 3 which is in agreement with Flory's assumptions. The same order was found by Ivanov[325] for the 1,10-decanediol/2-propylheptanedioic system.

Tang and Yao's works have been used as a reference by many authors who found orders of 2.5 or 2 for the whole reaction. This was the case for Vancso-Smercsanyi[9–12], Matsuzaki[268] and Kirivahk[269] who studied bulk reactions of diols with diacids.

Ueberreiter and Hager[19] rather surprisingly found an overall reaction order of 6 at the beginning of ethanediol esterification. They explained this with the existence of hydrogen-bonded dimers.

Recently, Lin[17, 270, 271] found a 3rd-order for the reaction of succinic and adipic acids with 1,2-ethanediol in the presence of an excess of a diol. Surprisingly, he found orders 1 with respect to acid and 2 with respect to alcohol. This was explained by the fact that acid dissociation depends on alcohol concentration, i.e.:

$$[H^\oplus] = k_h[R'OH] \tag{66}$$

Since the reaction rate is given by

$$r = k[H^\oplus][RCOOH][R'OH] \tag{67}$$

we obtain with Eq. (66):

$$r = k_h k[RCOOH][R'OH]^2 \tag{68}$$

When a large excess of diol is present the orders are 2 relative to acid and 0 relative to alcohol respectively. In this case Lin assumed that

$$[H^\oplus] = k'_h[RCOOH] \tag{69}$$

which is equivalent to the statement that $[H^\oplus]$ is too low for a complete saturation of the reaction medium in $H^\oplus$.

In the cases where a strong protonic acid is added as a catalyst orders 2 with respect to acid and 0 with respect to alcohol have been found. It is assumed that free ions are present which leads to:

$$K_a = \frac{[H^\oplus][RCOO^\ominus]}{[RCOOH]} \qquad (70)$$

and

$$r = k[RCOOH][R'OH][H^\oplus] \qquad (71)$$

$$r = kK_a[RCOOH]^2 \frac{[R'OH]}{[RCOO^\ominus]} \qquad (72)$$

with:

$$[H^\oplus] = [H^\oplus]_{cat} + [H^\oplus]_A \qquad (73)$$

where $[H^\oplus]_{cat}$ and $[H^\oplus]_A$ are the concentrations of $H^\oplus$ due to organic acid (reactant) and to added acid (catalyst) respectively.

In order to eliminate $[RCOO^\ominus]$, Lin assumed that:

$$[H^\oplus]_A = k_h[R'OH] \qquad (74)$$

and

$$[H^\oplus]_A = [RCOO^\ominus] \qquad (75)$$

which leads to

$$[RCOO^\ominus] = k_h[R'OH] \qquad (76)$$

and

$$r = \frac{kK_a}{k_h}[RCOOH]^2 \qquad (77)$$

Thus, according to this mechanism it would be possible to distinguish between catalyst and reactant protons which is impossible, at least when acids are not in an associated form. This mechanism is not consistant.

Moreover, these studies were carried out to very high conversions (99.9%) even when no catalyst was added. Evaporation of reactants (at least 0.1% of initial quantity) cannot be avoided, particularly in the case studied by Lin since the reaction temperature was 15 °C below the boiling point of the diol. Consequently, errors in experimental data obtained by Lin for the ultimate stages of the reaction can be as high as 50 to 100%.

6.1.3 Studies at High Temperatures (above 100 °C) in the Presence of Protonic Catalysts

With the exception of Lin's results all studies on the catalysis by strong protonic acids (mainly H_2SO_4, benzene-, toluene- and naphthalenesulfonic acids) fit an overall second-order reaction[1, 7, 13, 202, 229, 254, 289, 308, 313–315, 319, 321, 325, 327].

In the case of the esterification of terephthalic acid with an excess of 1,2-ethanediol a zero overall order is obtained; this is due to the control of the reaction rate by the dissolution of terephthalic acid.

When determined, orders in acid[202, 226, 229, 313, 314, 321], in alcohol[202, 226, 229, 313, 314, 321] and in catalysts[226, 229, 308, 319, 321, 325, 327] are 1.

This agreement between the various methods of determinations contrasts with the case of non-catalyzed reactions where the results differ widely. However, in spite of this concordance the various authors suggest mechanisms which involve either ion pairs or free ions.

6.1.4 Activation Parameters of Non-catalyzed and Protonic Acid-catalyzed Esterifications and Polyesterifications

6.1.4.1 Non-catalyzed Reactions

Table 3 shows that the activation enthalpies determined by various authors can be very different. These differences cannot be correlated to discrepancies in reaction orders since, even when these are the same, activation energies can vary. Since the theoretical difference between activation enthalpy and activation energy is low ($2\,RT = 3\,kJ \cdot mol^{-1}$) with regard to the differences found in experimental determinations, the values discussed below are either enthalpies or energies of activation (For more detailed information see Table 3).

Due to the differences in the values relative to any one system, conclusions cannot easily be drawn from the activation parameters listed in Table 3. However, an analysis of the results relative to 1,2-ethanediol, 2,2-dimethyl-1,3-propanediol, 1,5-pentanediol, 1,10-decanediol and diethylene glycol shows that a slight difference can be observed between aromatic and aliphatic acids: the activations enthalpies and entropies are in the ranges 70, 100 $kJ \cdot mol^{-1}$ and –80, –130 $J \cdot K^{-1} \cdot mol^{-1}$ for aromatic acids, and in the ranges 50, 70 $kJ \cdot mol^{-1}$ and –200, –100 $J \cdot K^{-1} \cdot mol^{-1}$ for the aliphatic acids.

Tang and Yao[308] found 59 and 67 $kJ \cdot mol^{-1}$ for the model reactions between decanedioic acid and respectively 1-octanol and isooctanol. Hamann and Solomon[13] reported 54 $kJ \cdot mol^{-1}$ for the reaction between 1-dodecanol and dodecanoic acid, Zajic and Buresova[310] 51 $kJ \cdot mol^{-1}$ for the system 1-hexadecanol/erucic acid, and Fradet and Maréchal[228, 229] 50 and 51 $kJ \cdot mol^{-1}$ for the reaction between 1-octadecanol and octadecanoic acid in, respectively, 1-octadecyl octadecanoate and benzophenone. In Fradet and Maréchal's study the activation entropy is –210 $J \cdot K^{-1} \cdot mol^{-1}$ in both cases. Similar values were found for the reactions between α,ω-dicarboxypolyester and α,ω-dihydroxypolyoxyethylene (60 $kJ \cdot mol^{-1}$, –200 $J \cdot K^{-1}mol^{-1}$)[232] and between monocarboxylic and monohydroxylic polyoxyethylenes (70 $kJ \cdot mol^{-1}$, –170 $J \cdot K^{-1} \cdot mol^{-1}$)[228].

It appears that the values obtained with models are in the same range as those resulting from polyesterification studies. This is evidence of the identity of the reactions which take place in macromolecular and model media.

Several authors studied the influence of substituents on activation parameters. Baddar et al.[315] who studied the polyesterification of γ-arylitaconic anhydrides and acids with 1,2-ethanediol found that in the non-catalyzed reaction a p-methoxy substituent decreases both the activation enthalpy and the entropy whereas an increase is observed with a p-chloro substituent. On the other hand, Huang et al., who studied the esterification of 2,2-dimethyl-1,3-propanediol with benzoic, butanedioic, hexanedioic, decanedioic and o-phthalic acid found the same values since the activation enthalpy is 64 kJ · mol^{-1} for the first reaction and 61 kJ · mol^{-1} for the others.

In spite of a rather surprising overall reaction order of 6 (see Sect. 6.1.2.2) and of E values noticeably above those generally found, Ueberreiter and Hager[19] reported some interesting variations of E in the early stages of the reaction of 1,2-ethanediol with the following dicarboxylic acids:

$$HOOC(CH_2)_n COOH \qquad 0 \leq n \leq 8$$

Alternating values of E and of ln A were found for n = 0 to 4 and increasing values above n = 5. This observation was correlated with the hydrogen-bonding ability of these acids: $E^{\neq}$ is proportional to the concentration of associated carboxy groups[19]. Their experimental observations are, to some extent, in agreement with the results from Matsuzaki and Mitani[268].

6.1.4.2 Strongly Acid-catalyzed Esterifications and Polyesterifications

These reactions have not been widely studied; however, most authors agree with a lowering of enthalpy or energy of activation. This applies to the system 2-ethylhexanol/phthalic anhydride[226] where $\Delta H^{\neq}$ is 68 kJ · mol^{-1} in the absence of a catalyst and 62 (or 61) kJ · mol^{-1} when H_2SO_4 (or PTS) is added. This is also true for the system 1-octadecanol/octadecanoic acid in octadecyl octadecanoate[229] where $\Delta H^{\neq}$ decreases from 50 to 18.5 kJ · mol^{-1} when PTS is added. In the case of the system 1,2-ethanediol/hexanedioic acid, the decrease is from 50 to 46 kJ · mol^{-1} when PTS is added[313].

On the other hand, Ignatov et al.[314] observed an increase of $\Delta H^{\neq}$ upon the addition of benzene-, p-toluene- or naphthalenesulfonic acid to the system maleic anhydride/1,2-ethanediol. The same observation was made by Ivanoff[325] with the system 1,10-decanediol/2-propylheptanedioic acid/benzenesulfonic acid (catalyst).

In the determination of the entropy of acitvation different trends [decrease[226, 315], increase[314, 315], roughly constant[229, 313]] have been observed by various authors.

Hence, many results are contradictory and much more studies are needed. These studies must be carried out using well-defined conditions and particularly low concentrations of the reactive groups.

6.2 Titanium Tetralkoxide-catalyzed Reactions

Many references relate to polyesterifications catalyzed by tetraalkoxy – particularly tetrabutoxy – titaniums, and the catalytic activity of these compounds is claimed in many patents (see Table 2). However, there is an almost complete lack of information on reactions catalyzed by titanium derivatives. During the last few years, Fradet and Maréchal[227, 230, 232] studied systematically the behaviour of tetrabutoxytitanium in polyesterification media and showed that it is very complex. Kinetic orders depend on the reaction medium and, even for the same reaction mixture, vary with changing experimental parameters (pressure in the reactor, residual humidity etc. ...).

6.2.1 Main Properties of Alkoxytitaniums

The thermal stability of alkoxytitaniums is sufficient, at least in the temperature range where polyesterifications take place.

One of the most important properties concerning their use as polyesterification catalysts is their sensitivity to hydrolysis. Hydrolysis leads to the formation of titanoxanes whose $\overline{DP}$ depends on R and the water concentration:

$$Ti(OR)_4 + H_2O \longrightarrow RO\underset{\underset{OR}{|}}{\overset{\overset{OR}{|}}{Ti}}-OH + ROH$$

$$RO\underset{\underset{OR}{|}}{\overset{\overset{OR}{|}}{Ti}}-OH + Ti(OR)_4 \longrightarrow RO\underset{\underset{OR}{|}}{\overset{\overset{OR}{|}}{Ti}}-O-\underset{\underset{OR}{|}}{\overset{\overset{OR}{|}}{Ti}}-OR + ROH$$

Resistance to hydrolysis increases with increasing length of group R.

Since the catalytical behaviour of titanium derivatives depends on their degree of condensation, these catalysts are used under conditions where the water concentration is very low, for instance in transesterifications or in the last steps of esterifications. However, the amounts of water required to hydrolyse these compounds are so low (less than 0.5 ppm for $Ti(OBu)_4$) that hydrolysis is probably the determining phenomenon in most studies. It seems that before Fradet and Maréchal[230] the contribution of this side effect has not been taken into consideration.

In polyesterification media, alkoxytitaniums do not react with water but can undergo exchange reactions with alcohols, acids and, in some cases, with esters, e.g.:

$$Ti(OR)_4 + 4R'OH \to Ti(OR')_4 + 4ROH$$

Fradet[227] studied this exchange reaction in the case of tetrabutoxytitanium and 1-octadecanol by determining the amount of butanol released. It appears that the true catalyst is not $Ti(OBu)_4$ but a compound containing both butoxy and 1-octadecyloxy groups.

Exchange reactions with esters are significant only if the resulting ester can be easily eliminated.

In the same way, exchange reactions occur with acids:

$$Ti(OR)_4 + n R'COOH \to (R'COO)_n Ti(OR)_{4-n} + n ROH$$

However, according to Mehrotra[209, 272, 294, 322] many side reactions take place: Thus intramolecular reactions are due to the fact that titanium diacylates are more stable than other acylates:

$$R(COO)_3TiOR' \rightarrow RCOO-\underset{\underset{O}{\|}}{Ti}-OOCR + RCOOR'$$

Intramolecular condensations may also occur:

$$Ti(OR)(OOCR')_3 + Ti(OR)_2(OOCR')_2 \longrightarrow O\underset{Ti(OR)(OOCR')_2}{\overset{Ti(OR)(OOCR')_2}{<}} + RCOOR'$$

Such reactions may be the reason why Fradet and Maréchal[230] observed a poor catalytic activity when performing model studies in the presence of an excess of acid.

Due to the electronic structure of the central atom, titanium compounds easily form chelates with the coordination number 6. These chelates have been obtained with donating compounds such as α-hydroxy acids, diacids, and diketones. They are much more resistant to hydrolysis than the alkoxy derivatives[337]. However, if such chelates contain alkoxy groups they can be hydrolyzed. Their infrared spectra exhibit no vibration corresponding to the carbonyl groups of free ligands but two bands at 1370 and 1580 cm^{-1} respectively (chelated carbonyl groups)[337]. Such vibrations were found in the spectra of the products resulting from the reaction of terephthalic acid with alkoxytitanium[302].

Fradet[227, 232], in an esterification study on models, examined the reaction of octadecanoic acid with tetrabutoxytitanium. He found that a small amount of butyl octadecanoate is formed (absorption of the ester carbonyl at 1740 cm^{-1}) and that the carboxy absorption at 1710 cm^{-1} disappears completely. Simultaneously, two bands appear at 1560 and 1450 cm^{-1}, which is in agreement with Yoshino[302]. The ratio of the intensity of each of these two peaks to the intensity of ester peak (1740 cm^{-1}) does not change when the concentration of the solution used in the spectroscopic study is varied; consequently, the interaction between carbonyl and titanium is most probably intramolecular:

If the interactions are very strong, an intermolecular chelate may be postulated:

Such interactions are probably very important in the catalytic action of titanium derivatives since they induce the formation of positive charges on the carbon atom of the acid carbonyl group, thus favoring the attack by the alcohol.

An important characteristic of alkoxytitaniums is their autoassociation in the crystalline state and even in solution[55, 95, 255]. Witters[55] supposed the existence of equilibria between monomer, dimer, trimer, and even tetramer in the case of tetrabutoxytitanium. However, Bradley[290] considered that in the case of tetrabutoxytitanium only the trimer exists. No information on titanium alkoxide associations under polyesterification conditions has been published. However, at room temperature, tetrabutoxytitanium is monomeric in benzene solution (0.05 mol kg^{-1})[322]. As a consequence, it is reasonable to assume that this compound is not autoassociated in polyesterification media where catalyst concentrations are low and the temperature lies in the range 150–200 °C. If hydrolysis occurs these conclusions no longer hold.

6.2.2 Catalytical Behaviour of Alkoxytitaniums

Several authors suggested mechanisms for esterifications catalyzed by titanium tetraalkoxides. Bolotina et al.[16, 221, 222] who studied the polyesterification of 2-ethylhexyl phthalate with 2-ethylhexanol found the same reaction order with respect to catalyst, acid and alcohol, namely 1; they suggested the following rate-determining step:

$$R-C\underset{OH}{\overset{O---Ti(OR')_4}{\diagup}} + R'OH \xrightarrow{slow} R-\underset{\overset{|}{\overset{\oplus}{O}-R'}}{\overset{\overset{OH}{|}}{C}}-O-Ti(OR')_4^{\ominus} \tag{I}$$
$$\overset{|}{H}$$

Studying the reaction of carboxylic oligoesters with monohydric alcohols, Sorokina and Barshtein[223] found the reaction orders 1 and 0.5 with respect to acid and alcohol. However, they did not determine the reaction order relative to catalyst. According to these authors, the most important steps are:

$$Ti(OR')_4 + 2\,R'OH \longrightarrow H_2Ti(OR')_6$$

$$H_2Ti(OR')_6 + RCOOH \longrightarrow H\,Ti(OR')_6^{\ominus} + RC(OH)_2^{\oplus}$$

$$RC(OH)_2^{\oplus} + R'OH \xrightarrow{slow} R-\underset{\overset{|}{OH}}{\overset{\overset{OH}{|}}{C}}-\overset{\oplus}{O}\diagdown_{H}^{R'}$$

According to Sapunov et al.[224, 226] the following intermediary complex is formed in the reaction of phthalic anhydride with various alkanols:

$$\begin{array}{c} O\diagdownR \\ R'\diagdown\overset{\|}{C}\diagdownH \\ OO \\ (R'O)_3Ti\downarrow H \\ \diagupO \\ R'OH\overset{|}{R'} \end{array}$$

These results and mechanisms differ from each other and from those obtained more recently by Fradet and Maréchal[230]. These authors studied the catalytic action of Ti(OBu)$_4$ in reaction of the model system 1-octadecanol/octadecanoic acid and with monofunctional oligomers. They showed that side reactions reported above (see Sect. 6.2.1) are responsible for the complex behaviour of this catalyst and found suitable experimental conditions for an accurate study of the catalytic mechanism. These are hydrophobic medium, low concentration of residual water and excess of alcohol.

Reactions between oligomers: ω-hydroxypolyoxyethylene and ω-carboxypolyoxyethylene ($\overline{M}_n \approx 1000$) did not lead to a satisfactory fit of experimental data with the established kinetic law[230]. This is presumably due to the hydrophilicity of polyoxyethylene which retains the reaction water and therefore favours the hydrolysis of the catalyst. Consequently, it is not surprising that only low values of rate constants were obtained. The best fit was found for an overall reaction order close to 2.5.

The reaction of 1-octadecanol with octadecanoic acid[230] was carried out in the corresponding ester, namely 1-octadecyl octadecanoate (C 36 ester). The authors observed the following influence of the experimental parameters:

Figure 2 reveals that the overall order changes continuously with conversion. Fradet and Maréchal[230] explained this change by a decrease of the catalytic activity with increasing conversion. From an analysis of the experimental results they concluded that this decrease of reactivity is due to a continuous change in the degree of condensation of the catalyst with conversion. Runs carried out with different catalyst concentrations showed that the reaction rate depends only slightly on this parameter. This is in agreement with an aggregation of tetrabutoxytitanium which makes active only the external part of the particles.

– When the pressure increases (14 to 100 torr) a marked decrease of the rate constant is observed; this is very probably due to the hydrolysis of the catalyst since the water concentration increases with increasing pressure.

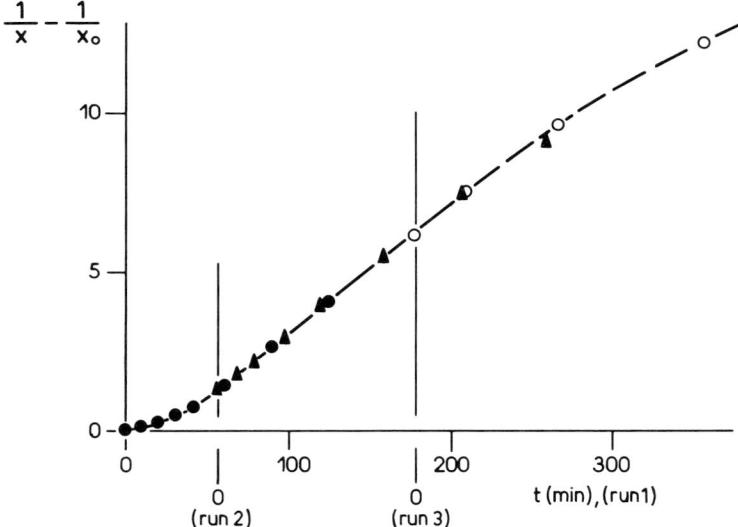

Fig. 2. Second-order plots for the stoichiometric reaction of 1-octadecanol with octadecanoic acid at 165.9 °C, 14 Torr, $[Ti(OBu)_4] = 2.69 \times 10^{-3}$ mol · kg^{-1} [230]
Run 1 (—●—): without solvent, $[COOH]_0 = 1.6$ eq · kg^{-1}.
Run 2 (—▲—): in octadecyl octadecanoate, $[COOH]_0 = 0.5$ eq · kg^{-1}.
Run 3 (—○—): in octadecyl octadecanoate, $[COOH]_0 = 0.15$ eq · kg^{-1}

Kinetics and Mechanisms of Polyesterifications

- The rate constant increases with rising temperature and the activation parameters are $\Delta H^{\neq} = 80$ kJ · mol^{-1} and $\Delta S^{\neq} = -100$ J · K^{-1} · mol^{-1}. According to these values the catalytic effect would result from an increase of the activation entropy.
- When α,ω-dimethoxypolyoxyethylene (10%) is added to the reaction mixture, a decrease (25%) of the initial rate is observed which very probably results from an increase of the hydrolysis of the catalyst due to a rise of the hydrophilicity of the reaction medium.
- When the reaction is carried out in the presence of an excess of acid, a strong decrease in the reaction rate occurs at high conversion. In contrast, the catalyst is stabilized when an excess of alcohol is present.

All these observations show that the catalytic activity of tetrabutoxytitanium depends on the composition of the medium used. When alcohol is used in excess, the catalyst is in a favourable neighbourhood since titanium can be complexed by two alcohol molecules.

$$\begin{array}{c}
R'\diagdown\quad\diagup H \\
O \\
R'O\diagdown\ |\ \diagup OR' \\
Ti \\
R'O\diagup\ |\ \diagdown OR' \\
O \\
R'\diagup\quad\diagdown H
\end{array}$$

In contrast, in the presence of an excess of acid formation of condensed species occurs:

$$\begin{array}{c}R'O\diagdown\\ \quad\quad Ti=O \\ R'O\diagup\end{array} \quad \text{or} \quad (R'O)_3Ti-O-Ti(OR')_3$$

The catalytic effect of these condensed species is probably lower than that of non-condensed species.

When water is present, hydrolysis takes place leading to the formation of aggregates. This would explain the low dependence of the reaction rate on the catalyst concentration.

6.3 Other Catalyzed Reactions

This section is concerned with elements other than titanium; however, it is sometimes difficult to treat this element separately from the others because several authors[49, 190] compared in the same work the relative reactivity of different catalysts. In most cases, a mechanism has not been suggested, nor has the efficiency of the catalyst been characterized. The order with respect to catalyst is almost never given.

Most compounds used as catalysts in esterifications contain elements belonging to groups III to VIII, mainly IV, V B, and VII A.

From the results of Malek et al.[49, 56, 190, 205, 206, 211] it may be concluded that the various metal derivatives used in esterifications exhibit many common catalytic characteristics. Thus, these authors[190] established relationships concerning the free energies of esterification of thirteen ortho-, meta- and para-substituted benzoic acids by 1,2-ethanediol. They obtained Hammett parameters which do not differ greatly from those found for the acid-catalyzed esterifications of the same acids with various alcohols. They concluded that free-energy relationships obtained for the Pb(II)-catalyzed esterification very likely hold for other catalytically active metal ions[1].

Other work by Malek et al.[56] is relative to the following system:

$$\underset{A}{\underset{HOOC}{\text{HOOC}}\text{-C}_6H_3(SO_3Na)\text{-COOH}} + \underset{B}{HO-CH_2-CH_2-OH} \rightleftharpoons \underset{C}{\underset{HOOC}{\text{HOOC}}\text{-C}_6H_3(SO_3Na)\text{-COO-CH}_2\text{-CH}_2\text{-OH}} + H_2O \quad \text{III}$$

$$C + B \rightleftharpoons HO-CH_2-CH_2-OOC\text{-C}_6H_3(SO_3Na)\text{-COO-CH}_2\text{-CH}_2\text{-OH} \quad \text{IV}$$

Both for reaction III and IV the order with respect to catalyst is 0.5. The activation enthalpies are 96.6 ± 3.4 and 97.6 ± 3.4 kJ · mol^{-1} respectively when Ti(OBu)$_4$ is used as the catalyst. This is not too far from the activation enthalpies[200] for the Sn(II)-catalyzed esterification of B with isophthalic acid (85.1 ± 4.9) and with 2-hydroxyethyl hydrogen isophthalate (85.8 ± 4.2). It is also close to the Ti(OBu)$_4$-catalyzed esterification of benzoic acid with B (85.8 ± 2.5)[49]. This is probably due to the formation of analogous intermediate complexes and similar catalytic mechanisms. On the other hand, the activation entropies of reactions III and IV are less negative than those of the reaction of benzoic or isophthalic acid with B. This probably corresponds to a stronger desolvation when the intermediary complex is formed and could be due to the presence of the sodium sulfonate group.

Habid and Malek[49] who studied the activity of metal derivatives in the catalyzed esterification of aromatic carboxylic acids with aliphatic glycols found a reaction order of 0.5 relative to the catalyst for Ti(OBu)$_4$, tin(II) oxalate and lead(II) oxide. As we have already mentioned in connection with other examples, it appears that the activation enthalpies of the esterifications carried out in the presence of Ti, Sn and Pb derivatives are very close to those reported by Hartman et al.[207, 208] for the acid-catalyzed esterification of benzoic and substituted benzoic acids with cyclohexanol. These enthalpies also approach those reported by Matsuzaki and Mitani[268] for the esterification of benzoic acids with 1,2-ethanediol in the absence of a catalyst. On the other hand, when activation entropies are considered, a difference exists between the esterification of benzoic acid with 1,2-ethanediol catalyzed by Ti, Sn and Pb derivatives and the non-catalyzed reaction[268]. Thus, activation enthalpies are nearly the same for metal ion-catalyzed and non-catalyzed reactions whereas the activation entropy of the metal ion-catalyzed reaction is much lower than that of the non-catalyzed reaction.

According to Habid and Malek[49] and to Basolo and Pearson[210], this highly negative value of $\Delta S^{\neq}$ for metal ion-catalyzed esterification is typical of bimolecular displacement

---

1 Malek et al.[49, 190] often use the terms "metal ion catalysis" and consider that metal ions play a very important role. According to our knowledge of esterification kinetics, this is only an assumption although these authors provided interesting arguments on esterification kinetics[49]

(SN 2) reactions where the formation of the transition state is characterized by a neat increase in bonding strength.

The efficiency of the various catalysts can be estimated from the pseudo half-rate constant $k^{49)}$. Their classification with respect to $k(mol^{1/2} \cdot kg^{-1/2} \cdot h^{-1})$ is as follows, the values of k being given in brackets:

$Ti(OBu)_4$ (925) < SnO (770) ≃ $Sn(COO)_2$ (767) < Bu_2SnO (570) < $Bi(OH)_3$ (515) < $Zn(CH_3COO)_2, 2H_2O$ (278) < $Pb(CH_3COO)_2, 3H_2O$ (260) < $Pb(C_6H_5COO)_2, H_2O$ (255) = PbO (255) < Sb_2O_3 (250) < $Al(CH_3COO)_3$ (215) ≃ $Mn(CH_3COO)_2, 4H_2O$ (214) < $Co(CH_3COO)_2, 4H_2O$ (186) ≃ $Cd(CH_3COO)_2$ (181) = $Cd(COO)_2$ (180).

This classification shows that tetrabutoxytitanium is by far the most efficient catalyst. Mg(II), Co(II), Cd(II), Zn(II), Pb(II), Al(III), Mn(II), and Sb(III) derivatives exhibit a rather low activity. Similarly, some of these compounds (Pb(II), Zn(II), Mn(II), Co(II)) display a low activity when used in the catalytic esterification of potassium 2,4-dicarboxybenzenesulfonate with 1,2-ethanediol[211]. Pb(II) derivatives fundamentally exhibit the same activity irrespective of the anion; the same holds for Cd(II) and Sn(II) derivatives. From these observations Habib and Malek[49] concluded that the catalytic activity of the metal ion plays a decisive role in the increase of the rate of the esterification of aromatic carboxylic acids with glycols. They also concluded that metal oxides, acetates, oxalates or benzoates are transformed into a complex under esterification conditions by an equilibrium reaction with the aromatic acid or the glycol. This complex is possibly that of the metal compound with the acid or more likely a complex of the acid with the metal glycolate such as $(RCOO)_xM[O(CH_2)_nOH]_y$. Since no induction period has been observed, this is a fast step; the rate-determining step would be the reaction of the complex with another glycol molecule to form the ester and regenerate the catalyst. According to Habid and Malek[49], this assumption is in agreement with the fact that tetrabutyltin, although fairly soluble in the reaction mixture, displays no catalytic activity in the esterification of benzoic acids with ethylene glycol. This compound fails to react with 1,2-ethanediol to give a glycolate of tetravalent tin even at high temperature, neither does it afford an activated complex with benzoic acid. This behaviour of tetrabutyltin differs significantly from that of lead (IV) oxide[212], dibutyltin oxide[213] and tetraalkoxytitanium[214] which react with glycol to give the corresponding complex glycoxides. The authors[49] conclude that the formation of an activated complex with the glycol or, more likely, with the carboxylic acid is a prerequisite to the catalytic activity of metal compounds. This complex should not be too stable. Thus, although they are resistant to poisoning by carboxylic derivatives[215], Sb(III) catalysts exhibit a low activity in the esterification of benzoic acid with a large excess of glycol. This could be due to the formation of stable complexes of Sb(III) with hydroxy ligands. The low activity of Zn(II) compounds in the esterification of benzoic acid[49] or of 2-hydroxyethyl hydrogen terephthalate[206] with 1,2-ethanediol, or in the first stage of the polyesterification of terephthalic acid[62], could be due to poisoning of the catalyst by COOH groups. According to Habib and Malek[49], this would explain the failure of the attempts to correlate the pseudo half-order rate constants (Table 3) with Pauling's metal ion electronegativities; however, it is important to mention that the catalytic efficiency of a series of metal ions often parallels their complexation constants with same chelating agent[216].

Some authors have used basic catalysts such as sulfonamides[217], primary[219] and tertiary[220] amines, but no mechanisms have been proposed. On the other hand, Kutepov and Skubin[27] who studied the synthesis of unsaturated polyesters in the presence of secondary amines suggested a possible mechanism but without any experimental support.

Caglioti et al.[201] suggested a mechanism for the action of hexachlorocyclotriphosphotriazene in the polyesterification of carboxylic acids with phenols. Higashi[291] catalyzed the reaction of various aromatic acids and alcohols by poly(ethyl phosphate). Both Caglioti[201] and Higashi[291] studied the influence of tertiary amines on the reactivity.

7 Concluding Remarks

From this critical analysis of the kinetics and mechanisms of linear polyesterifications by reaction of diols with diacids the following conclusions can be drawn:
– Non-catalyzed and protonic acid-catalyzed reactions are by far the best known. However, many studies have been carried out under conditions where kinetic values and relative mechanisms have no meaning.
– Patent literature shows that metal derivatives are increasingly used as catalysts in industrial preparations of polyesters. However, very little is known about the properties of these compounds. Moreover, experimental conditions obviously have a very strong influence on their catalytic activity. In fact, their behaviour in the reaction media and the actual effective catalytic species are not known.

8 References

1. Flory, P. J.: J. Am. Chem. Soc. *61*, 3334 (1939)
2. Flory, P. J.: Principles of Polymer Chemistry, Cornell University Press, Ithaca, New York 1953
3. Solomon, D. H.: Polyesterification, in: Kinetics and Mechanisms of Polymerization, vol. 3, p. 1, New York, Dekker 1972
4. Davies, M.: Research (London) *2*, 544 (1949)
5. Davies, M., Hill, D. R. J.: Trans. Faraday Soc. *49*, 395 (1953)
6. Robins, R. G.: Australian J. Appl. Sci. *5*, 187 (1954)
7. Tang, Au-Chin, Yao, Kuo-Sui: J. Polym. Sci. *35*, 219 (1959)
8. Huang, C. Y., Simono, Y., Onizuka, T.: Kobunshi Kagaku *23*, 408 (1966)
9. Vancso-Smercsanyi, I., Makay-Bodi, E., Szabo-Rethy, E.: J. Polym. Sci. A 1, *8*, 2861 (1970)
10. Vancso-Smercsanyi, I., Makay-Bodi, E.: J. Polym. Sci. *16*, 3709 (1968)
11. Vancso-Smercsanyi, I., Maros Greger, K., Makay-Bodi, E.: Eur. Polym. J. *5*, 155 (1969)
12. Makay-Bodi, E., Vancso-Smercsanyi, I.: Eur. Polym. J. *5*, 145 (1969)
13. Hamann, S. D., Solomon, D. H., Swift, J. D.: J. Macromol. Sci. Chem. A *2*, 153 (1968)
14. Sorokin, M. F., Kochnova, Z. A., Krovopalova, I. S.: Tr. Mosk. Khim. Tekhnol. Inst. *57*, 61 (1968)
15. Fang, Yao-Ren et al.: Sci. Sin. *18*, 72 (1975)
16. Bolotina, L. M., Maksimenko, E. G., Kutsenko, A. I.: Khim. Prom. St. *7*, 499 (1975)
17. Lin, Chen-Chong, Yu, Ping-Chang: J. Polym. Sci., Polym. Chem. Ed. *16*, 1005 (1978)
18. Amass, A. J.: Polymer *20*, 515 (1979)
19. Ueberreiter, K., Hager, W.: Makromol. Chem. *180*, 1697 (1979)
20. Bathia, J., Hussein, S. Z.: Chem. Eng. Sci. *28*, 337 (1973)
21. Campadelli, F., Nicora, C.: Eur. Polym. J. *8*, 1171 (1972)
22. Bhanumurty, J. V., Ayyanna, C., Chiranjivi, C.: Indian J. Technol. *14*, 438 (1976)
23. Charelishivili, B. L. et al.: Kinet. Katal. *19*, 899 (1978)
24. Charelishivili, B. L. et al.: React. Kin. Cat. Lett. *9*, 245 (1978)
25. Kreevoy, M. M., Kantner, S. S.: Croat. Chem. Acta *49*, 41 (1977)
26. Vinogradova, S. V. et al.: J. Polym. Sci. A 1, *9*, 3321 (1971)
27. Kutepov, D. F., Skubin, V. K.: Plast. Massy *2*, 18 (1972)

28. Seda De Barcelona: Span. Patent 451 776 (1977)
29. Kuroda, Y., Yamadera, R.: Jap. Kokai 73 103 587 (1973)
30. Cane, Ch., Yeomans, B.: Fr. Demande 2 362 63 (1978)
31. Murayama, K., Yamadera, R.: Jap. Kokai 73 57 945 (1973)
32. Takeo, S. et al.: Jap. Kokai 74 33 117 (1974)
33. Fujihara, K., Osokawa, T., Inamori, Y.: Jap. Kokai 77 151 112 (1977)
34. Erishev, B. Ya et al.: Z. Prikl. Khim. *51*, 953 (1978)
35. Kremer, E. B. et al.: USSR 334 820 (1976)
36. List, F., Wember, K.: Ger. Offen. 2 243 240 (1974)
37. Whang, K. Y., Lee, Y. K.: Taehan Hwahak Hoechi *22*, 184 (1978)
38. Ebisawa, K., Matsudaira, O.: Jap. Kokai 78 33 292 (1978)
39. Urbanska, H., Urbanski, J.: Polym. J. Chem. *52*, 523 (1978)
40. Ken, M., Reiso, Y.: Jap. Kokai 73 43 790 (1973)
41. Vizurraga, L. R.: US Patent 3 927 052 (1975)
42. Jackson, P. F., Morris, A., Ragg, P. L.: Ger. Offen. 2 255 268 (1973)
43. Koono, M. et al: Jap. Kokai 72 33 190 (1972)
44. Chimura, K. et al.: Jap. Kokai 73 32 836 (1973)
45. Chimura, K., Takahashi, S., Kawashima, M.: Jap. Kokai 74 05 936 (1974)
46. Knut, W., Gotintsewa, L.: GDR Patent 100 007 (1973)
47. Bartha, B.: Hung. Teljes 8012 (1974)
48. Chimura, K., Takishima, S., Kawashima, M.: Jap. Kokai 74 24 905 (1974)
49. Habid, O. M. O., Malek, J.: Collect. Czech. Chem. Commun. *41*, 2724 (1976)
50. Ishi, K., Kawashima, M., Terada, H.: Jap. Kokai 77 19 633 (1977)
51. Mitsubishi Chemical Industries Co. Ltd: Ger. Offen. 2 555 901 (1976)
52. Ikeuchi, H. et al.: Jap. Kokai 77 57 136 (1975)
53. Ganayem, I., Trevillyam, A. E.: US Patent 4 007 218 (1977)
54. Fukushima, S. et al.: Jap. Kokai 78 94 926 (1978)
55. Witters, R. D., Caughlan, C. N.: Nature (London) *205*, 1312 (1965)
56. Vejrosta, J., Zelena, E., Malek, J.: Collect. Czech. Chem. Commun. *43*, 424 (1978)
57. Chimura, K. et al.: Jap. Kokai 73 49 740 (1973)
58. Takeuchi, N., Tsuji, T.: Ger. Offen. 2 317 717 (1973)
59. Barkey, K. T., Brozek, C. T., May, D. C.: Fr. Demande 2 173 075 (1973); US Appl. 227 957 (1972)
60. Rubinstein, B. I. et al.: Nefetkhimiya *12*, 589 (1972)
61. Nosyreva, G. N., Ione, K., Zheivot, V. T.: Kinet. Katal. *20*, 270 (1979)
62. Krumpolc, M., Malek, J.: Makromol. Chem. *171*, 69 (1973)
63. Bhatia, J., Hussain, S. Z.: Indian Chem. J. *6*, 34 (1971)
64. Tanaka, M. et al.: Jap. Kokai 74 39 694 (1974)
65. Nomura, M., Yamamoto, T., Koyama, Y.: Jap. Patent 75 05 701 (1975)
66. Hayashi, M., Ikeuchi, H., Tanaka, M.: Jap. Kokai 75 82 196 (1975)
67. Ishibashi, Y., Matsubara, T.: Jap. Kokai 75 25 518 (1975)
68. Komoriya, T., Yamashita, G., Yamamoto, K.: Ger. Offen. 2 455 666 (1975)
69. Oka, I. et al.: Jap. Kokai 74 14 441 (1974)
70. Otake, M., Onoda, T.: Ger. Offen. 2 710 630 (1978)
71. Ezrielev, R. I., Arbuzova, I. A.: Zh. Obshch. Khim. *13*, 709 (1977)
72. Sergeev, V. F., Linchevskii, F. V., Maiorov, D. M.: Neftepererab. Neftekhim. *1977*, 37
73. Yatsevskaya, M. I., Ermolenko, N. F., Pavlyukevich, L. A.: Vestsi Akad. Navuk. Belarus. S.S.R., Ser. Khim. Navuk. *1972*, 11
74. Nomura, N., Fukamizu, K.: Jap. Patent 72 03 806 (1972)
75. Driscoll, L. G.: US Patent 3 671 572 (1972)
76. Ruzek, I., Pavelka, I., Sittlerova, H.: Czech. Patent 154 050 (1974)
77. Inoue, M., Oruichi, T., Nawada, K.: Jap. Kokai 77 136 294 (1977)
78. Sreeramulu, V., Rao, P. B.: Ind. Eng. Chem. Process., Design. Develop. *12*, 483 (1973)
79. Ruzek, I. et al.: Czech. Patent 138 959 (1970)
80. Kovacs, L., Szabos, L.: Brit. Patent 1 377 854 (1974)
81. Haken, J. K., Ho, D. K.: Australian Patent 454 792 (1974)
82. Gafurdzhanov, Ya. Yu., Rustamov, Kh. R.: Deposited. Publ., 1973, VINITI 5676-73

83. Chernysheva, D. A., Ostaeva, A. E., Polyanskii, N. G.: Tr. Tambovskogo Inst. Khim. Mashinostr. *4*, 172 (1970)
84. Lunin, A. F. et al.: USSR Patent 595 290 (1978)
85. Bhagade, S. S., Nageshwar, G. D.: Chem. Petro.-Chem. J. *9*, 3 (1978)
86. Thakur, D. S., Nageshwar, G. D.: J. Inst. Engl. (India), Part CH, *58*, 24 (1978)
87. Korovin, L. P. et al.: Izv. Vyssh. Uchebn. Zaved. Khim. Khim. Technol. *20*, 1847 (1977)
88. Libman, Z. G., Ul'Yankin, A. E.: Zh. Prikl. Khim. (Leningrad) *51*, 191 (1978)
89. Matyschok, H.: Chem. Stosow *21*, 155 (1977)
90. Matyschok, H.: Chem. Stosow *21*, 289 (1977)
91. Trofimenkova, T. K., Vesova, V. S., Reshetova, L. N.: Khim. Prom.-St. (Moscow) *1977*, 709
92. Abilova, Z. S., Mekhtiev, S. D.: Sb. Tr. Inst. Neftekhim. Protesessov AN Az SSR *1975*, 32
93. Vezely, V. et al.: Czech. Patent 165 254 (1976)
94. Karpov, O. N., Kazragis, A., Fedosyuk, L. G.: Liet. TSR Aukst. Mokyklu Mokslo Darb. Chem. Chem. Technol. *12*, 193 (1970)
95. Caughlan, C. N.: J. Am. Chem. Soc. *73*, 5652 (1951)
96. Sharma, O. M., Nageshwar, G. D., Mene, P. S.: Chem. Ind. Dev. *11*, 24 (1977)
97. Mallavarapu, G. R., Narasimhan, K.: Indian J. Chem. *16 B*, 725 (1978)
98. Rajkhova, P., Rao, M. G.: Ind. Chem. Eng. *17*, 41 (1975)
99. Nageshwar, G. D., Pandharpurkar, D. M., Mene, P. S.: Chem. Eng. World, *11*, 67 (1976)
100. Tomescu, M., Demian, N., Parausanu, V.: Rev. Chim. (Bucharest) *22*, 519 (1971)
101. Takamiya, N. et al.: Nippon Kagaku Kaishi *12*, 2069 (1975)
102. Parbuzina, I. L., Gordeev, L. S.: Khim. Farm. Zh. *8*, 39 (1974)
103. Kovacs, L., Szabo, L.: Hung. Teljes 4 927 (1972)
104. Strazhesko, D. N., Tovbina, Z. N., Stavitskaya, S. S.: Ukr. Khim. Zh. (Russ. edit.) *40*, 354 (1974)
105. Stavitskaya, S. S., Strazhesko, D. N.: Ukr. Khim. Zh. (russ. edit.) *41*, 1314 (1975)
106. Neises, B., Steglich, W.: Angew. Chem. *90*, 556 (1978)
107. Lazcano, R. L., Rivera, F. B. L.: Rev. Inst. Mex. Pet. *6*, 31 (1974)
108. Kubota, K., Kobayashi, D., Ohara, T.: Ger. Offen. 2 252 334 (1974)
109. Karpov, O. N.: Khim. Technol. (Kiev) *1974*, 61
110. Cheshko, F. F., Lugoyava, L. F.: Kinet. Katal. *18*, 1613 (1977)
111. Robin, M., Schulte, S. R.: US Patent 3 758 547 (1973)
112. Zienalov, B. K. et al.: Azerb. Neft. Khoz. *53*, 32 (1973)
113. Berlin, A. A., Rabia, A. M.: Vysokomol. Soedin. Ser. B. *15*, 336 (1973)
114. Rao, R. J.: Chem. Petro.-Chem. J. *7*, 3 (1976)
115. Nageshwar, C. D., Pandarpurkar, D. M., Mene, P. S.: Chem. Age India *26*, 365 (1975)
116. Yontanza, A., Kato, T., Shinofuji, K.: Jap. Patent 74 38 254 (1974)
117. Kimura, K., Ito, H.: Jap. Kokai 74 95 918 (1974)
118. Glushenkova, A. I.: USSR Patent 395 108 (1973)
119. Ratkevich, L. I., Mashkin, B. I., Roman'kov, V. A.: Neftepererab Neftekhim (Moscow) *7*, 34 (1978)
120. Volkova, I. A. et al.: Zh. Prikl. Khim. (Leningrad) *45*, 902 (1972)
121. Chernysheva, D. A., Ostaeva, A. E., Polyanskii, N. G.: Tr. Tambooskogo Inst. Khim. Mashinostr. *1970*, 172
122. Liu, H. J., Chan, W. H., Lee, S. P.: Tetrahedron Lett. *46*, 4461 (1978)
123. Shade, G., Vollkommer, N., Wolfes, W.: Ger. Offen. 2 349 115 (1975)
124. Safaev, A. S., Sultanov, A. S.: Tr. Tak. Politekh. Inst. *90*, 44 (1972)
125. Arifdzhanov, A., Karinov, K. G.: Mater. Resp. Nauchno-Tekh. Konf. Molodykh. Uch. Peperab Nefti Neftekhim *1*, 104 (1976)
126. Stojanovic, N., Keckarevic, P., Stojanovic, O.: Tecknika (Belgrade) *26*, 1993 (1971)
127. Rao, Y., Reddy, M. S., Rao, C. V.: Chem. Petro-Chem. J. *7*, 27 (1976)
128. Sinha, R., Raghavaiah, C. V., Chiranjivi, C.: Indian Chem. Manuf. *17*, 27 (1979)
129. Matsuzawa, K., Ohya, K.: Jap. Kokai 73 34 587 (1973)
130. Matsuzawa, K. et al.: Jap. Patent 73 26 748 (1973)
131. Wulf, H. D. et al.: Ger. Offen. 2 224 869 (1973)
132. Mashkova, V. V. et al.: Izv. Vyssh. Uchebn. Zaved. Khim. Khim. Tekhnol. *19*, 89 (1976)
133. Steinert, R., Sych, G., Mitchenko, L. N.: GDR Patent 104 307 (1974)

134. Maatman, R., Mahaffy, P., Mellema, R.: J. Catal. *35*, 44 (1974)
135. Abdurakhmanov, M. A. et al.: Katal. Pererab. Uglevodorod. Syr'ya *5*, 188 (1971)
136. Wulf, H. D. et al.: Ger. Offen. 2 224 869 (1973)
137. Bhatia, J., Hussain, S. Z.: Indian J. Technol. *13*, 348 (1975)
138. Senmura, K. et al.: Jap. Patent 71 42 492 (1971)
139. Standaert, R., Van Peteghem, A.: Belg. Patent 797 922 (1973)
140. Komoto, H., Toyomoto, K., Kobayashi, H.: Jap. Kokai 74 35 496 (1974)
141. Tsunawaki, K., Watanabe, K., Nawata, K.: Jap. Patent 74 20632 (1974)
142. Aitken, R. R., Trappe, G.: Brit. Patent 1 282 422 (1972)
143. Muryana, K., Yamadera, R.: Jap. Kokai 73 43 789 (1973)
144. Maiorano, G. et al.: Ger. Offen. 2 245 703 (1973)
145. Eguchi, H., Tsunoi, K., Takehisa, Y.: Jap. Patent 72 27 502 (1972)
146. Chimura, K., Takashima, S. Kawashima, M.: Jap. Patent 74 16 853 (1974)
147. Chimura, K. Takashima, S., Kawashima, M.: Jap. Patent 74 17 252 (1974)
148. Makimura, O., Yoshimura, M.: Jap. Kokai 74 48 632 (1974)
149. Okada, K. et al.: Jap. Kokai 73 68 540 (1973)
150. Okada, K., Tanaka, M., Kitagawa, H.: Jap. Kokai 73 68 539 (1973)
151. Senmura, K., Takashima, S., Kawashima, I.: Jap. Patent 75 05 698 (1975)
152. Kodaira, Y., Saito, S., Kobayashi, H.: Nippon Kagaku Kaishi *8*, 1503 (1972)
153. Nagai, Y. et al.: Jap. Kokai 77 39 644 (1977)
154. Ikeuchi, H. et al.: Jap. Kokai 77 83 424 (1977)
155. Chimura, K., Kakashima, S., Kawashima, M.: Jap. Kokai 74 14 741 (1974)
156. Takeda, S. et al.: Jap. Kokai 74 36 602 (1974)
157. Tanaka, M., Okada, K., Katagawa, H.: Jap. Kokai 74 35 495 (1974)
158. Eguchi, H., Tsunoi, K., Takehisa, Y.: Jap. Patent 72 27 502 (1972)
159. Nondek, L., Malek, J.: Collect. Czech. Chem. Commun. *43*, 1907 (1978)
160. Takashima, S., Kawashima, M., Ishii, K.: Jap. Kokai 77 128 342 (1977)
161. Galdecki, Z. et al.: Pol. Patent 98 272 (1978)
162. Sugiyama, I., Kasai, T.: Jap. Kokai 73 99 113 (1973)
163. Schumann, H. D., Bauereiss, G., Lehmann, P.: GDR Patent 98 285 (1973)
164. Erishev, B. Ya. et al.: Zh. Prikl. Khim. (Leningrad) *50*, 611 (1977)
165. Kazmierizak, A., Leo, J., Paryczak, T.: Zesz. Nauk. Polytechn. Lodz Chem. *30*, 97 (1974)
166. Itsikson, T. M. et al.: Khim. Technol. Topl. Masel *1975*, 10
167. Ichikawa, Y., Suzuki, N., Tanaka, Y.: Jap. Kokai 74 18 843 (1974)
168. Arifdzhanov, A., Safev, A., Sultanov, A. S.: Tr. Tashk. Politekh. Inst. *90*, 20 (1972)
169. Liu, H. J., Chan, W. H., Lee, S. P.: Tetrahedron Lett. *46*, 4461 (1978)
170. Casey, J. D., Gleckler, G. C.: Ger. Offen. 2 454 818 (1975)
171. Arifdzhanov, A., Gafurazhanov, Ya. Yu.: Tr. Tashk. Politekh. Inst. *90*, 45 (1972)
172. Rebhan, J., Matthies, H. G.: Ger. Offen. 2 449 162 (1976)
173. Izawa, N., Kosaka, N.: Jap. Kokai 73 05 736 (1973)
174. Onoda, T. et al.: Jap. Kokai 77 28 495 (1977)
175. Fricke, A. L., Alteper, R. J.: J. Catal. *25*, 33 (1972)
176. Mizabaeva, M., Glushenkova, A. I., Markman, A. L.: USSR Patent 408 646 (1973)
177. Moreschini, L., Petrini, G., Dalloro, L.: Fr. Demande 2 263 222 (1975)
178. Moreschini, L., Petrini, G., Dalloro, L.: Ger. Offen. 2 509 729 (1974)
179. Haas, Z., Alfred, F.: Czech. Patent 152 911 (1974)
180. Ikeuchi, H. et al.: Jap. Kokai 73 74 596 (1973)
181. Rod, V., Diwani, G. El., Minarik, M., Sir, Z.: Collect. Czech. Chem. Commun. *41*, 2339 (1976)
182. Eguchi, H., Kato, T.: Jap. Patent 71 22 460 (1971)
183. Wada, A., Miura, T., Shiraga, M.: Jap. Kokai 74 27 480 (1974)
184. Chimura, K., Takashima, S., Kawashima, I.: Jap. Patent 72 47013 (1972)
185. Massie, S. N.: US Patent 3 714 228 (1973)
186. Wuntke, K., Korovkina, T. N.: GDR Patent 104 506 (1974), USSR Patent 377 018 (1976)
187. Deleens, G., Foy, P., Maréchal, E.: Eur. Polym. J. *13*, 337 (1977)
188. Deleens, G., Foy, P., Maréchal, E.: Eur. Polym. J. *13*, 343 (1977)
189. Deleens, G., Foy, P., Maréchal, E.: Eur. Polym. J. *13*, 353 (1977)

190. Habib, O. M. O., Malek, J.: Collect. Czech. Chem. Commun. *41*, 3077 (1976)
191. Savicheva, O. I., Sedov, L. N.: Plast. Massy *9*, 5 (1972)
192. Rybachuk, V. P., Kadurina, T. I., Omel'chenko, S. I.: Vysokomol. Soedin. Ser. *A 15*, 1678 (1973)
193. Naudet, M., Cecchi, G.: Bull. Soc. Chim. France *1972*, 723
194. Kodaira, Y., Saito, S., Kobayashi, H.: Kogyo Kagaku Zasshi *74*, 2387 (1971)
195. Higashi, F., Kokubo, N., Gotto, M.: Polym. Prepr. Japan *28*, 946 (1979)
196. Sorokin, N. F., Kochnova, Z. A., Krivopalova, I. S.: Tr. Mosk. Khim. – Tekhnol. Inst. *61*, 100 (1969)
197. Beisebaev, M. Zh., Zhubanov, B. A., Mirfaizov, Kh. M.: Izv. Akad. Nauk. Kaz. SSR, Ser. Khim. *26*, 33 (1976)
198. Sorokin, M. F., Kochnova, Z. A., Krivopalova, I. S.: Plast. Massy *11*, 10 (1971)
199. Higashi, F., Kokubo, N., Gotto, M.: J. Polym. Sci., Polym. Lett. Ed. *18*, 385 (1980)
200. Nondek, L., Malek, J.: Makromol. Chem. *178*, 2211 (1977)
201. Caglioti, L., Poloni, M., Rosini, G.: J. Org. Chem. *33*, 2979 (1968)
202. Blaga, A. et al.: Rev. Chim. (Bucharest) *29*, 629 (1978)
203. Vladimirova, M. P. et al.: Khim. Volokna *1977*, 17
204. Mares, F., Bazant, V., Krupicka, J.: Collect. Czech. Chem. Commun. *34*, 2208 (1969)
205. Hradil, J., Malek, J.: Collect. Czech. Chem. Commun. *34*, 3959 (1969)
206. Krumpolc, M., Malek, J.: Makromol. Chem. *168*, 119 (1973)
207. Hartman, R. J., Storms, L. B., Gassmann, A. G.: J. Am. Chem. Soc. *61*, 2167 (1939)
208. Hartman, R. J., Hoogsteen, H. M., Moede, J. A.: J. Am. Chem. Soc. *66*, 1714 (1944)
209. Pande, K. C., Mehrotra, R. C.: Z. Anorg. Allg. Chem. *290*, 87, 95 (1957)
210. Basolo, F., Pearson, R. G.: Mechanisms of inorganic reactions, 2nd edit., Chapter 3, New York, Wiley 1967
211. Malek, J., Rerichova, M.: Chem. Prumysl. *25*, 407 (1975)
212. Torraca, G., Turriziani, R.: Rass. Chim. *14*, 100 (1962)
213. Bornstein, J. et al.: J. Org. Chem. *24*, 886 (1959)
214. Puri, D., Mehrotra, R.: Indian J. Chem. *5*, 448 (1967)
215. Hovenkamp, S. G.: J. Polym. Sci. Al, *9*, 3617 (1971)
216. Eichhorn, G. L.: Adv. Chem. Ser. *37*, 37 (1963)
217. Rueggeberg, W. H. C.: US Patent 3 008 928 (1958)
218. Ratkevich, L. I., Mashkin, B. I., Roman'kov, V. A.: Nefteperobad Neftekhim. (Moscow) *7*, 34 (1978)
219. Wilfong, R. E.: J. Polym. Sci. *54*, 385 (1961)
220. Fodor, G.: Szerves Kemia (Budapest) *1*, 630 (1960)
221. Bolotina, L. M., Kutsenko, A. I., Maksimenko, E. G.: Plast. Massy *7*, 13 (1973)
222. Kutsenko, A. I., Bolotina, L. M., Soinov, S. D.: Plast. Massy *1*, 23 (1971)
223. Sorokina, I. A., Barshtein, R. S.: Plast. massy *9*, 24 (1975)
224. Lebedev, N. N., Sapunov, V. N., Lemman, G. M.: Izv. Vyssh. Ucheb. Zaved. Khim., Khim. Teknol. *16*, 1762 (1973)
225. Sapunov, V. N., Lemman, G. M., Lebedev, N. N.: Kratk. Tezisy-Vses. Soveshch. Probl., Mekh. Geteroliticheskikh. Reakts. *1974*, 141
226. Sapunov, V. N., Lemman, G. M., Lebedev, N. N.: Izv. Vyssh. Ucheb. Zaved. Khim. Khim. Teknol. *19*, 696 (1976)
227. Fradet, A.: Thèse d'Etat, Paris 1980
228. Fradet, A., Maréchal, E.: J. Polym. Sci., Polym. Chem. Ed., in press
229. Fradet, A., Maréchal, E.: J. Macromol. Sci. Chem., in press
230. Fradet, A., Maréchal, E.: J. Macromol. Sci. Chem., in press
231. Korshak, V. V., Vinogradova, S. V.: Polyesters, Oxford, Pergamon Press 1965
232. Fradet, A., Maréchal, E.: Eur. Polym. J. *14*, 761 (1978)
233. Berthelot, M., Péan de Saint Gilles, L.: Ann. Chim. Phys. *65*, 385 (1862); *68*, 225 (1863)
234. Berthelot, M.: C. R. Acad. Sci. *85*, 883 (1877); *86*, 1296 (1878)
235. Goldschmidt, H. A.: Ber. Dtsch. Chem. Ges. *29*, 2208 (1896)
236. Goldschmidt, H. A., Udby, O.: Z. Physik. Chem. *60*, 728 (1907); *70*, 627 (1910)
237. Goldschmidt, H. A., Thuesen, A.: Z. Physik. Chem. *81*, 30 (1913)
238. Goldschmidt, H. A., Melbye, R. S.: Z. Physik. Chem. *143*, 139 (1929)

239. Goldschmidt, H. A., Haaland, H., Melbye, R. S.: Z. Physik. Chem. *143*, 278 (1929)
240. Hinshelwood, C. N., Legard, A. R.: J. Chem. Soc. *1935*, 587
241. Fairclough, R. A., Hinshelwood, C. N.: J. Chem. Soc. *1939*, 593
242. Smith, H. A.: J. Am. Chem. Soc. *61*, 254 (1939); *62*, 1136 (1940)
243. Smith, H. A., Reichardt, C. H.: J. Am. Chem. Soc. *63*, 605 (1941)
244. Van der Zeeuw, A. J.: Chem. Ind. (London) *22*, 978 (1969)
245. Leyes, C. E., Othmer, D. F.: Ind. Eng. Chem. *37*, 968 (1945)
246. Berman, S., Melnychuk, A. A., Othmer, D. F.: Ind. Eng. Chem. *40*, 1312 (1948)
247. Shlechter, N., Othmer, D. F., Marshak, S.: Ind. Eng. Chem. *37*, 900 (1945)
248. Ingold, C. K.: Structure and mechanisms in organic chemistry, 2nd edit., p. 1129, Ithaca, New York, Cornell University Press 1969
249. Tripathi, S. R., Singh, B., Krishna, B.: Chim. Anal. (Paris) *51*, 70 (1969)
250. March, J.: Advanced organic chemistry, 2nd edit., p. 351, New York, Mc Graw Hill 1977
251. Rolfe, A. C., Hinshelwood, C. N.: Trans. Faraday Soc. *30*, 935 (1934)
252. Flory, P. J.: J. Am. Chem. Soc. *58*, 1877 (1936)
253. Flory, P. J.: J. Am. Chem. Soc. *59*, 466 (1937)
254. Flory, P. J.: J. Am. Chem. Soc. *62*, 2261 (1940)
255. Martin, R. L., Winter, G.: Nature (London) *197*, 687 (1963)
256. Kemkes, J. F.: J. Polym. Sci. *C22*, 713 (1967)
257. Mares, F., Bazant, V., Kurpicka, K.: Collect. Czech. Commun. *34*, 2208 (1969)
258. Gordon, M., Scantlebury, G. R.: J. Chem. Soc. Part B, *1967*, 1
259. Davies, M.: Trans. Faraday Soc. *34*, 410 (1938)
260. Szabo Sarkadi, D., De Boer, J. H.: Rec. Trav. Chim. Pays Bas *76*, 628 (1957)
261. Haurie, M., Novak, A.: C. R. Acad. Sci. Ser. *B 264*, 694 (1967)
262. Madec, P. J., Maréchal, E.: J. Macromol. Sci. Chem. *A 12*, 1091 (1978)
263. Saigo, K. et al.: Bull. Chem. Soc. Japan *50*, 1863 (1977)
264. Rafikov, S. R., Korshak, V. V.: Dokl. Acad. Nauk. SSSR *64*, 211 (1949)
265. Korshak, V. V., Frunze, T. M., Li, I.: Vysokomolekul. Soedin. *3*, 665 (1961)
266. Korshak, V. V., Vinogradova, Sz. V.: Z. Obshch. Khim. *22*, 1176 (1952)
267. Gordon, M., Leonis, C. G.: J. Chem. Soc. Fraday Trans. *71*, 161 (1975)
268. Matsuzaki, K., Mitani, K.: Kogyo Kagaku Zasshi *70*, 470 (1967)
269. Kivirahk, S. Fomina, A. S.: Eesti NSV Tead. Akad. Toim. Keem. Geol. *22*, 118 (1973)
270. Lin, Chen Chong, Hsieh, Kuo Huang: J. Appl. Polym. Sci. *21*, 2711 (1977)
271. Chen Chong Lin, Ping Chang Yu: J. Appl. Polym. Sci. *22*, 1797 (1978)
272. Pande, K. C., Mehrotra, R. C.: Z. Anorg. Allg. Chem. *291*, 97 (1957)
273. Patai, S.: The chemistry of the hydroxyl group, New York, Interscience Publishers 1971
274. Vogel, A. I.: Elementary practical organic chemistry, Part III: Quantitative organic analysis, p. 676, London, Longman 1970
275. Reed, D. H., Critchfield, F. E., Elder, D. K.: Anal. Chem. *35*, 571 (1963)
276. Burns, E. A., Muraka, R. F.: Anal. Chem. *31*, 397 (1959)
277. Breitmaier, E. et al.: Angew. Chem. *82*, 82 (1970); Angew. Chem. Intern. Ed. Engl. *9*, 75 (1970) and references cited therein
278. Ho, Fl. F. L.: Anal. Chem. *45*, 603 (1973)
279. Hoppesch, C. W., Van Atta, R. E., Musulin, B.: Trans. Illinois State Acad. Sci. *54*, 96 (1961)
280. Kaduji, I. I., Rees, J. H.: Analyst (London) *99*, 435 (1974)
281. Fritz, D. F. et al.: Anal. Chem. *51*, 7 (1979)
282. Dellacherie, J., Foucaut, J. F., Scacchi, G.: Actualite Chimique *1980*, 35
283. Dufek, E. J., Butterfield, R. O., Frankel, E. N.: J. Amer. Oil. Chem. Soc. *49*, 302 (1972)
284. Vejrosta, J., Zelena, E., Malek, J.: Collect. Czech. Chem. Commun. *43*, 424 (1978)
285. Gordon, M., Leonis, C. G.: J. Chem. Soc. Faraday Trans. *71*, 178 (1975)
286. Schumann, H. D.: Faserforsch. Textiltechn. *23*, 80 (1972)
287. Deme, A., Alksnis, A., Surna, J.: Latv. PSR Zinat. Akad. Vestis Khim. Ser. *1974*, 214
288. Scipioni, A., Zanetti, A.: Tecnica Ital. *26*, 237 (1961)
289. Gareev, G. A., Men'shutin, V. P.: Kinet. Katal. *8*, 1369 (1967)
290. Bradley, D. C., Holloway, C. E.: Inorg. Chem. *3*, 1163 (1964)
291. Higashi, F., Kubota, K., Sekizuka, M.: Makromol. Chem. Rapid Commun. *1*, 461 (1980)
292. Pope, M. T., Williams, R. J. P.: J. Chem. Soc. *1959*, 3579

293. Szabo-Rethy, E.: Eur. Polym. J. 7, 1485 (1971)
294. Pande, K. C., Mehrotra, R. C.: J. Prakt. Chem. 5, 101 (1957)
295. Rybachuk, V. P., Kadurina, T. I., Omel'chenko, S. I.: Dopov Akad. Nauk. Ukr. RSR, Ser. B 35, 435 (1973); Vysokomol. Soedin. Ser. A 15, 1678 (1973)
296. Veibel, S., in: The Analysis of Organic Materials (ed.) Belcher, R., Anderson, D. M. N., Determination of OH groups, Vol. 1, New York, Academic Press 1972
297. Kuroda, Y., Yamadera, R.: Jap. Kokai 73 103 537 (1973)
298. Pancholy, M., Saksena, T. K.: J. Phys. Soc. Jap. 22, 1110 (1967)
299. Oehme, F.: Chem. Tech. (Berlin) 3, 171 (1951)
300. Axtmann, R. C.: J. Am. Chem. Soc. 73, 5367 (1951)
301. Davies, M., Hill, D. R. J.: Trans. Faraday Soc. 49, 395 (1953)
302. Yoshino, N., Yoshino, T.: Kogyo Kagaku Zasshi, 74, 1673 (1971)
303. Fradet, A., Maréchal, E.: Polym. Bull. 3, 441 (1980)
304. Margerison, D.: Comprehensive chemical kinetics (ed.) Bamford, C. H., Tipper, C. F. H., vol. 1, p. 343, Amsterdam, Elsevier 1969
305. Adamov, A. A. et al.: Zh. Obshch. Khim. 47, 1900 (1977)
306. Reimschuessel, H. K., Debona, B. T.: J. Polym. Sci., Polym. Chem. Ed. 17, 3241 (1979)
307. Manakov, M. N., Hasan, F.: Izv. Vyssh. Uchebn. Zaved. Khim. Khim. Teknol. 13, 705 (1970)
308. Tang, A. C. et al.: Ko Hsueb Chi Lu 3, 7 (1959)
309. Maeda, H. et al.: Shikizai Kyokaishi 39, 489 (1966)
310. Zajic, J., Buresova, M.: Sb. Vysokeskoly. Chem. Technol. Praze, Oddil. Fak. Potravenareske Technol. 4, 275 (1969)
311. Reimschuesel, H. K., Debona, B. T., Murthy, K. K. S.: J. Polym. Sci., Polym. Chem. Ed. 17, 321 (1979)
312. Dostal, H., Raff, R.: Monatsh. Chem. 68, 188, 247 (1936)
313. Sumoto, M., Hasegawa, Y.: Kogyo Kagaku Zasshi 68, 1900 (1965)
314. Ignatov, V. A. et al.: Izv. Vyssh. Uchebn. Zaved. Khim. Khim. Teknol. 21, 419 (1978)
315. Baddar, F. G. et al.: Eur. Polym. J. 7, 1621 (1971)
316. Shkol'man, E. E., Zeidler, I. I.: Zhur. Priklad. Khim. 26, 1205 (1953)
317. Kokochashvili, V. I., Sepiashvili, L. M., Bagdadlishivi, M. K.: Tr. Triblis. Gos. Univ. 126, 335 (1968)
318. Repina, L. P., Kremer, E. B., Aizenshtein, E. M.: Khim. Volokna 6, 9 (1969)
319. Pope, M. T., Williams, R. J. P.: J. Chem. Soc. 1959, 3579
320. Siggia, S., Hanna, J. G., Culme, R.: Anal. Chem. 33, 900 (1961)
321. Gumenchuk, L. M. et al.: Izv. Vyssh. Uchebn. Zaved., Khim. Khim. Teknol. 21, 844 (1978)
322. Puri, D. M., Mehrotra, R. C.: J. Ind. Chem. Soc. 39, 447 (1962)
323. Nerdel, F., Remmets, Th.: Z. Elektrochem. 60, 377 (1956)
324. Shlechter, N., Othmer, D. F., Marshak, S.: Ind. Eng. Chem. 37, 900 (1945)
325. Ivanoff, N.: Bull. Soc. Chim. France 1950, 347
326. Vinogradova, S. V. et al.: Vysokomol. Soedin. Ser. A, 9, 2152 (1967)
327. Bawn, C. E. H., Huglin, M. B.: Polym. 3, 257 (1962)
328. Robinson, W. T., Cundiff, R. H., Markunas, P. C.: Anal. Chem. 33, 1030 (1961)
329. Hase, A., Hase, T.: Analyst. 97, 998 (1972)
330. Burnnett, J. F.: Investigation of rates and mechanisms of reactions, in: Technics of Chemistry (ed.) E. S. Lewis, 3rd edit., p. 101, New York, Wiley Interscience 1974
331. Murayama, K., Yamadera, R.: Jap. Patent 73 43 789 (1973)
332. Vladimirova, M. P., Malykh, V. A., Chegolya, A. S.: Khim. Volokna 15, 20 (1973)
333. Kutsenko, A. I. et al.: Khim. Prom. St. (Moscow) 48, 182 (1972)
334. Yatsevskaya, M. I., Ermolenko, N. F., Pavlyukevich, L. A.: Vesti Akad. Nauk. Belarus SSR, Ser. Khim. Navuk 4, 11 (1972)
335. Lofquist, L. A.: US Patent 3 734 892 (1973)
336. Okuneva, A. G. et al.: Plast. Massy 7, 23 (1975)
337. Yamamoto, A., Kambara, S.: J. Am. Chem. Soc. 79, 4344 (1977)
338. Red'ko, V. P. et al.: Plast. Massy 5, 11 (1976)

Received July 7, 1981
H.-J. Cantow (Ed.)

Table 3. Kinetic parameters of esterification and polyesterification reactions. – The listed values relate to the following cases: Reactions where at least one of the two reactants is difunctional and reactions between two monofunctional reactants having more than 4 carbon atoms in their molecule. Table is in two parts: Part 1 (page 99 to 122) for experimental conditions. Part 2 (page 122 to 142) for results and references.

$(T-ED)_n$ = H(OOC—⟨C$_6$H$_4$⟩—COO—CH$_2$CH$_2$)$_n$OH

$ED(T-ED)_n$ = HO—CH$_2$CH$_2$(OOC—⟨C$_6$H$_4$⟩—COO—CH$_2$CH$_2$)$_n$OH

$T(ED-T)_n$ = HOOC—⟨C$_6$H$_4$⟩—COO(CH$_2$CH$_2$—OOC—⟨C$_6$H$_4$⟩—COO)$_n$H

PTS = p-Toluenesulfonic acid
r = initial molar ratio of alcohol (or diol) to acid (or diacid); when only r is given (third column) no solvent is present; % (W) means wt-% catalyst/weight reaction mixture.
Orders with quotation marks mean that the rate constants given by the authors are complex parameters depending on the acid concentration. (In this case see corresponding reference.) The overall reaction order does not include the order with respect to catalyst.
k values refer to reactions carried out without a catalyst. In the case of reactions performed in the presence of an added catalyst, two rate constants are given: either k^0 with: rate = $k^0[Cat.]^n[Reactants]^n$ or k_{cat} with: rate = $k_{cat}[Reactants]^n$.
If two esterifications take place (e.g. monoalcohol with diacid), two rate constants are given: $k^{(1)}$ corresponds to the formation of the monoester and $k^{(2)}$ to the formation of the diester.
If esterifications are carried out over a temperature range, the listed k values correspond to the lowest and to the highest temperature. When k is given with no specification, r = 1 for this value; k (homogeneous or heterogeneous) means that during esterification the medium changes from a heterogeneous to a homogeneous one.
k, $\Delta S^{\neq}$ and $\Delta H^{\neq}$ values are given in all cases with 3 significant figures at the most. E is the Arrhenius activation energy. [COOH] and [OH] are the concentrations of acid and alcohol end groups, respectively

Table 3 Part 1

No.	Alcohol	Acid	Solvent	Catalyst	Temperature Range (°C)
1	Methanol	Hexanedioic	r = 2	No	200
2	Ethanol	Butanedioic	r = 2	No	200
3	Ethanol	Pentanedioic	r = 2	No	200

Table 3 Part 1 (continued)

No.	Alcohol	Acid	Solvent	Catalyst	Temperature Range (°C)
4	Ethanol	Hexanedioic	r = 2	No	200
5	Ethanol	Heptanedioic	r = 2	No	200
6	Ethanol	Octanedioic	r = 2	No	200
7	Ethanol	Nonanedioic	r = 2	No	200
8	Ethanol	Decanedioic	r = 2	No	200
9	2-Butoxyethanol	Poly(diethylene glycol adipate), [Acid end group] = 2.52 meq · g^{-1}	r = 1.1	$Ti(OBu)_4$ 0.27% (W)	170
10	2-Butoxyethanol	Poly(diethylene glycol adipate), [Acid end group] = 2.52 meq · g^{-1}	r = 1.1	No	200
11	2-(2-Methoxyethoxy)-ethanol	Terephthalic	Kinetic parameters are deduced from studies of the following equilibria (no removal of water): – Esterification of alcohol with acid $[Acid]_0 = 0.279$ mol · kg^{-1}; $[Alcohol]_0 = 7.94$ mol · kg^{-1} – Esterification of monoester with alcohol $[Monoester]_0 = 0.266$ mol · kg^{-1}; $[Alcohol]_0 = 7.73$ mol · kg^{-1} – Transesterification of monoester $[Monoester]_0 = 3.73$ mol · kg^{-1}. No catalyst		230
12	2-(2-Methoxyethoxy)-ethanol	Hydrogen-2(2-methoxyethoxy)-ethyl terephthalate	Studied as a concurrent reaction to 11.		
13	2-(2-Butoxyethoxy)-ethanol	Poly(diethylene glycol adipate), [Acid end group] = 2.52 meq · g^{-1}	r = 1.1	$Ti(OBu)_4$ 0.27% (W)	170

#	Alcohol	Acid	Solvent	Catalyst	T (°C)
14	2-(2-Butoxyethoxy)-ethanol	Poly(diethylene glycol adipate), [Acid end group] = 2.52 meq · g^{-1}	r = 1.1	No	200
15	1-Propanol	Hexanedioic	r = 2	No	200
16	1-Butanol	Hexanedioic	r = 2	No	200
17	1-Butanol	Hydrogen butyl phthalate	r = 11	H$_2$SO$_4$ 0.011 mol · l^{-1}	80–120
18	1-Butanol	Hydrogen butyl phthalate	r = 3.03 to 30.21	H$_2$SO$_4$ 0.09 to 3.59% (W)	80–150
19	1-Butanol	Hydrogen butyl maleate	r = 5 to 8	PTS 0.5 to 4% (W)	85–115
20	2-Ethylbutanol	Phthalic anhydride	Decane r = 10, [Anhydride]$_0$ = 0.1 mol · l^{-1}	Ti(OBu)$_4$ 0.149 mol · l^{-1}	130
21	1-Pentanol	Hexanedioic	r = 2	No	200
22	1-Pentanol	Hydrogen pentyl phthalate	r = 13.5	H$_2$SO$_4$ 0.017 mol · l^{-1}	80–120
23	1-Hexanol	Hexanedioic	r = 2	No	200
24	1-Hexanol	Hydrogen hexyl phthalate	r = 12.5	H$_2$SO$_4$ 0.012 mol · l^{-1}	120
25	2-Ethylhexanol	Phthalic anhydride	Decane r = 3 to 15, [Anhydride]$_0$ = 0.1 mol · l^{-1}	Ti(O-2-Ethylhexyl)$_4$ 0.0149 mol · l^{-1}	130
26	2-Ethylhexanol	Phthalic anhydride	Decane, r = 10 [Anhydride]$_0$ = 0.1 mol · l^{-1}	Ti(OBu)$_4$ 0.0149 mol · l^{-1}	130
27	2-Ethylhexanol	Phthalic anhydride	Xylene, r = 10 [Anhydride]$_0$ = 0.1 mol · l^{-1}	Ti(O-2-Ethylhexyl)$_4$ 0.0149 mol · l^{-1}	130
28	2-Ethylhexanol	Phthalic anhydride	Ethylbenzene, r = 10 [Anhydride]$_0$ = 0.1 mol · l^{-1}	Ti(O-2-Ethylhexyl)$_4$ 0.0149 mol · l^{-1}	130
29	2-Ethylhexanol	Phthalic anhydride	Chlorobenzene, r = 10 [Anhydride]$_0$ = 0.1 mol · l^{-1}	Ti(O-2-Ethylhexyl)$_4$ 0.0149 mol · l^{-1}	130
30	2-Ethylhexanol	Phthalic anhydride	Dichlorobenzene, r = 10 [Anhydride]$_0$ = 0.1 mol · l^{-1}	Ti(O-2-Ethylhexyl)$_4$ 0.0149 mol · l^{-1}	130

Table 3 Part 1 (continued)

No.	Alcohol	Acid	Solvent	Catalyst	Temperature Range (°C)
31	2-Ethylhexanol	Phthalic anhydride	Nitrobenzene, r = 10 [Anhydride]$_0$ = 0.1 mol · l^{-1}	Ti(O-2-Ethylhexyl)$_4$ 0.0149 mol · l^{-1}	130
32	2-Ethylhexanol	Phthalic anhydride	Anisole, r = 10 [Anhydride]$_0$ = 0.1 mol · l^{-1}	Ti(O-2-Ethylhexyl)$_4$ 0.0149 mol · l^{-1}	130
33	2-Ethylhexanol	Phthalic anhydride	Bis(2-ethylhexyl) phthalate [Anhydride]$_0$ = 0.1 mol · l^{-1}	Ti(O-2-Ethylhexyl)$_4$ 0.0149 mol · l^{-1}	130
34	2-Ethylhexanol	Phthalic anhydride	excess 2-ethylhexanol [Anhydride]$_0$ = 0.1 mol · l^{-1}	Ti(O-2-Ethylhexyl)$_4$ 0.0149 mol · l^{-1}	130–165
35	2-Ethylhexanol	Phthalic anhydride	excess 2-ethylhexanol [Anhydride]$_0$ = 0.1 mol · l^{-1}	H$_2$SO$_4$, no [ ] given	135–165
36	2-Ethylhexanol	Phthalic anhydride	excess 2-ethylhexanol [Anhydride]$_0$ = 0.1 mol · l^{-1}	PTS, no [ ] given	135–165
37	2-Ethylhexanol	Hydrogen 2-ethylhexyl phthalate	r = 1 to 9	Ti(OBu)$_4$ 0.1 to 5% (W)	≥ 165
38	2-Ethylhexanol	Hydrogen 2-ethylhexyl phthalate	Bis(2-ethylhexyl) phthalate, r = 1 to 3 [Acid]$_0$ = 0.086 (mol/mol diester)	No	175
39	2-Ethylhexanol	Hydrogen 2-ethylhexyl phthalate	Bis(2-ethylhexyl) phthalate, r = 1 to 3 [Acid]$_0$ = 0.86 (mol/mol diester)	Ti(OBu)$_4$, 0.1 to 1% (W/W phthalic anhydride)	155–185
40	2-Ethylhexanol	Hydrogen 2-ethylhexyl phthalate	r = 1.86 to 133	Ti(O-2-ethylhexyl)$_4$ 0.98 to 5.55 · 10^{-3} mol · l^{-1}	135–187
41	2-Ethylhexanol	Poly(diethylene glycol adipate), [Acid end group] = 2.52 meq · g^{-1}	r = 1 to 12	Ti(OBu)$_4$ 0.27% (W)	160–170

#	Alcohol	Acid	Conditions	Catalyst	T (°C)
42	2-Ethylhexanol	Poly(diethylene glycol adipate), [Acid end group] = 2.52 meq · g^{-1}	r = 1.1 to 6	No	200
43	1-Heptanol	Phthalic anhydride	Decane, r = 10 [Anhydride]$_0$ = 0.1 mol · l^{-1}	Ti(OBu)$_4$ 0.0149 mol · l^{-1}	130
44	1-Heptanol	Poly(diethylene glycol adipate), [Acid end group] = 2.52 meq · g^{-1}	r = 1.1	Ti(OBu)$_4$ 0.27% (W)	170
45	1-Heptanol	Poly(diethylene glycol adipate), [Acid end group] = 2.52 meq · g^{-1}	r = 1.1	No	200
46	1-Heptanol	Hydrogen heptyl phthalate	r = 11	H$_2$SO$_4$ 0.12 mol · l^{-1}	120
47	1-Heptanol	Pentanoic	Decane, r = 1.68 to 2.09 [Acid]$_0$ = 0.383 to 0.238 mol · l^{-1}	No	140
48	1-Heptanol	Pentanoic	Decane, r = 1.16 to 1.30 [Acid]$_0$ = 0.420 to 0.236 mol · l^{-1}	Co(stearate)$_2$ 0.32 × 10^{-3} mol · l^{-1}	140
49	1-Octanol	Decanedioic	No	No	170
50	1-Octanol	Decanedioic	No	H$_2$SO$_4$	150
51	1-Octanol	Poly(diethylene glycol adipate), [Acid end group] = 2.52 meq · g^{-1}	r = 1.1	Ti(OBu)$_4$ 0.27% (W)	170
52	1-Octanol	Poly(diethylene glycol adipate), [Acid end group] = 2.52 meq · g^{-1}	r = 1.1	No	200
53	1-Octanol	Phthalic anhydride	Decane, r = 10 [Anhydride]$_0$ = 0.1	Ti(OBu)$_4$ 0.0149 mol · l^{-1}	130
54	1-Octanol	Hydrogen octyl phthalate	r = 14	H$_2$SO$_4$ 0.009 mol · l^{-1}	120

Table 3 Part 1 (continued)

No.	Alcohol	Acid	Solvent	Catalyst	Temperature Range (°C)
55	Isooctanol	Decanedioic	No	No	170
56	Isooctanol	Decanedioic	No	H_2SO_4	160
57	1-Nonanol	Phthalic anhydride	decane, r = 10 [Anhydride]$_0$ = 0.1 mol · l^{-1}	Ti(OBu)$_4$ 0.0149 mol · l^{-1}	130
58	1-Nonanol	Poly(diethylene glycol adipate), [Acid end group] = 2.52 meq · g^{-1}	r = 1.1	Ti(OBu)$_4$ 0.27% (W)	170
59	1-Nonanol	Poly(diethylene glycol adipate), [Acid end group] = 2.52 meq · g^{-1}	r = 1.1	No	200
60	1-Nonanol	Hydrogen nonyl phthalate	r = 11	H_2SO_4 0.012 mol · l^{-1}	120
61	1 Decanol	Phthalic anhydride	Decane, r = 10 [Anhydride]$_0$ = 0.1 mol · l^{-1}	Ti(OBu)$_4$ = 0.0149 mol · l^{-1}	130
62	1 Decanol	Poly(diethylene glycol adipate), [Acid end group] = 2.52 meq · g^{-1}	r = 1.1	Ti(OBu)$_4$ 0.27% (W)	170
63	1 Decanol	Poly(diethylene glycol adipate), [Acid end group] = 2.52 meq · g^{-1}	r = 1.1	No	200
64	1 Decanol	Hydrogen decyl phthalate	r = 10.7	H_2SO_4 0.012 mol · l^{-1}	80–120
65	1-Dodecanol	Dodecanoic	Dodecyl dodecanoate r = 0.5 to 2 [Acid]$_0$ = 0.537 mol · kg^{-1}	No	163–195
66	1-Dodecanol	Dodecanoic	r = 8.93	No	195

#	Alcohol	Acid	Conditions	Catalyst	Temp (°C)
67	1-Dodecanol	Dodecanoic	Dodecyl dodecanoate $r = 1$; $[Acid]_0 = 0.537$ mol·kg^{-1}	PTS 1.07×10^{-2} eq·kg^{-1}	163
68	1-Dodecanol	Dodecanoic	$r = 1$	No	202
69	1-Dodecanol	Dodecanoic	$r = 1.1$	Ti(OBu)$_4$ 0.27% (W)	170
70	1-Dodecanol	Poly(diethylene glycol adipate), [Acid end group] = 2.52 meq·g^{-1}	$r = 1.1$	No	200
71	1-Dodecanol	Phthalic	$r = 1$	No	170–210
72	1-Dodecanol	Hexanedioic	$r = 2$	No	202
73	1-Hexadecanol	Poly(diethylene glycol adipate), [Acid end group] = 2.52 meq·g^{-1}	$r = 1.1$	Ti(OBu)$_4$ 0.27% (W)	170
74	1-Hexadecanol	Poly(diethylene glycol adipate), [Acid end group] = 2.52 meq·g^{-1}	$r = 1.1$		200
75	1-Hexadecanol	cis 13-Docosenoic (erucic)	$r = 1$	No	160–240
76	1-Octadecanol	Octadecanoic	$r = 1$	No	165–175
77	1-Octadecanol	Octadecanoic	Octadecyl octadecanoate $r = 0.5$ to 4 $[Acid]_0 = 0.178$ to 0.614 eq·kg^{-1}	No	166.0–202.8
78	1-Octadecanol	Octadecanoic	Octadecyl octadecanoate $r = 0.5$ to 2 $[Acid]_0 = 0.288$ to 0.570 eq·kg^{-1}	PTS 1.15 to 10.5×10^{-3} eq·kg^{-1}	166.0 to 202.8
79	1-Octadecanol	Octadecanoic	Excess of alcohol $[Acid]_0 = 0.552$ eq·kg^{-1}	Ti(OBu)$_4$ 0.587×10^{-3} mol·kg^{-1}	166.0

Table 3 Part 1 (continued)

No.	Alcohol	Acid	Solvent	Catalyst	Temperature Range (°C)
80	1-Octadecanol	Octadecanoic	Benzophenone – r = 0.5 to 4 [Acid]$_0$ = 0.241 to 0.496 eq · kg^{-1}	No	166.0–204.0
81	1-Octadecanol	Octadecanoic	Excess of alcohol [Acid]$_0$ = 0.452 eq · kg^{-1}	No	166.0
82	Octadecanol	13-Docosenoic(cis) (Erucic)	r = 1	No	240
83	cis 9-Octadecen-1-ol	13-Docosenoic(cis) (Erucic)	r = 1	No	240
84	Benzyl alcohol	Phthalic anhydride	[Decane] – r = 10 [Anhydride]$_0$ = 0.1 mol · l^{-1}	Ti(OBu)$_4$ 0.0149 mol · l^{-1}	130
85	ω-Hydroxyundecanoic		Benzene (0.02 to 0.5 mol · l^{-1})	PTS (2 × 10^{-3} to 10^{-2} mol · l^{-1})	59 to 80.2
86	ω-Hydroxyundecanoic		CHCl$_3$ (0.1 to 0.5 mol · l^{-1})	PTS (2 × 10^{-3} to 10^{-2} mol · l^{-1})	40.84 to 84.20
87	ω-Hydroxyundecanoic		Decaline (0.0444 to 0.399 mol · l^{-1})	No	129 to 189
88	1,2-Ethanediol	Butanoic	r = 1 to 3	H$_2$SO$_4$ 0.25–0.75% (W)	55–70
89	1,2-Ethanediol	Heptanoic	Excess alcohol [Acid]$_0$ = 1.2 mol · l^{-1}	No	160
90	1,2-Ethanediol	Heptanoic	Excess alcohol [Acid]$_0$ = 4 mol · l^{-1}	No	160
91	1,2-Ethanediol	Heptanoic	Excess alcohol [Acid]$_0$ = 0.3 mol · l^{-1}	No	160
92	1,2-Ethanediol	Heptanoic	Diphenyl oxide – r = 4; [Acid]$_0$ = 1.0 mol · l^{-1}	No	160

Kinetics and Mechanisms of Polyesterifications

#	Diol	Acid	r	Catalyst	T (°C)
93	1,2-Ethanediol	Phenylacetic	r = 0.5	No	160–180
94	1,2-Ethanediol	3-Phenylpropionic	r = 0.5	No	160–180
95	1,2-Ethanediol	4-Phenylbutyric	r = 0.5	No	160–180
96	1,2-Ethanediol	Cyclohexanecarboxylic	r = 0.5	No	180–190
97	1,2-Ethanediol	3-Cyclohexylpropionic	r = 0.5	No	160–190
98	1,2-Ethanediol	Acrylic	Benzene; r = 1 $[\text{Acid}]_0 = 5$ mol·l^{-1}	H_2SO_4; 1.5% (W/W of reactants)	Reflux benzene
99	1,2-Ethanediol	Methacrylic	Benzene; r = 1 $[\text{Acid}]_0 = 5$ mol·l^{-1}	H_2SO_4; 1.5% (W/W of reactants)	Reflux benzene
100	1,2-Ethanediol	Methacrylic	r = 0.5 – Benzene; same quantity of solvent and of reactants	PTS 0.1 mol·l^{-1}	86–87
101	1,2-Ethanediol	Benzoic	r = 30	No	187–197
102	1,2-Ethanediol	Benzoic	r = 30	MgO 0.002 mol·kg^{-1}	197
103	1,2-Ethanediol	Benzoic	r = 30	Zn(OAc)$_2$, 2H$_2$O 0.002 mol·kg^{-1}	197
104	1,2-Ethanediol	Benzoic	r = 30	Cd(OAc)$_2$ 0.002 mol·kg^{-1}	197
105	1,2-Ethanediol	Benzoic	r = 30	CdC$_2$O$_4$ 0.002 mol·kg^{-1}	197
106	1,2-Ethanediol	Benzoic	r = 30	Al(OAc)$_3$ 0.002 mol·kg^{-1}	197
107	1,2-Ethanediol	Benzoic	r = 30	Ti(OBu)$_4$ 5×10^{-4} to 2×10^{-3} mol·kg^{-1}	167–197
108	1,2-Ethanediol	Benzoic	r = 30	PbO 5×10^{-4} to 2×10^{-3} mol·kg^{-1}	167–197
109	1,2-Ethanediol	Benzoic	r = 30	Pb(OAc)$_2$, 3H$_2$O 0.002 mol·kg^{-1}	197
110	1,2-Ethanediol	Benzoic	r = 30	Pb(OOCC$_6$H$_5$)$_2$ 0.002 mol·kg^{-1}	197
111	1,2-Ethanediol	Benzoic	r = 30	SnO, 0.002 mol·kg^{-1}	197

Table 3 Part 1 (continued)

No.	Alcohol	Acid	Solvent	Catalyst	Temperature Range (°C)
112	1,2-Ethanediol	Benzoic	r = 30	SnC_2O_4 5×10^{-4} to 3×10^{-3} mol · kg^{-1}	167–197
113	1,2-Ethanediol	Benzoic	r = 30	$(C_4H_9)_2SnO$ 0.002 mol · kg^{-1}	197
114	1,2-Ethanediol	Benzoic	r = 30	$(C_4H_9)_4Sn$ 0.002 mol · kg^{-1}	197
115	1,2-Ethanediol	Benzoic	r = 30	Sb_2O_3 0.002 mol · kg^{-1}	197
116	1,2-Ethanediol	Benzoic	r = 30	$Bi(OH)_3$ 0.002 mol · kg^{-1}	197
117	1,2-Ethanediol	Benzoic	r = 30	$Mn(OAc)_2$, $4H_2O$ 0.002 mol · kg^{-1}	197
118	1,2-Ethanediol	Benzoic	r = 30	$Co(OAc)_2$, $4H_2O$ 0.002 mol · kg^{-1}	197
119	1,2-Ethanediol	Benzoic	r = 8.542 to 257.8	ZnO 8.14×10^{-4} to 2.44×10^{-2} mol · kg^{-1}	175–197
120	1,2-Ethanediol	Benzoic	r = 13.02 to 126.9	No	197
121	1,2-Ethanediol	Benzoic	r = 1	PTS 1% (W)	140
122	1,2-Ethanediol	Benzoic	r = 0.5	No	140–190
123	1,2-Ethanediol	Benzoic	Values of rate constants and of activation parameters are obtained from the study of the following reactions (No removal of water). – Hydrolysis of 1,2-ethanediol monobenzoate ([H$_2$O]/[monoester] = 1/1(mol)) – Hydrolysis of 1,2-ethanediol dibenzoate ([H$_2$O]/[diester] = 1/1 (mol)) – condensation of 1,2-ethanediol monobenzoate [monoester] = 5.97 to 6.15 mol · kg^{-1} [benzoic acid] = 0 to 0.6 mol · kg^{-1}	No catalyst	202–226
124	1,2-Ethanediol	Benzoic	"	Sb_2O_3 1.09 to 4.41×10^{-3} mol · kg^{-1}	202–226

Kinetics and Mechanisms of Polyesterifications

125	1,2-Ethanediol	o-Methylbenzoic	r = 30	No	197
126	1,2-Ethanediol	o-Methylbenzoic	r = 30	PbO 0.002 mol · kg^{-1}	197
127	1,2-Ethanediol	m-Methylbenzoic	r = 30	No	197
128	1,2-Ethanediol	m-Methylbenzoic	r = 30	PbO 0.002 mol · kg^{-1}	197
129	1,2-Ethanediol	p-Methylbenzoic	r = 30	No	197
130	1,2-Ethanediol	p-Methylbenzoic	r = 30	PbO 0.002 mol · kg^{-1}	197
131	1,2-Ethanediol	o-Methoxybenzoic	r = 30	No	197
132	1,2-Ethanediol	o-Methoxybenzoic	r = 30	PbO 0.002 mol · kg^{-1}	197
133	1,2-Ethanediol	p-Methoxybenzoic	r = 30	No	197
134	1,2-Ethanediol	p-Methoxybenzoic	r = 30	PbO 0.002 mol · kg^{-1}	197
135	1,2-Ethanediol	o-Chlorobenzoic	r = 30	No	197
136	1,2-Ethanediol	o-Chlorobenzoic	r = 30	PbO 0.002 mol · kg^{-1}	197
137	1,2-Ethanediol	p-Chlorobenzoic	r = 30	No	197
138	1,2-Ethanediol	p-Chlorobenzoic	r = 30	PbO 0.002 mol · kg^{-1}	197
139	1,2-Ethanediol	o-Bromobenzoic	r = 30	No	197
140	1,2-Ethanediol	o-Bromobenzoic	r = 30	PbO 0.002 mol · kg^{-1}	197
141	1,2-Ethanediol	p-Bromobenzoic	r = 30	No	197
142	1,2-Ethanediol	p-Bromobenzoic	r = 30	PbO 0.002 mol · kg^{-1}	197
143	1,2-Ethanediol	m-Trifluoromethylbenzoic	r = 30	No	197
144	1,2-Ethanediol	m-Trifluoromethylbenzoic	r = 30	PbO 0.002 mol · kg^{-1}	197
145	1,2-Ethanediol	o-Nitrobenzoic	r = 30	No	197
146	1,2-Ethanediol	o-Nitrobenzoic	r = 30	PbO 0.002 mol · kg^{-1}	197
147	1,2-Ethanediol	m-Nitrobenzoic	r = 30	No	197
148	1,2-Ethanediol	m-Nitrobenzoic	r = 30	PbO 0.002 mol · kg^{-1}	197
149	1,2-Ethanediol	p-Nitrobenzoic	r = 30	No	197

Table 3 Part 1 (continued)

No.	Alcohol	Acid	Solvent	Catalyst	Temperature Range (°C)
150	1,2-Ethanediol	p-Nitrobenzoic	r = 30	PbO 0.002 mol · kg^{-1}	197
151	1,2-Ethanediol	Sodium 3,5-dicarboxybenzenesulfonate	r = 20 + Et$_3$N (4.3% (W))	Ti(OBu)$_4$ 1.1 to 3.10^{-3} mol · kg^{-1}	162.5 to 190
152	1,2-Ethanediol	Sodium 3,5-dicarboxybenzenesulfonate	r = 20 + Et$_3$N (4.3% (W))	SnC$_2$O$_4$ 1.1 to 3.10^{-3} mol · kg^{-1}	180
153	1,2-Ethanediol	Potassium 3,5-dicarboxybenzenesulfonate	r = 20	SnC$_2$O$_4$ 0.015 mol · kg^{-1}	180
154	1,2-Ethanediol	Sodium 3-(2-hydroxyethyl carboxylate)-5-carboxybenzenesulfonate	Studied as concurrent reaction to 151		
155	1,2-Ethanediol	Sodium 3-(2-hydroxyethyl carboxylate)-5-carboxybenzenesulfonate	Studied as concurrent reaction to 152		
156	1,2-Ethanediol	Potassium 3-(2-hydroxyethyl carboxylate)-5-carboxybenzenesulfonate	Studied as concurrent reaction to 153		
157	1,2-Ethanediol	Potassium 4-carboxybenzenesulfonate	r = 20	SnC$_2$O$_4$ 0.01 mol · kg^{-1}	180
158	1,2-Ethanediol	Potassium 3-carboxy-6-methylbenzenesulfonate	r = 20	SnC$_2$O$_4$ 0.01 mol · kg^{-1}	180
159	1,2-Ethanediol	Potassium 4-t-butyl-2,6-dicarboxybenzenesulfonate	r = 20	SnC$_2$O$_4$ 0.015 mol · kg^{-1}	180
160	1,2-Ethanediol	Potassium 2-(2-hydroxyethyl carboxylate)-4-t-butyl-6-carboxybenzene sulfonate	Studied as concurrent reaction to 159		

Kinetics and Mechanisms of Polyesterifications

#	Diol	Diacid	Conditions	Catalyst	Temp
161	1,2-Ethanediol	Ethanedioic	r = 1	No	120–130
162	1,2-Ethanediol	Propanedioic	r = 1	No	140–150
163	1,2-Ethanediol	Butanedioic	r = 1	No	173–188
164	1,2-Ethanediol	Butanedioic	r = 1	No	170–190
165	1,2-Ethanediol	Butanedioic	Dioxane r = 1 $[\text{Acid}]_0 = 0.1$ to $0.5 \text{ mol} \cdot l^{-1}$	No	110 to 150
166	1,2-Ethanediol	Butanedioic	r = 1	No	123
167	1,2-Ethanediol	Butanedioic	Excess alcohol – $[\text{Acid}]_0 = 0.5 \text{ mol} \cdot l^{-1}$	No	140
168	1,2-Ethanediol	Butanedioic	Experimental data of 167 replotted		
169	1,2-Ethanediol	Butanedioic	Experimental data of 166 replotted		
170	1,2-Ethanediol	Butanedioic	Experimental data of 165 replotted		
171	1,2-Ethanediol	Butanedioic	r = 0.833 to 25	No	130–160
172	1,2-Ethanediol	Butanedioic	r = 1	No	160
173	1,2-Ethanediol	Butanedioic	r = 1 to 5	No	195
174	1,2-Ethanediol	Butanedioic	r = 1 to 1.6	PTS (0.099 to 0.0128 eq $\cdot$ kg^{-1})	195
175	1,2-Ethanediol	Butanedioic	r = 1	H_3PO_4 (1.28×10^{-3} eq/eq COOH at time 0)	140
176	1,2-Ethanediol	Pentanedioic	r = 1	No	178–188
177	1,2-Ethanediol	Hexanedioic	r = 1	No	160–180
178	1,2-Ethanediol	Hexanedioic	r = 1	No	177–188
179	1,2-Ethanediol	Hexanedioic	r = 1	No	160
180	1,2-Ethanediol	Hexanedioic	r = 0.091 to 30	No	130–160
181	1,2-Ethanediol	Hexanedioic	r = 1	No	126–160
182	1,2-Ethanediol	Hexanedioic	r = 0.44 to 3	No	150–235
183	1,2-Ethanediol	Hexanedioic	r = 1.2 to 2.2	No	160

Table 3 Part 1 (continued)

No.	Alcohol	Acid	Solvent	Catalyst	Temperature Range (°C)
184	1,2-Ethanediol	Hexanedioic	r = 4	No	180
185	1,2-Ethanediol	Hexanedioic	r = 2	PTS 0.01 to 0.2% (mol/W)	130–180
186	1,2-Ethanediol	Hexanedioic	r = 1.2 to 2.0	PTS 0.0116 eq · kg^{-1}	151–180
187	1,2-Ethanediol	Hexanedioic	r = 1	H_2SO_4 (0.214 × 10^{-3} eq/eq COOH at time 0)	140
188	1,2-Ethanediol	Hexanedioic	r = 1	H_3PO_4 (0.172 × 10^{-3} eq/eq COOH at time 0)	140
189	1,2-Ethanediol	Heptanedioic	r = 1	No	178–188
190	1,2-Ethanediol	Octanedioic	r = 1	No	173–188
191	1,2-Ethanediol	Octadecanoic	r = 0.5 to 1.5	NaOH 0.1% W/(W acid)	180
192	1,2-Ethanediol	Nonanedioic	r = 1	No	173–188
193	1,2-Ethanediol	Decanedioic	r = 1	No	180–188
194	1,2-Ethanediol	Decanedioic	r = 1	No	140–180
195	1,2-Ethanediol	Decanedioic	r = 1	No	160
196	1,2-Ethanediol	Maleic anhydride	r = 10	No	109–140
197	1,2-Ethanediol	Maleic anhydride	r = 10	Benzenesulfonic (0.2–0.8% mol/mol anhydride)	109–140
198	1,2-Ethanediol	Maleic anhydride	r = 10	PTS (0.2 to 0.4% mol/mol anhydride)	109–140
199	1,2-Ethanediol	Maleic anhydride	r = 10	β-Naphthalenesulfonic (0.2 to 0.4% mol/mol anhydride)	109–140
200	1,2-Ethanediol	Maleic anhydride	r = 10	Aluminumbenzenesulfonate (0.2 to 0.4% mol/mol anhydride)	109–140

#	Diol	Acid	Conditions	Catalyst	Temp (°C)
201	1,2-Ethanediol	Maleic anhydride	r = 10	Titaniumbenzenesulfonate (0.2 to 0.4% mol/mol anhydride)	109–140
202	1,2-Ethanediol	Maleic acid	r = 1	No	160–170
203	1,2-Ethanediol	Fumaric	r = 1	No	170–180
204	1,2-Ethanediol	γ-Phenylitaconic anhydride	r = 1 (2nd-step of esterification)	No	160–200
205	1,2-Ethanediol	γ-(p-methoxyphenyl)itaconic anhydride	r = 1 (2nd-step of esterification)	No	160–200
206	1,2-Ethanediol	γ-(p-chlorophenyl)itaconic anhydride	r = 1 (2nd-step of esterification)	No	160–200
207	1,2-Ethanediol	γ-phenylitaconic anhydride	r = 1 (2nd-step of esterification)	PTS 0.004 mol/mol diol	160–200
208	1,2-Ethanediol	γ-(p-methoxyphenyl)itaconic anhydride	r = 1 (2nd-step of esterification)	PTS 0.004 mol/mol diol	160–200
209	1,2-Ethanediol	γ-(p-chlorophenyl)itaconic anhydride	r = 1 (2nd-step of esterification)	PTS 0.004 mol/mol diol	160–200
210	1,2-Ethanediol	Phthalic anhydride	r = 0.67 to 1.25	No	150–170
211	1,2-Ethanediol	Phthalic anhydride	r = 1	No	140–178
212	1,2-Ethanediol	Phthalic	r = 1	No	150–190
213	1,2-Ethanediol	Phthalic anhydride	r = 1	No	160–190
214	1,2-Ethanediol	Isophthalic	r = 2 to 8	No	130–235
215	1,2-Ethanediol	Isophthalic	r = 25, $[Acid]_0 = 0.569$ mol·kg^{-1}	No	180
216	1,2-Ethanediol	Isophthalic	r = 25, $[Acid]_0 = 0.569$ mol·kg^{-1}	SnC_2O_4 0.01 mol·kg^{-1}	171–193
217	1,2-Ethanediol	Isophthalic	r = 25, $[Acid]_0 = 0.569$ mol·kg^{-1}	$Zn(OAc)_2$, $2H_2O$ 0.01 mol·kg^{-1}	180
218	1,2-Ethanediol	Isophthalic	r = 25, $[Acid]_0 = 0.569$ mol·kg^{-1}	$Cd(OAc)_2$, $2H_2O$ 0.01 mol·kg^{-1}	180

Table 3 Part 1 (continued)

No.	Alcohol	Acid	Solvent	Catalyst	Temperature Range (°C)
219	1,2-Ethanediol	Isophthalic	$r = 25$, $[Acid]_0 = 0.569$ mol · kg^{-1}	Ti(OBu)$_4$ 0.01 mol · kg^{-1}	180
220	1,2-Ethanediol	Isophthalic	$r = 25$, $[Acid]_0 = 0.569$ mol · kg^{-1}	Pb(OAc)$_2$, 3 H$_2$O 0.01 mol · kg^{-1}	180
221	1,2-Ethanediol	Isophthalic	$r = 25$, $[Acid]_0 = 0.569$ mol · kg^{-1}	Co(OAc)$_2$, 4 H$_2$O 0.01 mol · kg^{-1}	180
222	1,2-Ethanediol	2-Hydroxyethyl isophthalate	Studied as concurrent reaction to 215		
223	1,2-Ethanediol	2-Hydroxyethyl isophthalate	Studied as concurrent reaction to 216		
224	1,2-Ethanediol	2-Hydroxyethyl isophthalate	Studied as concurrent reaction to 217		
225	1,2-Ethanediol	2-Hydroxyethyl isophthalate	Studied as concurrent reaction to 218		
226	1,2-Ethanediol	2-Hydroxyethyl isophthalate	Studied as concurrent reaction to 219		
227	1,2-Ethanediol	2-Hydroxyethyl isophthalate	Studied as concurrent reaction to 220		
228	1,2-Ethanediol	2-Hydroxyethyl isophthalate	Studied as concurrent reaction to 221		
229	1,2-Ethanediol	Terephthalic acid	Excess of undissolved acid $r_{total} = 2$ $r_{dissolved} =$ 36 to 56.4 $[COOH]_0 = 0.241$ to 0.273 mol · kg^{-1}	No	212–240
230	1,2-Ethanediol	Terephthalic acid	Excess of alcohol; $[Acid]_0 = 0.0427$ to 0.1913 mol · kg^{-1}	No	180–196.7
231	1,2-Ethanediol	Terephthalic acid	$r = 4$ (excess of undissolved acid)	No	200
232	1,2-Ethanediol	Terephthalic acid	$r = 4$ (excess of undissolved acid)	H$_2$SO$_4$ (96%) : 0.3% (W/W acid)	200

#	Diol	Acid	Conditions	Catalyst	T (°C)
233	1,2-Ethanediol	Terephthalic acid	r = 4 (excess of undissolved acid)	Antimonic acid (0.5% W/W acid)	200
234	1,2-Ethanediol	Terephthalic acid	r = 4 (excess of undissolved acid)	Ti(OBu)$_4$ 0.4 to 0.5% (W/W acid)	200
235	1,2-Ethanediol	Terephthalic acid	r = 4 (excess of undissolved acid)	Tin dibutyl phthalate 0.5% (W/W acid)	200
236	1,2-Ethanediol	Terephthalic	r = 10	Co(OAc)$_2$, 4H$_2$O or Mn(OAc)$_2$, 4H$_2$O 0.03% to 0.3% (W) + KCl (0 to 0.3% (W total))	240
237	1,2-Ethanediol	Terephthalic	r = 10	No	240
238	1,2-Ethanediol	Terephthalic	r = 10 to 77	ZnO 0.002 mol · kg^{-1}	180–196
239	1,2-Ethanediol	Terephthalic	Oligomeric poly(ethylene-terephthalate) [Acid]$_0$ = 0.435 to 0.639 mol · kg^{-1} r = 7.01 to 15.24	No	230–260
240	1,2-Ethanediol	Hydrogen 2-Hydroxyethyl terephthalate (T–ED)	Studied as a concurrent reaction to 238		
241	1,2-Ethanediol	Hydrogen 2-Hydroxyethyl terephthalate (T–ED)	Excess of alcohol [Acid]$_0$ = 0.120 mol · kg^{-1}	No	196.7
242	1,2-Ethanediol	Hydrogen 2-Hydroxyethyl terephthalate (T–ED)	Studied as a side reaction to 236		
243	1,2-Ethanediol	Hydrogen 2-Hydroxyethyl terephthalate (T–ED)	Studied as a side reaction to 237		
244	1,2-Ethanediol	Hydrogen 2-Hydroxyethyl terephthalate (T–ED)	r = 13 to 25	ZnO 2 × 10^{-3} mol · kg^{-1}	170–190
245	1,2-Ethanediol	Hydrogen 2-Hydroxyethyl terephthalate (T–ED)	Studied as a side reaction to 229		
246	1,2-Ethanediol	T–ED–T–ED	Studied as a side reaction to 236		
247	1,2-Ethanediol	T–ED–T–ED	Studied as a side reaction to 237		

Table 3 Part 1 (continued)

No.	Alcohol	Acid	Solvent	Catalyst	Temperature Range (°C)
248	1,2-Ethanediol	T–ED–T–ED	Studied as a side reaction to 244		
249	1,2-Ethanediol	T–ED–T–ED	Studied as a side reaction to 238		
250	2-Hydroxyethyl heptanoate	1-Heptanoic	Diphenyl oxide $[Acid]_0 = 0.25$ to 1.25 mol $\cdot$ l^{-1} $[Alcohol]_0 = 0.45$ to 1.2 mol $\cdot$ l^{-1}	No	160
251	2-Hydroxyethyl heptanoate	1-Heptanoic	Excess alcohol $[Acid]_0 = 0.3$ mol $\cdot$ l^{-1}	No	160
252	2-Hydroxyethyl heptanoate	1-Heptanoic	Excess alcohol, $[Acid]_0 = 0.25$ to 0.3 mol $\cdot$ l^{-1}	No	160
253	2-Hydroxyethyl methacrylate	Methacrylic acid	Studied as a concurrent to 100		
254	2-Hydroxyethyl benzoate	Benzoic	$r = 1$	PTS ≈ 2.5 to 10×10^{-2} mol $\cdot$ kg^{-1}	140–175
255	2-Hydroxyethyl benzoate	Benzoic	Studied as a concurrent reaction to 123		
256	2-Hydroxyethyl benzoate	Benzoic	Studied as a concurrent reaction to 124		
257	ED–T–ED	Terephthalic	Studied as a concurrent reaction to 238		
258	T–ED	T–ED	Studied as a concurrent reaction to 238		
259	ED–T–ED	T–ED	Studied as a concurrent reaction to 238		
260	ED–T–ED	T–ED	Studied as a concurrent reaction to 244		
261	T–ED	T–ED	Studied as a concurrent reaction to 244		
262	ED–T–ED	Terephthalic	Studied as a concurrent reaction to 236		
263	ED–T–ED	Terephthalic	Studied as a concurrent reaction to 237		
264	T–ED	T–ED	Studied as a concurrent reaction to 236		

Kinetics and Mechanisms of Polyesterifications

#	Diol	Acid	Conditions	T
265	T-ED		Studied as a concurrent reaction to 237	
266	ED-T-ED		Studied as a concurrent reaction to 236	
267	ED-T-ED		Studied as a concurrent reaction to 237	
268	ED-T-ED-T-ED		Studied as a concurrent reaction to 236	
269	ED-T-ED-T-ED		Studied as a concurrent reaction to 237	
270	ED-T-ED-T-ED		Studied as a concurrent reaction to 236	
271	ED-T-ED-T-ED-T-ED		Studied as a concurrent reaction to 237	
272	$HOCH_2CH_2OCH_2CH_2OH$ (Diethylene glycol)	Hexanoic	$r = 1$	No	166
273	$HOCH_2CH_2OCH_2CH_2OH$ (Diethylene glycol)	Acrylic	Benzene $r = 1$ [Acid] = 0.5 mol · l^{-1}	H_2SO_4 1.5% (W/W reactants)	Reflux
274	$HOCH_2CH_2OCH_2CH_2OH$ (Diethylene glycol)	Methacrylic	Benzene $r = 1$ [Acid] = 0.5 mol · l^{-1}	H_2SO_4 1.5% (W/W reactants)	Reflux
275	$HOCH_2CH_2OCH_2CH_2OH$ (Diethylene glycol)	Methacrylic	As in 100	PTS 0.1 mol · l^{-1}	86–87
276	$HOCH_2CH_2OCH_2CH_2OH$ (Diethylene glycol)	Succinic	$r = 1$	No	160
277	$HOCH_2CH_2OCH_2CH_2OH$ (Diethylene glycol)	Succinic	$r = 1$	No	160–180
278	$HOCH_2CH_2OCH_2CH_2OH$ (Diethylene glycol)	Hexanedioic	$r = 1$	No	160
279	$HOCH_2CH_2OCH_2CH_2OH$ (Diethylene glycol)	Hexanedioic	$r = 1$; $p > 0.80$	No	161–190
280	$HOCH_2CH_2OCH_2CH_2OH$ (Diethylene glycol)	Hexanedioic	$r = 1$	PTS 0.4% mol/mol glycol	109
281	$HOCH_2CH_2OCH_2CH_2OH$ (Diethylene glycol)	Hexanedioic	$r = 1$	No	166–202
282	$HOCH_2CH_2OCH_2CH_2OH$ (Diethylene glycol)	Hexanedioic	$r = 1$	No	160–193

Table 3. Part 1 (continued)

No.	Alcohol	Acid	Solvent	Catalyst	Temperature Range (°C)
283	$HOCH_2CH_2OCH_2CH_2OH$ (Diethylene glycol)	Hexanedioic	$r = 1$	PTS No [ ] given	109–140
284	$HOCH_2CH_2OCH_2CH_2OH$ (Diethylene glycol)	Maleic	$r = 1$	No	140–170
285	$HOCH_2CH_2OCH_2CH_2OH$ (Diethylene glycol)	Fumaric	$r = 1$	No	150–200
286	$H(OCH_2CH_2)_3OH$ (Triethylene glycol)	Acrylic	Benzene $r = 1$ $[COOH]_0 = 5$ mol $\cdot l^{-1}$	H_2SO_4 1.5% (W/W reactants)	Reflux
287	$H(OCH_2CH_2)_3OH$ (Triethylene glycol)	Methacrylic	Benzene $r = 1$ $[COOH]_0 = 5$ mol $\cdot l^{-1}$	H_2SO_4 1.5 (W/W reactants)	Reflux
288	$H(OCH_2CH_2)_3OH$ (Triethylene glycol)	Methacrylic	As in 100	As in 100	86–87
289	$H(OCH_2CH_2)_3OH$ (Triethylene glycol)	Methacrylic	$r = 10$	H_2SO_4 0.087 to 0.39 mol $\cdot l^{-1}$	90–120
290	$H(OCH_2CH_2)_3OH$ (Triethylene glycol)	Butanedioic	$r = 1$	No	160
291	$H(OCH_2CH_2)_3OH$ (Triethylene glycol)	Decanedioic	$r = 1$	No	160
292	$CH_2=C(CH_3)-COO(CH_2CH_2O)_3H$ Triethylene glycol methacrylate	Methacrylic	Studied as a concurrent reaction to 289		
293	$H(OCH_2CH_2)_n OH$ $\overline{M}_n = 975$	$HOOC(CH_2)_4CO[NH(CH_2)_{10}CO]_n OH$ $\overline{M}_n = 2090$ α,ω-dicarboxy-poly-(hexamethylene adipate) [Acid end group] = 2.04 eq $\cdot$ kg^{-1}	$r = 1$	$Ti(OBu)_4$ 0.3% (W)	250–280
294	$H(OCH_2CH_2)_n OH$ [Hydroxy end group] = 2.085 eq $\cdot$ kg^{-1}		$r = 1$ to 2	No	160–180

#	Alcohol	Acid	Conditions	Catalyst	Temp (°C)
295	H+(OCH$_2$CH$_2$)$_n$OH [Hydroxy end group] = 2.085 eq·kg^{-1}	α,ω-dicarboxy-poly-(hexamethylene adipate) [Acid end group] = 2.04 eq·kg^{-1}	r = 0.5 to 2	Ti(OBu)$_4$ 0.1% (W)	160–180
296	HO+(CH$_2$CH$_2$O)$_n$CH$_3$ [Hydroxy end group] = 1.03 eq·kg^{-1}	HOOCCH$_2$O+(CH$_2$CH$_2$O)$_n$CH$_3$ [Acid end group] = 1.09 eq·kg^{-1}	r = 0.5 to 2	No	160.6–202.8
297	H+(OCH$_2$CH$_2$)$_n$OCH$_3$ $\overline{M}_n$ = 750	HOOC(CH$_2$)$_4$CO[NH(CH$_2$)$_{10}$CO]$_n$OH $\overline{M}_n$ = 2140	r = 1	Ti(OBu)$_4$ 0.3% W	260–280
298	1,2-Propanediol	Octadecanoic	r = 0.5 to 2	NaOH 0.1% (W/W acid)	180
299	1,2-Propanediol	Acrylic	Benzene r = 1 [Acid]$_0$ = 5 mol·l^{-1}	H$_2$SO$_4$ 1.5% (W/W reactants)	Reflux
300	1,2-Propanediol	Benzoic	r = 1	PTS 1% (W)	140
301	1,2-Propanediol	Benzoic	r = 30	No	187
302	1,2-Propanediol	Benzoic	r = 30	PbO 0.002 mol·kg^{-1}	187
303	1,2-Propanediol monomaleate	—	—	No	200–210
304	1,2-Propanediol monobenzoate	Benzoic	r = 1	PTS ≃ 0.03 to 0.1 mol·kg^{-1}	140–175
305	1,3-Propanediol	Acrylic	Benzene r = 1 [Acid]$_0$ = 5 mol·l^{-1}	H$_2$SO$_4$ 1.5% (W/W reactants)	Reflux
306	1,3-Propanediol	Methacrylic	Benzene r = 1 [Acid]$_0$ = 5 mol·l^{-1}	H$_2$SO$_4$ 1.5% (W/W reactants)	Reflux
307	1,3-Propanediol	Benzoic	r = 30	No	187
308	1,3-Propanediol	Benzoic	r = 30	PbO 0.002 mol·kg^{-1}	187
309	1,3-Propanediol	Hexanedioic	r = 3	No	235
310	2,2-dimethyl-1,3-propanediol	Benzoic	r = 30	No	187
311	2,2-dimethyl-1,3-propanediol	Benzoic	r = 30	PbO 0.002 mol·kg^{-1}	187
312	2,2-dimethyl-1,3-propanediol	Maleic anhydride	r = 1.05 Study of the 2nd step of esterification	No	161.6–190.7

Table 3 Part 1 (continued)

No.	Alcohol	Acid	Solvent	Catalyst	Temperature Range (°C)
313	2,2-dimethyl-1,3-propanediol	Butanedioic	r = 1.05 Study of the 2nd step of esterification	No	160.2–189.2
314	2,2-dimethyl-1,3-propanediol	Hexanedioic	r = 1.05 Study of the 2nd step of esterification	No	152.2–182.6
315	2,2-dimethyl-1,3-propanediol	Hexanedioic	r = 3	No	235
316	2,2-dimethyl-1,3-propanediol	Decanedioic	r = 1.05	No	160.5–189.9
317	2,2-dimethyl-1,3-propanediol	o-Phthalic	r = 1.05	No	171.6–200.1
318	2,2-bis-(chloromethyl)-1,3-propanediol	Hexanedioic	r = 3	No	235
319	1,3-Butanediol	Acetic	$[Alcohol]_0 = 9.17$; $[Acid]_0 = 0.764$; $[H_2O] = 9.17$ mol · l^{-1}	PTS 0.1 mol · l^{-1}	50
320	2,3-Butanediol	Acetic	r = 0.5 to 1	H_2SO_4 0.0162 mol/mol diol	62–120
321	1,4-Butanediol	Acetic	As in 319	As in 319	50
322	1,4-Butanediol	Acrylic	Benzene r = 1 $[Acid]_0 = 5$ mol · l^{-1}	H_2SO_4 1.5% (W/W reactants)	Reflux
323	1,4-Butanediol	Methacrylic	Benzene r = 1 $[Acid]_0 = 5$ mol · l^{-1}	H_2SO_4 1.5% (W/W reactants)	Reflux
324	1,4-Butanediol	Benzoic	r = 30	No	187
325	1,4-Butanediol	Benzoic	r = 30	PbO 0.002 mol · kg^{-1}	187
326	1,4-Butanediol	Butanedioic	r = 1	No	160
327	1,4-Butanediol	Butanedioic	Dioxane r = 1 $[COOH]_0 = 0.5$ mol · l^{-1}	No	140
328	1,4-Butanediol	Butanedioic	Experimental data of 327 are replotted		
329	1,4-Butanediol	Hexanedioic	r = 1	No	150–180

#	Diol	Diacid	r	Catalyst	Temp
330	1,4-Butanediol	Hexanedioic	r = 3	No	235
331	1,4-Butanediol	Decanedioic	r = 3	No	160
332	1,5-Pentanediol	Hexanedioic	Diphenyl oxide r = 0.1 to 9.9 $[COOH]_0$ = 0.11 to 1.79 eq · l^{-1}	No	166.5–233.5
333	1,5-Pentanediol	Hexanedioic	Diethylaniline r = 1 $[COOH]_0$ = 0.245 to 1.128 eq · l^{-1}	No	178.5–208.5
334	1,5-Pentanediol	Hexanedioic	r = 1	No	208.5
335	1,5-Pentanediol	Hexanedioic	r = 3	No	235
336	1,6-Hexanediol	Maleic acid	r = 1	No	160–170
337	1,6-Hexanediol	Fumaric	r = 1	No	170
338	1,6-Hexanediol	Butanedioic	r = 1	No	160
339	1,6-Hexanediol	Hexanedioic	r = 1	No	160
340	1,6-Hexanediol	Decanedioic	r = 1	No	160
341	1,10-Decanediol	Hexanedioic	r = 3	No	235
342	1,10-Decanediol	Hexanedioic	r = 1	No	138–180
343	1,10-Decanediol	Hexanedioic	r = 1 (p > 0.82)	PTS 6.25 × 10^{-3} eq · kg^{-1}	161
344	1,10-Decanediol	Hexanedioic	r = 1 (p > 0.82)	No	202
345	1,10-Decanediol	Hexanedioic	r = 1	PTS 0.007 mol · kg^{-1}	82.8–109.1
346	1,10-Decanediol	Hexanedioic	r = 1 (p > 0.82)	No	161–190
347	1,10-Decanediol	Hexanedioic	r = 1	Benzenesulfonic acid 0.002 to 0.5 eq/eq acid	100–118
348	1,10-Decanediol	2-Methylheptanedioic	r = 1	Benzenesulfonic acid 2 × 10^{-3} eq/eq acid	78–118
349	1,10-Decanediol	2-Ethylheptanedioic	r = 1	Benzenesulfonic acid 2 × 10^{-3} eq/eq acid	78–109

Table 3 Part 1 (continued)

No.	Alcohol	Acid	Solvent	Catalyst	Temperature Range (°C)
350	1,10-Decanediol	2-Propylheptanedioic	r = 1	Benzenesulfonic acid 2×10^{-3} eq/eq acid	100–118
351	1,10-Decanediol	2-Propylheptanedioic	r = 1	No	109–118
352	1,10-Decanediol	2-Butylheptanedioic	r = 1	Benzenefulfonic acid 2×10^{-3} eq/eq acid	100
353	1,10-Decanediol	Decanedioic	r = 1	Benzenesulfonic acid 2×10^{-3} eq/eq acid	126–160
354	1,4-bis(2-hydroxy-propoxy)benzene	Fumaric	r = 1	No	200–230
355	1,3-bis(2-hydroxy-propoxy)benzene	Fumaric	r = 1	No	200–230
356	12-Hydroxyoctadecanoic acid		Melt (10^{-2} torr)	PTS 0.0354 mol · l^{-1}	116.6–160.5

Table 3 Part 2

No.	Overall order	Order in alcohol	Order in acid	Order in catalyst	k	$\Delta H^{\neq}$ kJ · mol^{-1}	$\Delta S^{\neq}$ J · K^{-} · mol^{-1}	Ref.
1	2.5	–	–	–	k = 4.5×10^{-5} eq$^{-1.5}$kg$^{1.5}$s^{-1}	–	–	305)
2	2.5	–	–	–	k = 11.5×10^{-5} eq$^{-1.5}$kg$^{1.5}$s^{-1}	–	–	305)
3	2.5	–	–	–	k = 7.97×10^{-5} eq$^{-1.5}$kg$^{1.5}$s^{-1}	–	–	305)
4	2.5	–	–	–	k = 6.37×10^{-5} eq$^{-1.5}$kg$^{1.5}$s^{-1}	–	–	305)
5	2.5	–	–	–	–	–	–	305)
6	2.5	–	–	–	k = 5.75×10^{-5} eq$^{-1.5}$kg$^{1.5}$s^{-1}	–	–	305)
7	2.5	–	–	–	k = 6.95×10^{-5} eq$^{-1.5}$kg$^{1.5}$s^{-1}	–	–	305)

8	2.5	—	—	—	$k = 7.86 \times 10^{-5}$ eq$^{-1.5}$kg$^{1.5}$s^{-1}	—	—	305
9	1.5	0.5	—	1	$k_{cat} = 18.7 \times 10^{-5}$ eq$^{-0.5}$kg$^{0.5}$s^{-1}	—	—	223
10	2	0.5	—	1.5	$k = 7.17 \times 10^{-5}$ eq^{-1}kg s^{-1}	—	—	223
11	"2"	1	—	"1"	$k = k^a + k^b$[COOH] (see Ref.)	—	—	306
12	"2"	1	—	"1"	— (see Ref.)	—	—	306
13	1.5	0.5	—	1	$k_{cat} = 18.2 \times 10^{-5}$ eq$^{-0.5}$kg$^{0.5}$s^{-1}	—	—	223
14	2	0.5	—	1.5	$k = 11.3 \times 10^{-5}$ eq^{-1}kg s^{-1}	—	—	223
15	2.5	—	—	—	$k = 5.06 \times 10^{-5}$ eq$^{-1.5}$kg$^{1.5}$s^{-1}	—	—	305
16	2.5	—	—	—	$k = 6.74 \times 10^{-5}$ eq$^{-1.5}$kg$^{1.5}$s^{-1}	—	—	305
17	—	—	—	—	$r_G = 0.97$ log $k_G = 1.37$; (120 °C; k_G unit l·mol^{-1}h^{-1}); k_G and r_G refer to Goldschmidt's equation[235–239]	E = 65	—	244
18	—	—	—	2	$k_{cat} = a + bC + (C/r)c$ mol^{-1} l min^{-1}; $a = 0.000021$; $b = 0.000889642$; $c = 0.00122829$; $C = \%$ (W) of H_2SO_4	E = 37	—	246
19	—	—	—	2	$k_{cat} = (-0.09498 + 10.89\,W + 0.03063\,r) \exp\left(16.121 - \dfrac{6714}{T}\right)$ mol^{-1}·l·min^{-1}; W = [Cat]	—	—	310
20	2	—	—	1	$k^0 = 2.46 \times 10^{-3}$ mol^{-2}l^{2}s^{-1}	E = 56	—	226
21	2.5	—	—	—	$k = 10.4 \times 10^{-5}$ eq$^{-1.5}$kg$^{1.5}$s^{-1}	—	—	305
22	see 17	1	—	—	$r_G = 0.55$, log $k_G = 1.47$ see 17	—	—	244
23	2.5	—	—	—	$k = 13.2 \times 10^{-5}$ eq$^{-1.5}$kg$^{1.5}$s^{-1}	E = 64	—	305
24	see 17	—	—	—	$r_G = 0.31$, log $k_G = 1.48$	—	—	244
25	2	1	—	1	$k^0 = 2.8 \times 10^{-3}$ mol^{-2}l^{2}s^{-1} ($r = 10$)	—	—	226
26	2	—	—	1	$k^0 = 2.8 \times 10^{-3}$ mol^{-2}l^{2}s^{-1}	—	—	226
27	2	—	—	1	$k^0 = 11 \times 10^{-3}$ mol^{-2}l^{2}s^{-1}	—	—	226

Table 3 Part 2 (continued)

No.	Overall order	Order in alcohol	Order in acid	Order in catalyst	k	$\Delta H^{\#}$ kJ·mol^{-1}	$\Delta S^{\#}$ J·K·mol^{-1}	Ref.
28	2	–	–	1	$k^0 = 11.5 \times 10^{-3}$ mol^{-2}l^{2}s^{-1}	–	–	226)
29	2	–	–	1	$k^0 = 15.2 \times 10^{-3}$ mol^{-2}l^{2}s^{-1}	–	–	226)
30	2	–	–	1	$k^0 = 11.4 \times 10^{-3}$ mol^{-2}l^{2}s^{-1}	–	–	226)
31	2	–	–	1	$k^0 = 6.4 \times 10^{-3}$ mol^{-2}l^{2}s^{-1}	–	–	226)
32	2	–	–	1	$k^0 = 7.0 \times 10^{-3}$ mol^{-2}l^{2}s^{-1}	–	–	226)
33	2	–	–	1	$k^0 = 6.0 \times 10^{-3}$ mol^{-2}l^{2}s^{-1} ($r = 10$)	–	–	226)
34	2	–	–	1	$k^0 = 3.2 \times 10^{-3}$ mol^{-2}l^{2}s^{-1} ($r = 10$) (130 °C)	68	–127	226)
35		–	1	1	–	62	–146	226)
36		–	–	1	–	61	–153	226)
37	2	1	1	1	($r = 1$) $k = 37.5 \times 10^{-5}$ mol^{-2}kg^{2}s^{-1}	–	–	221)
38	3	2	1	–	($r = 1.4$) $k_{cat} = (21.7$ to $91.7) \times$ (0.5% cat) 10^{-5} mol^{-1}kg s^{-1}	–	–	16)
39	2	1	–	1	$k^0 = (0.4$ to $2.64) \times 10^{-2}$ mol^{-2}l^{2}s^{-1} ([cat] $= 5.35 \times 10^{-2}$ mol·l^{-1} $r = 62.5$)	$E = 74$	–	16)
40	2	1	1	1		$E = 71$	$A = 4.75 \times 10^6$ mol^{-1}·l·s^{-1}	224)
41	1.5	0.5	1	–	$k_{cat} = 16.6 \times 10^{-5}$ ($r = 1$; 160 °C) to 46.2×10^{-5} ($r = 6$; 170 °C) eq$^{-0.5}$·kg$^{0.5}$·s^{-1}	–	–	223)
42	2	0.5	1.5	–	$k = 11.7 \times 10^{-5}$ eq^{-1}·kg·s^{-1} ($r = 1.1$)	–	–	223)
43	2	–	–	1	$k^0 = 1.35 \times 10^{-3}$ mol^{-2}·l^2·s^{-1}	–	–	226)
44	1.5	0.5	1	–	$k_{cat} = 31.3 \times 10^{-5}$ eq$^{-0.5}$·kg$^{0.5}$·s^{-1}	–	–	223)
45	2	0.5	1.5	–	$k = 11 \times 10^{-5}$ mol^{-1}·kg·s^{-1}	–	–	223)
46	see 17	–	–	–	$r_G = 0.15$ $\log k_G = 1.71$	–	–	244)
47	3	–	–	–	$k = 0.81 \times 10^{-4}$ mol^{-2}·l^2·s^{-1} (Ar)	–	–	307)

No.					Expression			Ref.
48	3	—	—	—	$k_{cat} = 1.26 \times 10^{-4}$ mol$^{-2} \cdot$ l$^2 \cdot$ s^{-1} (Ar)	—	—	307)
49	2.5	—	—	—	—	E = 59	—	308)
50	2	—	—	1	—	—	—	308)
51	1.5	0.5	1	—	$k_{cat} = 30.7 \times 10^{-5}$ mol$^{-0.5} \cdot$ kg$^{0.5} \cdot$ s^{-1}	—	—	223)
52	2	0.5	1.5	—	$k = 11.3 \times 10^{-5}$ eq^{-1}kg s^{-1}	—	—	223)
53	2	—	—	1	$k^0 = 1.57 \times 10^{-3}$ mol^{-2}l^{2}s^{-1}	—	—	226)
54	see 17	—	—	—	$r_G = 0.53$ $\log k_G = 1.53$	—	—	244)
55	2.5	—	—	1	—	E = 67	—	308)
56	2	—	—	1	—	—	—	308)
57	2	—	—	—	$k^0 = 2.00 \times 10^{-3}$ mol$^{-2} \cdot$ l$^2 \cdot$ s^{-1}	—	—	226)
58	1.5	0.5	1	—	$k_{cat} = 31.5 \times 10^{-5}$ eq$^{-0.5} \cdot$ kg$^{0.5} \cdot$ s^{-1}	—	—	223)
59	2	0.5	1.5	—	$k = 12.7 \times 10^{-5}$ eq$^{-1} \cdot$ kg $\cdot$ s^{-1}	—	—	223)
60	see 17	—	—	—	$r_G = 0.36$ $\log k_G = 1.67$	—	—	244)
61	2	—	—	1	$k^0 = 2.30 \times 10^{-3}$ mol$^{-2} \cdot$ kg$^2 \cdot$ s^{-1}	—	—	226)
62	1.5	0.5	1	—	$k_{cat} = 26.5 \times 10^{-5}$ eq$^{-0.5} \cdot$ kg$^{0.5} \cdot$ s^{-1}	—	—	223)
63	2	0.5	1.5	—	$k = 13.8 \times 10^{-5}$ eq$^{-1} \cdot$ kg $\cdot$ s^{-1}	—	—	223)
64	see 17	—	—	—	$r_G = 0.22$ $\log k_G = 1.84$ (120 °C)	E = 63	—	244)
65	3	2	1	—	$k = (10.7$ to $30.5) \times 10^{-5}$ eq$^{-2} \cdot$ kg$^2 \cdot$ s^{-1} (r = 1)	E = 54	—	13)
66	3	2	1	—	$k = 48 \times 10^{-5}$ eq$^{-2} \cdot$ kg$^2 \cdot$ s^{-1} (r = 1)	—	—	13)
67	2	—	—	—	$k_{cat} = 4.53 \times 10^{-3}$ eq$^{-1} \cdot$ kg$^1 \cdot$ s^{-1}	—	—	13)
68	3	—	—	—	$k = 41.7 \times 10^{-5}$ mol$^{-2} \cdot$ kg$^2 \cdot$ s^{-1} (p > 0.8)	—	—	1)
69	1.5	0.5	1	—	$k_{cat} = 26.7 \times 10^{-5}$ eq$^{-0.5} \cdot$ kg$^{0.5} \cdot$ s^{-1}	—	—	223)
70	2	0.5	1.5	—	$k = 15.2 \times 10^{-5}$ eq$^{-1} \cdot$ kg $\cdot$ s^{-1}	—	—	223)
71	2	—	—	—	$k = 2.0 \times 10^{-4}$ eq$^{-1} \cdot$ l $\cdot$ s^{-1} (190 °C)	E = 31	—	309)
72	3	—	—	—	$k = 26.2 \times 10^{-5}$ mol$^{-2} \cdot$ kg$^2 \cdot$ s^{-1} (p > 0.8)	—	—	1)

Table 3 Part 2 (continued)

No.	Overall order	Order in alcohol	Order in acid	Order in catalyst	k	$\Delta H^{\#}$ kJ·mol^{-1}	$\Delta S^{\#}$ J·K·mol^{-1}	Ref.
73	1.5	0.5	1	–	$k_{cat} = 28.7 \times 10^{-5}$ eq$^{-0.5}$·kg$^{0.5}$·s^{-1}	–	–	223)
74	2	0.5	1.5	–	$k = 16.5 \times 10^{-5}$ eq^{-1}·kg·s^{-1}	–	–	223)
75	2	–	–	–	$k = (0.39 \text{ to } 3.69) \times 10^{-5}$ No unit given. Probably (mg KOH)$^{-1}$·kg·s^{-1}	$E = 51$	–	310)
76	2 to 2.5 (pressure-dependent)				$k_{cat} = (12.7 \text{ to } 14.7) \times 10^{-5}$ eq^{-1}·kg·s^{-1}	24	–	232)
77	2.5	1	1.5	–	$k = (10.6 \text{ to } 33.4) \times 10^{-5}$ eq$^{-1.5}$·kg$^{1.5}$·s^{-1} (r = 1)	50	–210	229)
78	2	1	1	1	$k^0 = 0.829$ to 1.33 eq^{-2}·kg^2·s^{-1} (r = 1)	18.5	–210	229)
79	–	–	0	–	$k_{cat} = 1.04 \times 10^{-2}$ eq·kg^{-1}·s^{-1}	–	–	230)
80	3	1	2	–	$k = (13.4 \text{ to } 45.9) \times 10^{-5}$ eq^{-2}·kg^2·s^{-1} (r = 1)	51	–210	228)
81	3	1	2	–	$k = 0.779 \times 10^{-4}$ eq^{-2}·kg^2·s^{-1}	–	–	229)
82	2	–	–	–	$k = 2.86 \times 10^{-5}$ Unit see 75	–	–	310)
83	2	–	–	–	$k = 1.56 \times 10^{-5}$ Unit see 75	–	–	310)
84	3	–	–	–	$k_{cat} = 1.50 \times 10^{-3}$ mol^{-2}·l^2·s^{-1}	–	–	226)
85	1	–	–	1	$k^0 = (1.27 \text{ to } 2.60) \times 10^{-2}$ mol^{-1}·l·s^{-1} [cat] = 0.01 mol·l^{-1}	–	–	259)
86	1	–	–	1	$k^0 = (0.883 \text{ to } 4.2) \times 10^{-2}$ mol^{-1}·l·s^{-1} [cat] = 0.002 mol·l^{-1}	–	–	259)
87	2	–	–	1	–	$E = 43$	–	259)
88	2	0	2	1	$k'_T = (0.4613 \text{rW} - 0.14586\text{W} - 0.06758) \cdot \frac{1}{0.1390} \times 10^{\left(5.6281 - \frac{2160}{T}\right)}$ mol^{-1}·l·s^{-1} W = W % H_2SO_4	–	–	22)

Kinetics and Mechanisms of Polyesterifications

						E	ΔS	Ref.
89	2	0	—	—	$k = 0.56 \times 10^{-4}\ mol^{-1} \cdot l \cdot s^{-1}$	—	—	198)
90	—	1	—	—	—	—	—	198)
91	—	—	2	—	—	—	—	198)
92	—	—	2	—	—	—	—	198)
93	2	—	—	—	$k = (6.56\ to\ 12.1) \times 10^{-5}\ eq^{-1} \cdot kg \cdot s^{-1}$	E = 51	−154	268)
94	2.25	—	—	—	$k = (192\ to\ 346) \times 10^{-3}\ eq^{-1.25} \cdot g^{1.25} \cdot s^{-1}$	E = 48	−150	268)
95	2.5	—	—	—	$k = (569\ to\ 1200) \times 10^{-3}\ eq^{-1.5} \cdot g^{1.5} \cdot s^{-1}$	E = 61	−111	268)
96	2.25	—	—	—	$k = (202\ to\ 300) \times 10^{-3}\ eq^{-1.25} \cdot g^{1.25} \cdot s^{-1}$	E = 69	−110	268)
97	2.5	—	—	—	$k = (590\ to\ 2410) \times 10^{-3}\ eq^{-1.5} \cdot g^{1.5} \cdot s^{-1}$	E = 79	−66	268)
98	2	—	—	—	$k_{cat}^{(1)} = 81.7 \times 10^{-5}\ mol^{-1} \cdot l \cdot s^{-1}$ $k_{cat}^{(2)} = 3.6 \times 10^{-5}\ mol^{-1} \cdot l \cdot s^{-1}$	—	—	289)
99	2	—	—	—	$k_{cat}^{(1)} = 13.3 \times 10^{-5}\ mol^{-1} \cdot l \cdot s^{-1}$ $k_{cat}^{(2)} = 0.63 \times 10^{-5}\ mol^{-1} \cdot l \cdot s^{-1}$	—	—	289)
100	2	—	—	—	$k_{cat}^{(1)} = (8.83 \pm 0.83) \times 10^{-5}\ mol^{-1} \cdot l \cdot s^{-1}$ $k_{cat}^{(1)}/k_{cat}^{(2)} \approx 2$	—	—	24)
101	—	—	1	—	$k = (5.47\ to\ 6.39) \times 10^{-5}\ s^{-1}$	—	—	49)
102	—	—	0.5	—	$k_{cat} = 4.17 \times 10^{-5}\ mol^{0.5} \cdot kg^{-0.5} \cdot s^{-1}$	—	—	49)
103	—	—	0.5	—	$k_{cat} = 7.72 \times 10^{-5}\ mol^{0.5} \cdot kg^{-0.5} \cdot s^{-1}$	—	—	49)
104	—	—	0.5	—	$k_{cat} = 5.03 \times 10^{-5}\ mol^{0.5} \cdot kg^{-0.5} \cdot s^{-1}$	—	—	49)
105	—	—	0.5	—	$k_{cat} = 5.00 \times 10^{-5}\ mol^{0.5} \cdot kg^{-0.5} \cdot s^{-1}$	—	—	49)
106	—	—	0.5	—	$k_{cat} = 5.97 \times 10^{-5}\ mol^{0.5} \cdot kg^{-0.5} \cdot s^{-1}$	—	—	49)
107	—	—	0.5	0.5	$k_{cat} = (5.61\ to\ 25.7) \times 10^{-5}\ mol^{0.5} \cdot kg^{-0.5} \cdot s^{-1}$ $[cat] = 0.002\ mol \cdot kg^{-1}$	82	−143	49)
108	—	—	0.5	0.5	$k_{cat} = (1.61\ to\ 7.08) \times 10^{-5}\ mol^{0.5} \cdot kg^{-0.5} \cdot s^{-1}$ $[cat] = 0.002\ mol \cdot kg^{-1}$	82	−154	49)
109	—	—	0.5	—	$k_{cat} = 7.22 \times 10^{-5}\ mol^{0.5} \cdot kg^{-0.5} \cdot s^{-1}$	—	—	49)
110	—	—	0.5	—	$k_{cat} = 7.08 \times 10^{-5}\ mol^{0.5} \cdot kg^{-0.5} \cdot s^{-1}$	—	—	49)

Table 3 Part 2 (continued)

No.	Overall order	Order in alcohol	Order in acid	Order in catalyst	k	$\Delta H^{\#}$ kJ·mol^{-1}	$\Delta S^{\#}$ J·K·mol^{-1}	Ref.
111	–	–	0.5	–	$k_{cat} = 21.4 \times 10^{-5}$ mol$^{0.5}$·kg$^{-0.5}$·s^{-1}	–	–	49)
112	–	–	0.5	0.5	$k_{cat} = (4.94$ to $21.3) \times 10^{-5}$ mol$^{0.5}$·kg$^{-0.5}$·s^{-1} [cat] = 0.002	77	–153	49)
113	–	–	0.5	–	$k_{cat} = 15.83 \times 10^{-5}$ mol$^{0.5}$·kg$^{-0.5}$·s^{-1}	–	–	49)
114	–	–	0.5	–	No catalytic effect on esterification	–	–	49)
115	–	–	0.5	–	$k_{cat} = 6.94 \times 10^{-5}$ mol$^{0.5}$kg$^{-0.5}$s^{-1}	–	–	49)
116	–	–	0.5	–	$k_{cat} = 14.3 \times 10^{-5}$ mol$^{0.5}$kg$^{-0.5}$s^{-1}	–	–	49)
117	–	–	0.5	–	$k_{cat} = 5.94 \times 10^{-5}$ mol$^{0.5}$kg$^{-0.5}$s^{-1}	–	–	49)
118	–	–	0.5	–	$k_{cat} = 5.17 \times 10^{-5}$ mol$^{0.5}$kg$^{-0.5}$s^{-1}	–	–	49)
119	–	–	0.5	0.5	$k^0 = 1.27 \times 10^{-3}$ s^{-1} (r = 13.02; 197 °C)	E = 83	$A = 6.3 \times 10^8$ h^{-1}	205)
120	–	–	1	–	$k = 9.82 \times 10^{-5}$ s^{-1} (r = 13.02)	–	–	205)
121	2	1	1	–	$k_{cat} = 2.70 \times 10^{-4}$ mol^{-1}·kg·s^{-1}	–	–	202)
122	2	–	–	–	$k = (2.02$ to $32.7) \times 10^{-3}$ eq^{-1}·g·s^{-1}	E = 89	–86	268)
123	"2"	1	"1"	–	$k = k^a + k^b$[COOH] k^a: (see reference) k^b:	0 86	$A = 2.55 \times 10^{-5}$ kg·mol^{-1}·min^{-1} $A = 8.672 \times 10^5$ kg^2·mol^{-2}·min^{-1}	311) 311)
124	–	–	–	–	No catalytic effect on esterification	–	–	190)
125	–	–	1	–	$k = 2.22 \times 10^{-5}$ s^{-1}	–	–	190)
126	–	–	0.5	–	$k_{cat} = 2.44 \times 10^{-5}$ mol$^{0.5}$·kg$^{-0.5}$·s^{-1}	–	–	190)
127	–	–	1	–	$k = 5.28 \times 10^{-5}$ s^{-1}	–	–	190)

Kinetics and Mechanisms of Polyesterifications

128	—	—	0.5	—	$k_{cat} = 5.97 \times 10^{-5}\ mol^{0.5} kg^{-0.5} s^{-1}$	190)
129	—	—	1	—	$k = 4.83 \times 10^{-5} s^{-1}$	190)
130	—	—	0.5	—	$k_{cat} = 5.4 \times 10^{-5}\ mol^{0.5} kg^{-0.5} s^{-1}$	190)
131	—	—	1	—	$k = 7.58 \times 10^{-5} s^{-1}$	190)
132	—	—	0.5	—	$k_{cat} = 6.81 \times 10^{-5}\ mol^{0.5} \cdot kg^{-0.5} \cdot s^{-1}$	190)
133	—	—	1	—	$k = 3.14 \times 10^{-5} s^{-1}$	190)
134	—	—	0.5	—	$k_{cat} = 4.03 \times 10^{-5}\ mol^{0.5} \cdot kg^{-0.5} \cdot s^{-1}$	190)
135	—	—	1	—	$k = 7.11 \times 10^{-5} s^{-1}$	190)
136	—	—	0.5	—	$k_{cat} = 5.36 \times 10^{-5}\ mol^{0.5} \cdot kg^{-0.5} \cdot s^{-1}$	190)
137	—	—	1	—	$k = 7.78 \times 10^{-5} s^{-1}$	190)
138	—	—	0.5	—	$k_{cat} = 8.06 \times 10^{-5}\ mol^{0.5} \cdot kg^{-0.5} \cdot s^{-1}$	190)
139	—	—	1	—	$k = 5.72 \times 10^{-5} s^{-1}$	190)
140	—	—	0.5	—	$k_{cat} = 4.25 \times 10^{-5}\ mol^{0.5} \cdot kg^{-0.5} \cdot s^{-1}$	190)
141	—	—	1	—	$k = 8.53 \times 10^{-5} s^{-1}$	190)
142	—	—	0.5	—	$k_{cat} = 8.39\ mol^{0.5} \cdot kg^{-0.5} \cdot s^{-1}$	190)
143	—	—	1	—	$k = 9.22 \times 10^{-5} s^{-1}$	190)
144	—	—	0.5	—	$k_{cat} = 10.1 \times 10^{-5}\ mol^{0.5} \cdot kg^{-0.5} \cdot s^{-1}$	190)
145	—	—	1	—	$k = 2.56 \times 10^{-5} s^{-1}$	190)
146	—	—	0.5	—	$k_{cat} = 2.89\ mol^{0.5} \cdot kg^{-0.5} \cdot s^{-1}$	190)
147	—	—	1	—	$k = 12.0 \times 10^{-5} s^{-1}$	190)
148	—	—	0.5	—	$k_{cat} = 13.6 \times 10^{-5}\ mol^{0.5} \cdot kg^{-0.5} \cdot s^{-1}$	190)
149	—	—	1	—	$k = 14.1 \times 10^{-5} s^{-1}$	190)
150	—	—	0.5	—	$k_{cat} = 14.4 \times 10^{-5}\ mol^{0.5} \cdot kg^{-0.5} \cdot s^{-1}$	190)
151	—	—	1	0.5	$k_{cat} = (19.9\ to\ 97.4) \times 10^{-5} s^{-1}$ $[cat] = 3.0 \times 10^{-3}\ mol \cdot kg^{-1}$ $\quad E = 97\quad -113$	56)

Table 3 Part 2 (continued)

No.	Overall order	Order in alcohol	Order in acid	Order in catalyst	k	$\Delta H^{\neq}$ kJ · mol^{-1}	$\Delta S^{\neq}$ J · K · mol^{-1}	Ref.
152	–	–	1	0.5	$k_{cat} = 37.3 \times 10^{-5} s^{-1}$ [cat] $= 3.0 \times 10^{-3}$ mol · kg^{-1}	–	–	56)
153	–	–	1	–	$k_{cat} = 177 \times 10^{-5} s^{-1}$	–	–	159)
154	–	–	1	0.5	$k_{cat} = (10.5$ to $51.6) \times 10^{-5} s^{-1}$	E = 98	–117	56)
155	–	–	1	0.5	$k_{cat} = 18.5 \times 10^{-5} s^{-1}$	–	–	56)
156	–	–	1	–	$k_{cat} = 107 \times 10^{-5} s^{-1}$	–	–	159)
157	–	–	1	–	$k_{cat} = 74.8 \times 10^{-5} s^{-1}$	–	–	159)
158	–	–	1	–	$k_{cat} = 81.0 \times 10^{-5} s^{-1}$	–	–	159)
159	–	–	1	–	$k_{cat} = 48.7 \times 10^{-5} s^{-1}$	–	–	159)
160	–	–	1	–	$k_{cat} = 18.8 \times 10^{-5} s^{-1}$	–	–	159)
161	6	–	–	–	$k = (5.47$ to $7.50) \times 10^{-3}$ mol^{-5} · l^5 · s^{-1}	71	ln(A/min^{-1}) = 11	19)
162	6	–	–	–	$k = (2.50$ to $11.7) \times 10^{-3}$ mol^{-5} · l^5 · s^{-1}	226	ln(A/min^{-1}) = 64	19)
163	6	–	–	–	$k = (2.83$ to $10.0) \times 10^{-3}$ mol^{-5} · l^5 · s^{-1}	142	ln(A/min^{-1}) = 37	19)
164	2.5	–	–	–	$k = (656$ to $1390) \times 10^{-3}$ eq$^{-1.5}$ · g$^{1.5}$ · s^{-1}	E = 64	–107	268)
165	2	–	–	–	–	–	–	312)
166	2	–	–	–	–	–	–	312)
167	2	–	–	–	–	–	–	312)
168	3	1	2	–	$k = 5.25 \times 10^{-6}$ eq^{-2} · l^2 · s^{-1}	–	–	253)
169	3	–	–	–	$k = 8.8 \times 10^{-7}$ eq^{-2} · l^2 · s^{-1}	–	–	253)
170	3	–	–	–	$k = (0.94$ to $4.58) \times 10^{-6}$ eq^{-2} · l^2 · s^{-1} (120 to 140 °C)	105	–	253)

Kinetics and Mechanisms of Polyesterifications

#			Rate expression				Ref.
171	2.5	—	Rate = k[COOH]$^{3/2}$[OH] $\dfrac{1}{1+\dfrac{K'[OH]}{[COOH]}}$	$k \approx (0.45 \pm 0.22) \times 10^{-5}$ mol$^{-1.5}$ · kg$^{1.5}$ · s^{-1} (140 °C); $K' = 2.3 \times 10^{-2}$	59	—	7)
172	2.5	—	—	$k = 1.75 \times 10^{-5}$ eq$^{-1.5}$ · kg$^{1.5}$ · s^{-1}	—	—	10)
173	3	1	—	$k = 3.0 \times 10^{-5}$ eq^{-2} · kg^2 · s^{-1} (r = 1)	—	—	17, 271)
174	2	2	—	$k_{cat} = 82 \times 10^{-5}$ eq^{-1} · kg · s^{-1} ([cat] = 0.0099 eq · kg^{-1}) (r = 1.6)	—	—	17, 271)
175	2	—	Same rate equation as in 187		—	—	7)
176	6	—	—	$k = (2.13$ to $8.07) \times 10^{-3}$ mol^{-5} · l^5 · s^{-1}	234	ln(A/min^{-1}) = 58	19)
177	2.5	—	—	$k = (381$ to $909) \times 10^{-3}$ eq$^{-1.5}$ · g$^{1.5}$ · s^{-1}	—	-92	268)
178	6	—	—	$k = (3.28$ to $7.99) \times 10^{-3}$ mol^{-5} · l^5 · s^{-1}	138	ln(A/min^{-1}) = 36	19)
179	2.5	—	—	$k = 1.25 \times 10^{-5}$ eq$^{1.5}$ · kg$^{-1.5}$ · s^{-1}	—	—	12)
180	2.5	—	Same rate equation as in 171	$k_{140} \approx (0.45 \pm 0.22) \times 10^{-5}$ mol$^{-1.5}$ · kg$^{1.5}$ · s^{-1}; $K' = 2.3 \times 10^{-2}$	59	—	7)
181	2	—	—	$k = (1.2$ to $4.3) \times 10^{-5}$ eq · kg^{-1} · s^{-1} (doubtful unit)	50	—	264)
182	2.5	1.5	—	$k = (0.896$ to $4.40) \times 10^{-5}$ eq$^{-1.5}$ · l$^{1.5}$ · s^{-1}; (r = 2; 150 to 200 °C) $k = 9.22 \times 10^{-5}$ (r = 3; 235 °C)	50	-226	313)
183	3	1	—	$k = 1.03 \times 10^{-5}$ eq^{-2} · kg^2 · s^{-1} (r = 1.2)	—	—	270, 271)
184	2	2	—	—	—	—	270, 271)
185	2	1	—	$k_{cat} = (8.25$ to $36.0) \times 10^{-5}$ eq^{-1} · l · s^{-1}	46	-220	313)
186	2	2	—	$k_{cat} = 38.2 \times 10^{-5}$ eq^{-1} · kg · s^{-1} $\left(\begin{array}{c}r = 1.2\\160\,°C\end{array}\right)$	—	—	270, 271)

Table 3 Part 2 (continued)

No.	Overall order	Order in alcohol	Order in acid	Order in catalyst	k	$\Delta H^{\#}$ kJ·mol^{-1}	$\Delta S^{\#}$ J·K·mol^{-1}	Ref.
187	2	Rate = k[H$^+$][COOH][OH] $\dfrac{1}{1+k'\dfrac{[OH]}{[COOH]}}$ No value given				–	–	7)
188	2	Same rate equation as in 187			No value given	–	–	7)
189	6	–	–	–	k = (3.70 to 7.58) × 10^{-3} mol^{-5}·l^5·s^{-1}	121	ln(A/min^{-1}) = 30	19)
190	6	–	–	–	k = (2.28 to 7.47) × 10^{-3} mol^{-5}·l^5·s^{-1}	134	ln(A/min^{-1}) = 34	19)
191	2	1	1	–	k$_{cat}$ = 11.7 × 10^{-5} mol^{-1}·s^{-1} (r = 1) (unit: see reference)	–	–	193)
192	6	–	–	–	k = (1.89 to 7.49) × 10^{-3} mol^{-5}·l^5·s^{-1}	155	ln(A/min^{-1}) = 37	19)
193	6	–	–	–	k = (1.76 to 9.00) × 10^{-3} mol^{-5}·l^5·s^{-1}	184	ln(A/min^{-1}) = 43	19)
194	2.5	–	–	–	k = (215 to 949) × 10^{-3} eq$^{-1.5}$·g$^{1.5}$·s^{-1}	E = 58	–120	268)
195	2.5	–	–	–	k = 1.22 × 10^{-5} eq$^{-1.5}$·kg$^{1.5}$·s^{-1}	–	–	12)
196	2	1	1	–	k = (0.325 to 1.14) × 10^{-5} mol^{-1}·l·s^{-1}	52	–181	314)
197	2	1	1	–	k$_{cat}$ = (0.917 to 3.77) × 10^{-5} mol^{-1}·l·s^{-1} Extrapolated for [Cat] = 1%	59	–155	314)
198	2	1	1	–	k$_{cat}$ = (0.715 to 3.00) × 10^{-5} mol^{-1}·l·s^{-1} Extrapolated for [Cat] = 1%	60	–154	314)
199	2	1	1	–	k$_{cat}$ = (0.783 to 2.90) × 10^{-5} mol^{-1}·l·s^{-1} Extrapolated for [Cat] = 1%	55	–166	314)
200	2	1	1	–	k$_{cat}$ = (1.50 to 5.52) × 10^{-5} mol^{-1}·l·s^{-1} Extrapolated for [Cat] = 1%	57	–157	314)

#				Rate expression	E	ΔS	Ref
201	2	1	1	$k_{cat} = (0.683 \text{ to } 2.64) \times 10^{-5}$ mol$^{-1} \cdot$l$\cdot$s^{-1}. Extrapolated for [Cat] = 1%	54	−170	314
202	2	−	−	$k = (3.88 \text{ to } 5.78) \times 10^{-5}$ eq$^{-1} \cdot$kg$\cdot$s^{-1}	−	−	10
203	2	−	−	$k = (5.35 \text{ to } 7.65) \times 10^{-5}$ eq$^{-1} \cdot$kg$\cdot$s^{-1}	−	−	10
204	2	−	−	$k = (0.285 \text{ to } 2.61) \times 10^{-5}$ eq$^{-1} \cdot$kg$\cdot$s^{-1}	96	−109	315
205	2	−	−	$k = (0.210 \text{ to } 0.658) \times 10^{-5}$ eq$^{-1} \cdot$kg$\cdot$s^{-1}	50	−216	315
206	2	−	−	$k = (0.297 \text{ to } 3.95) \times 10^{-5}$ eq$^{-1} \cdot$kg$\cdot$s^{-1}	109	−73	315
207	2	−	−	$k_{cat} = (0.349 \text{ to } 2.68) \times 10^{-5}$ eq$^{-1} \cdot$kg$\cdot$s^{-1}	89	−120	315
208	2	−	−	$k_{cat} = (0.262 \text{ to } 0.891) \times 10^{-5}$ eq$^{-1} \cdot$kg$\cdot$s^{-1}	59	−192	315
209	2	−	−	$k_{cat} = (0.626 \text{ to } 8.99) \times 10^{-5}$ eq$^{-1} \cdot$kg$\cdot$s^{-1}	125	−44	315
210	2	−	1	$k = (4.9 \text{ to } 12.6) \times 10^{-5}$ mol$^{-1} \cdot$l$\cdot$s^{-1}	73	−	316
211	2	−	−	$k = (0.13 \text{ to } 0.645) \times 10^{-5}$ mol$^{-1} \cdot$l$\cdot$s^{-1}	E = 68	−	317
212	2	−	−	$k = (0.661 \text{ to } 4.22) \times 10^{-3}$ eq^{-1}kg s^{-1}	E = 82	−97	268
213	2	−	−	$k = (0.778 \text{ to } 3.49) \times 10^{-5}$ eq^{-1}kg s^{-1}	E = 87	−86	268
214	2.5	1	1.5	$k = (3.52 \text{ to } 9.22) \times 10^{-5}$ eq$^{-1.5} \cdot$l$^{1.5} \cdot$s^{-1}	−	−	313
215	−	−	1	$k = 14.3 \times 10^{-5}$ s^{-1}	105	−126	200
216	−	−	1	$k_{cat} = (67.9 \text{ to } 198) \times 10^{-5}$ s^{-1}	85	−134	200
217	−	−	1	$k_{cat} = 23.6 \times 10^{-5}$ s^{-1}	−	−	200
218	−	−	1	$k_{cat} = 10.7 \times 10^{-5}$ s^{-1}	−	−	200
219	−	−	1	$k_{cat} = 88.0 \times 10^{-5}$ s^{-1}	−	−	200
220	−	−	1	$k_{cat} = 20.5 \times 10^{-5}$ s^{-1}	−	−	200
221	−	−	1	$k_{cat} = 17.9 \times 10^{-5}$ s^{-1}	−	−	200
222	−	−	0.5	$k = 3.27 \times 10^{-5}$ mol$^{0.5} \cdot$kg$^{-0.5} \cdot$s^{-1}	−	−	200
223	−	−	0.5	$k_{cat} = (20.5 \text{ to } 60.5) \times 10^{-5}$ s^{-1}	83	−149	200
224	−	−	0.5	$k_{cat} = 6.50 \times 10^{-5}$ s^{-1}	−	−	200
225	−	−	0.5	$k_{cat} = 3.70 \times 10^{-5}$ s^{-1}	−	−	200

Table 3 Part 2 (continued)

No.	Overall order	Order in alcohol	Order in acid	Order in catalyst	k	$\Delta H^{\#}$ kJ·mol^{-1}	$\Delta S^{\#}$ J·K·mol^{-1}	Ref.
226	–	–	0.5	–	$k_{cat} = 22.53 \times 10^{-5} \cdot s^{-1}$	–	–	200)
227	–	–	0.5	–	$k_{cat} = 5.23 \times 10^{-5} \cdot s^{-1}$	–	–	200)
228	–	–	0.5	–	$k_{cat} = 3.85 \times 10^{-5} \cdot s^{-1}$	–	–	200)
229	2	0	2	–	$k = (1.52 \text{ to } 4.0) \times 10^{-3}$ mol^{-1}·kg·s^{-1} $p < 0.15$	68	–	203)
230	–	–	2	–	$k = 0.957 \times 10^{-5}$ mol^{-1}·kg·s^{-1} (196.7 °C) [acid]$_0$ = 0.191 mol·kg^{-1} 1st step	–	–	204)
231	0	–	–	–	$k = 7.8 \times 10^{-5}$ eq·kg^{-1}·s^{-1} [acid] = C^t (dissolution of acid as esterification progress)	–	–	318)
232	0	–	–	–	$k_{cat} = 263.0 \times 10^{-5}$ eq·kg^{-1}·s^{-1} [acid] = C^t (dissolution of acid as esterification progress)	–	–	318)
233	0	–	–	–	$k_{cat} = 25.6 \times 10^{-5}$ eq·kg^{-1}·s^{-1} [acid] = C^t (dissolution of acid as esterification progress)	–	–	318)
234	0	–	–	–	$k_{cat} = 30.2 \times 10^{-5}$ eq·kg^{-1}·s^{-1} [acid] = C^t (dissolution of acid as esterification progress)	–	–	318)
235	0	–	–	–	$k_{cat} = 22.6 \times 10^{-5}$ eq·kg^{-1}·s^{-1} [acid] = C^t (dissolution of acid as esterification progress)	–	–	318)
236	2	1	1	–	$k_{cat} = 46.4 \times 10^{-5}$ mol^{-1}·l·s^{-1} (Mean value) – First step of T + ED esterification [Cat] = 0.3%	–	–	181)

#					Description		Ref
237	2	1	—	—	$k = 21.2 \times 10^{-5}$ mol$^{-1} \cdot$ l $\cdot$ s^{-1} (Mean value) First step of T + ED esterification [Cat] = 0.3%	—	181
238	2	1	1	—	k_{cat} (heterogeneous)/k_{cat} (homogeneous) = $1.53 \times 10^{-5}/1.67 \times 10^{-5}$ mol $\cdot$ kg$^{-1} \cdot$ s^{-1}. Doubtful unit. More probably: mol$^{-1} \cdot$ kg $\cdot$ s^{-1}	—	62
239	2	—	1	—	$k = (16.3$ to $43.1) \times 10^{-5}$ eq $\cdot$ kg$^{-1} \cdot$ s^{-1}	$E = 72$	$A = \exp(16.50)$ h^{-1}	256
240	2	1	1	—	See 238 $0.92 \times 10^{-5}/1.19 \times 10^{-5}$ eq $\cdot$ kg$^{-1} \cdot$ s^{-1}	—	62
241	—	—	1	—	$k = 1.25 \times 10^{-5}$ s^{-1}	—	204
242	2	1	1	—	$k_{cat} = 5.36 \times 10^{-5}$ mol$^{-1} \cdot$ l $\cdot$ s^{-1} Second step of T + ED esterification See 236	—	181
243	2	1	1	—	$k = 3.72 \times 10^{-5}$ mol$^{-1} \cdot$ l $\cdot$ s^{-1} Second step of T + ED esterification See 237	—	181
244	2	1	1	—	$k_{cat}^{(1)} = 0.861 \times 10^{-5}$ mol $\cdot$ kg$^{-1} \cdot$ s^{-1} ($r = 19.52; 190 \,°C$) First step of T + ED esterification Doubtful unit. More probably mol$^{-1} \cdot$ kg $\cdot$ s^{-1}	—	206
245	—	—	1	—	$k = 3.34 \times 10^{-4}$ to 1.45×10^{-3} mol$^{-1} \cdot$ kg $\cdot$ s^{-1} Doubtful unit. More probably s^{-1}	108	203
246	2	1	1	—	$k_{cat} = 3.86 \times 10^{-4}$ mol$^{-1} \cdot$ l $\cdot$ s^{-1}	—	181
247	2	1	1	—	$k = 9.7 \times 10^{-5}$ mol$^{-1} \cdot$ l $\cdot$ s^{-1}	—	181
248	2	1	1	—	$k_{cat}^{(6)} = 3.06 \times 10^{-5}$ mol $\cdot$ kg$^{-1} \cdot$ s^{-1}. Doubtful unit. Step of the reaction of 1,2-ethane diol (ED) with terephthalic acid (T) giving ED–T–ED–T–ED	—	206

Table 3 Part 2 (continued)

No.	Overall order	Order in alcohol	Order in acid	Order in catalyst	k	$\Delta H^{\neq}$ kJ·mol^{-1}	$\Delta S^{\neq}$ J·K·mol^{-1}	Ref.
249	2	1	1	–	k_{cat} (homogeneous)/k_{cat} (heterogeneous) = $1.56 \times 10^{-5}/3.89 \times 10^{-5}$ mol^{-1}·kg·s^{-1} (see 238)	–	–	62)
250	2	1	1	–	–	–	–	14)
251	–	1 to 0	–	–	–	–	–	198)
252	–	–	1	–	–	–	–	198)
253	2	–	1	–	$k_{cat} = 4.33 \times 10^{-5}$ mol^{-1}·l·s^{-1}	–	–	24)
254	2	1	1	–	$k_{cat} = (2.98 \text{ to } 6.33) \times 10^{-4}$ mol^{-1}·kg·s^{-1} [Catalyst] = 1% (W)	33	–	202)
255	"2"	1	"1"	–	$k = k^a + k^b$ [COOH] (see reference) k^a: k^b:	124 43	$A = 8.426 \times 10^8$ kg·mol^{-1}·min^{-1} $A = 7.564 \times 10^1$ kg·mol^{-1}·min^{-1}	311)
256	–	–	–	–	No catalytic effect on esterification	–	–	311)
257	2	1	1	–	$4.72 \times 10^{-5}/8.06 \times 10^{-5}$ mol^{-1}·kg·s^{-1} (see 238) Step of T + ED esterification leading to T-ED-T-ED	–	–	62)
258	2	–	–	–	$11.1 \times 10^{-5}/4.44 \times 10^{-5}$ mol^{-1}·kg·s^{-1} (see 238) Step of T + ED esterification leading to T-ED-T-ED	–	–	62)
259	2	1	1	–	$7.5 \times 10^{-5}/3.89 \times 10^{-5}$ mol^{-1}·kg·s^{-1} (see 238) Step of T + ED esterification leading to ED-T-ED-T-ED	–	–	62)

260	2	1	—	$k_{cat}^{(4)} = 5.0 \times 10^{-5}$ mol · kg^{-1} · s^{-1} (see 244) Step of T + ED esterification leading to ED–T–ED–T–ED. Doubtful unit	206)
261	2	—	—	$k_{cat}^{(5)} = 5.28 \times 10^{-5}$ mol · kg^{-1} · s^{-1} (see 244) Step of T + ED esterification leading to ED–T–ED–T. Doubtful unit.	206)
262	2	1	—	$k = 6.3 \times 10^{-4}$ mol^{-1} · l · s^{-1} (see 236) Step of T + ED esterification leading to ED–T–ED–T	181)
263	2	1	—	$k = 7 \times 10^{-4}$ mol^{-1} · l · s^{-1} (see 237) Step of T + ED esterification leading to ED–T–ED–T	181)
264	2	—	—	$k_{cat} = 6.3 \times 10^{-4}$ mol^{-1} · l · s^{-1} (see 236) T–ED–T–ED	181)
265	2	—	—	$k = 7 \times 10^{-4}$ mol^{-1} · l · s^{-1} (see 237) Step of T + ED esterification leading to ED–T–ED–T	181)
266	2	1	—	$k_{cat} = 9.2 \times 10^{-4}$ mol^{-1} · l · s^{-1} (see 236) Step of T + ED esterification leading to ED–T–ED–T–ED	181)
267	2	1	—	$k = 3.5 \times 10^{-4}$ mol^{-1} · l · s^{-1} (see 237) Step of T + ED esterification leading to ED–T–ED–T–ED	181)
268	2	1	—	$k_{cat} = 1.58 \times 10^{-4}$ mol^{-1} · l · s^{-1} (see 236) Step of T + ED esterification leading to ED–T–ED–T–ED–T–ED	181)
269	2	1	—	$k = 1.97 \times 10^{-4}$ mol^{-1} · l · s^{-1} (see 237) Step of T + ED esterification leading to ED–T–ED–T–ED–T–ED	181)
270	2	1	—	$k_{cat} = 3.5 \times 10^{-4}$ mol^{-1} · l · s^{-1} (see 236) Step of T + ED esterification leading to ED–T–ED–T–ED–T–ED–T–ED	181)

Table 3 Part 2 (continued)

No.	Overall order	Order in alcohol	Order in acid	Order in catalyst	k	$\Delta H^{\neq}$ kJ·mol^{-1}	$\Delta S^{\neq}$ J·K·mol^{-1}	Ref.
271	2	1	1	–	$k = 6.89 \times 10^{-4}$ mol^{-1}·l·s^{-1} (see 237) Step of T + ED esterification leading to ED–T–ED–T–ED–T–ED–T–ED	–	–	181)
272	3	–	–	–	$k = 5.83 \times 10^{-5}$ mol^{-2}·kg^2·s^{-1} $p > 0.8$	–	–	1)
273	2	–	–	–	$k_{cat}^{(1)} = 71.7 \times 10^{-5}$ $k_{cat}^{(2)} = 3.55 \times 10^{-5}$ (mol^{-1}·l·s^{-1})	–	–	289)
274	2	–	–	–	$k_{cat}^{(1)} = 12 \times 10^{-5}$ $k_{cat}^{(2)} = 0.58 \times 10^{-5}$ (mol^{-1}·l·s^{-1})	–	–	289)
275	2	–	–	–	$k = 10^{-4}$ mol^{-1}·l·s^{-1} $k_{cat}^{(1)}/k_{cat}^{(2)} \simeq 2$	–	–	24)
276	2.5	–	–	–	$k = 2.42 \times 10^{-5}$ eq$^{-1.5}$·kg$^{1.5}$·s^{-1}	–	–	12)
277	2	–	–	–	$k = (2.67$ to $5.58) \times 10^{-5}$ eq^{-1}·kg·s^{-1}	E = 57	–	192)
278	2.5	–	–	–	$k = 2.05 \times 10^{-5}$ eq$^{-1.5}$·kg$^{1.5}$·s^{-1}	–	–	10)
279	3	–	–	–	$k = (2.0$ to $3.5) \times 10^{-5}$ eq^{-2}·kg^2·s^{-1}	E = 32	–	13)
280	2	–	–	–	–	–	–	1)
281	3	–	–	–	$k = 6.83 \times 10^{-5}$ mol^{-2}·kg^2·s^{-1} (202 °C) (p > 0.8)	–	–	1)
282	2.5	Same rate equation as in 171			–	59	–	7)
283	2	–	–	1	$k_{cat} = 0.95$ mol^{-1}·l·s^{-1} (140 °C)	–	–	319)
284	2	–	–	–	$k = (1.58$ to $5.05) \times 10^{-5}$ eq^{-1}·kg·s^{-1}	E = 54	–	10)
285	2	–	–	–	$k = (2.58$ to $15.5) \times 10^{-5}$ eq^{-1}·kg·s^{-1}	E = 57	–	10)
286	2	–	–	–	$k_{cat}^{(1)} = 83 \times 10^{-5}$ $k_{cat}^{(2)} = 4.4 \times 10^{-5}$ (mol^{-1}·l·s^{-1})	–	–	289)
287	2	–	–	–	$k_{cat}^{(1)} = 11.7 \times 10^{-5}$ $k_{cat}^{(2)} = 0.62 \times 10^{-5}$ (mol^{-1}·l·s^{-1})	–	–	289)

Kinetics and Mechanisms of Polyesterifications

288	2	–	–	$k_{cat}^{(1)} = 11.8 \times 10^{-5}$ $k_{cat}^{(1)}/k_{cat}^{(2)} \approx 2$ mol$^{-1} \cdot$ l $\cdot$ s^{-1}	–	–	24)
289	2	1	1	$k_{cat} = (14.1$ to $67.5) \times 10^{-5}$ mol$^{-1} \cdot$ l $\cdot$ s^{-1} [cat] = 0.26 mol $\cdot$ l^{-1}	87	$A = 1.01 \times 10^{11}$ mol$^{-1} \cdot$ min^{-1}	321)
290	2.5	–	–	$k = 2.82 \times 10^{-5}$ eq$^{-1.5} \cdot$ kg$^{1.5} \cdot$ s^{-1}	–	–	12)
291	2.5	–	–	$k = 2.03 \times 10^{-5}$ eq$^{-1.5} \cdot$ kg$^{1.5} \cdot$ s^{-1}	–	–	12)
292	2	1	1	$k_{cat} = (3.67$ to $25.5) \times 10^{-5}$ mol$^{-1} \cdot$ l $\cdot$ s^{-1} [cat] = 0.26 mol $\cdot$ l^{-1}	74	$A = 4.16 \times 10^{8}$ mol$^{-1} \cdot$ min^{-1}	321)
293	2	–	–	$k_{cat} = (2.4$ to $5.6) \times 10^{-3}$ eq$^{-1} \cdot$ kg $\cdot$ s^{-1}	59	–	189)
294	3	1	2	$k_{cat} = (6.37$ to $14.1) \times 10^{-5}$ eq$^{-2} \cdot$ kg$^{2} \cdot$ s^{-1} (r = 1)	60	–200	232)
295	1	–	–	$k_{cat} = (11.2$ to $24.2) \times 10^{-5}$ s^{-1} (r = 1)	60	–165	232)
296	3	1	2	$k_{cat} = (5.61$ to $34.9) \times 10^{-5}$ eq$^{-2} \cdot$ kg$^{2} \cdot$ s^{-1} (r = 1)	71	–170	228)
297	2	–	–	$k_{cat} = (5.3$ to $6.4) \times 10^{-3}$ eq$^{-1} \cdot$ kg $\cdot$ s^{-1}	–	–	189)
298	2	1	1	$k_{cat} = 9.44 \times 10^{-5}$ mol$^{-1} \cdot$ s^{-1} (unit: see reference)	–	–	193)
299	2	–	–	$k_{cat}^{(1)} = 21.7 \times 10^{-5}$ $k_{cat}^{(2)} = 0.87 \times 10^{-5}$ mol$^{-1} \cdot$ l $\cdot$ s^{-1}	–	–	289)
300	2	1	1	$k_{cat} = 2.8 \times 10^{-4}$ mol$^{-1} \cdot$ kg $\cdot$ s^{-1}	–	–	202)
301	–	–	1	$k = 1.83 \times 10^{-5}$ s^{-1}	–	–	49)
302	–	–	0.5	$k_{cat} = 2.00 \times 10^{-5}$ mol$^{0.5} \cdot$ kg$^{-0.5} \cdot$ s^{-1}	–	–	49)
303	3	–	–	$k = (2.23$ to $4.03) \times 10^{-52}$ (reactive group)$^{-2} \cdot$ s^{-1} (unit: see reference)	–	–	6)
304	2	1	1	$k_{cat} = (1.15$ to $3.2) \times 10^{-4}$ mol$^{-1} \cdot$ kg $\cdot$ s^{-1}	55	–	202)

Table 3 Part 2 (continued)

No.	Overall order	Order in alcohol	Order in acid	Order in catalyst	k	$\Delta H^{\neq}$ kJ · mol^{-1}	$\Delta S^{\neq}$ J · K · mol^{-1}	Ref.
305	2	–	–	–	$k_{cat}^{(1)} = 76.7 \times 10^{-5}$ $k_{cat}^{(2)} = 3.72 \times 10^{-5}$ mol^{-1} · l · s^{-1}	–	–	289)
306	2	–	–	–	$k_{cat}^{(1)} = 8.3 \times 10^{-5}$ $k_{cat}^{(2)} = 0.43 \times 10^{-5}$ mol^{-1} · l · s^{-1}	–	–	289)
307	–	–	1	–	$k = 4.61 \times 10^{-5}$ s^{-1}	–	–	49)
308	–	–	0.5	–	$k_{cat} = 4.31$ mol$^{0.5}$ · kg$^{-0.5}$ · s^{-1}	–	–	49)
309	2.5	1	1.5	–	$k = 9.15 \times 10^{-5}$ eq$^{-1.5}$ · l$^{1.5}$ · s^{-1}	–	–	313)
310	–	–	1	–	$k = 1.56 \times 10^{-5}$ s^{-1}	–	–	49)
311	–	–	0.5	–	$k_{cat} = 1.19 \times 10^{-5}$ mol$^{0.5}$ · kg$^{-0.5}$ · s^{-1}	–	–	49)
312	2.5	Same rate equation as in 171			$k = (3.20 \text{ to } 10.6) \times 10^{-5}$ eq$^{-1.5}$ · kg$^{1.5}$ · s^{-1} $K' = 4.8 \times 10^{-2}$	E = 64	–	8)
313	2.5	Same rate equation as in 171			$k = (2.40 \text{ to } 7.05) \times 10^{-5}$ eq$^{-1.5}$ · kg$^{1.5}$ · s^{-1} $K' = 4.8 \times 10^{-2}$	E = 61	–	8)
314	2.5	Same rate equation as in 171			$k = (1.38 \text{ to } 4.43) \times 10^{-5}$ eq$^{-1.5}$ · kg$^{1.5}$ · s^{-1} $K' = 4.8 \times 10^{-2}$	E = 61	–	8)
315	2.5	1	1.5	–	$k = 8.48 \times 10^{-5}$ eq$^{1.5}$ · l$^{-1.5}$ · s^{-1}	–	–	313)
316	2.5	Same rate equation as in 171			$k = (2.05 \text{ to } 5.55) \times 10^{-5}$ eq$^{-1.5}$ · kg$^{1.5}$ · s^{-1} $K' = 4.8 \times 10^{-2}$	E = 61	–	8)
317	2.5	Same rate equation as in 171			$k = (1.72 \text{ to } 4.62) \times 10^{-5}$ eq$^{-1.5}$ · kg$^{1.5}$ · s^{-1} $K' = 4.8 \times 10^{-2}$	E = 61	–	8)
318	2.5	1	1.5	–	$k \simeq 0$	–	–	313)
319	2	–	–	–	$k_{cat}^{(1)} = 0.80 \times 10^{-6}$ $k_{cat}^{(2)} = 0.467 \times 10^{-6}$ mol^{-1} · l · s^{-1}	–	–	323)
320	Do not conform to a simple-order equation					–	–	324)

No.								Ref.
321	2	–	–	–	$k_{cat}^{(1)} = 1.70 \times 10^{-5}$ $k_{cat}^{(2)} = 0.833 \times 10^{-5}$ mol^{-1}·l·s^{-1}	–	–	323
322	2	–	–	–	$k_{cat}^{(1)} = 70 \times 10^{-5}$ $k_{cat}^{(2)} = 3.5 \times 10^{-5}$ mol^{-1}·l·s^{-1}	–	–	289
323	2	–	–	–	$k_{cat}^{(1)} = 8 \times 10^{-5}$ $k_{cat}^{(2)} = 0.37 \times 10^{-5}$ mol^{-1}·l·s^{-1}	–	–	289
324	–	–	1	–	$k = 2.33 \times 10^{-5}$ s^{-1}	–	–	49
325	–	–	0.5	–	$k_{cat} = 2.69 \times 10^{-5}$ mol$^{0.5}$·kg$^{-0.5}$·s^{-1}	–	–	49
326	2.5	–	–	–	$k = 4.42 \times 10^{-5}$ eq$^{-1.5}$·kg$^{1.5}$·s^{-1}	–	–	12
327	2	–	–	–	–	–	–	312
328	3	–	–	–	$k = 2.83 \times 10^{-6}$ eq^{-2}·l^2·s^{-1}	–	–	253
329	2.5	–	–	–	$k = (2.22$ to $4.9) \times 10^{-5}$ eq$^{-1.5}$·kg$^{1.5}$·s^{-1}	$E = 47$	–	10
330	2.5	1	1.5	–	$k = 10.01 \times 10^{-5}$ eq$^{-1.5}$·l$^{1.5}$·s^{-1}	–	–	313
331	2.5	–	–	–	$k = 3.22 \times 10^{-5}$ eq$^{-1.5}$·kg$^{1.5}$·s^{-1}	–	–	10
332	2	–	–	–	$k = (10.2$ to $66.7) \times 10^{-5}$ eq^{-1}·kg·s^{-1} $(r = 1)$ [COOH]$_0 = 0.703$ to 1.10 eq·l^{-1} (166.5 °C to 233.5 °C)	$E = 50$	–	5
333	2	–	–	–	$k = (14.17$ to $32.67) \times 10^{-5}$ $(r = 1)$ [COOH]$_0 = 1.128$ to 1.125 eq·l^{-1} (178.5 °C to 208.5 °C)	–	–	5
334	2	–	–	–	$k = 29.17 \times 10^{-5}$ eq·l^{-1}	–	–	5
335	2.5	1	1.5	–	$k = 10.97 \times 10^{-5}$ eq$^{-1.5}$·l$^{1.5}$·s^{-1}	–	–	313
336	2	–	–	–	$k = (7.38$ to $10.6) \times 10^{-5}$ eq^{-1}·kg·s^{-1}	–	–	10
337	2	–	–	–	$k = 10.6 \times 10^{-5}$ eq^{-1}·kg·s^{-1}	–	–	10
338	2.5	–	–	–	$k = 5.38 \times 10^{-5}$ eq$^{-1.5}$·kg$^{1.5}$·s^{-1}	–	–	10
339	2.5	–	–	–	$k = 3.67 \times 10^{-5}$ eq$^{-1.5}$·kg$^{1.5}$·s^{-1}	–	–	12
340	2.5	–	–	–	$k = 4.10 \times 10^{-5}$ eq$^{-1.5}$·kg$^{1.5}$·s^{-1}	–	–	12

Table 3 Part 2 (continued)

No.	Overall order	Order in alcohol	Order in acid	Order in catalyst	k	$\Delta H^{\neq}$ kJ·mol^{-1}	$\Delta S^{\neq}$ J·K·mol^{-1}	Ref.
341	2.5	1	1.5	–	$k = 20.0 \times 10^{-5}$ eq$^{-1.5}$l$^{1.5}$·s^{-1}	–	–	313)
342	2	–	–	–	$k = (5.5$ to $19.8) \times 10^{-5}$ eq·kg^{-1}·s^{-1} (doubtful unit)	E = 50	–	264)
343	2	–	–	–	$k_{cat} = 1.62 \times 10^{-3}$ eq^{-1}·kg·s^{-1}	–	–	13)
344	3	–	–	–	$k = 21.2 \times 10^{-5}$ mol^{-2}·kg^2·s^{-1} (202 °C) Only for p > 0.8	–	–	1)
345	2	–	–	.1	$k^0 = (1.42$ to $4.25) \times 10^{-2}$ eq^{-2}·kg^2·s^{-1}	E = 47	–	254)
346	3	–	–	–	$k^0 = (7.50$ to $21.8) \times 10^{-5}$ eq^{-2}·kg^2·s^{-1}	E = 59	–	13)
347	2	–	–	1	$k^0 = (1.95$ to $4.07) \times 10^{-2}$ mol^{-2}·l^2·s^{-1}	49	–151	325)
348	2	–	–	1	$k^0 = (0.27$ to $1.17) \times 10^{-2}$ mol^{-2}·l^2·s^{-1}	49	–161	325)
349	2	–	–	1	$k^0 = (0.038$ to $0.147) \times 10^{-2}$ mol^{-2}·l^2·s^{-1}	50	–172	325)
350	2	–	–	1	$k^0 = (1.3$ to $2.17) \times 10^{-4}$ mol^{-2}·l^2·s^{-1}	50	–192	325)
351	3	–	–	–	$k = (3.5$ to $5.5) \times 10^{-5}$ mol^{-2}·l^2·s^{-1}	58	–	325)
352	2	–	–	1	$k^0 = 1.3 \times 10^{-4}$ mol^{-2}·l^2·s^{-1}	50	–192	325)
353	2	–	–	1	$k^0 = (2.42$ to $5.25) \times 10^{-4}$ mol^{-2}·l^2·s^{-1}	–	–	325)
354	2	–	–	–	$k = (10.28$ to $23.3) \times 10^{-5}$ eq^{-1}·kg^1·s^{-1}	E = 51	A = 0.26 × 10^{-4} min^{-1}	326)
355	2	–	–	–	$k = (8.63$ to $17.3) \times 10^{-5}$ eq^{-1}·kg^1·s^{-1}	E = 50	A = 0.161 × 10^{-4} min^{-1}	326)
356	2	–	–	1	$k^0 = (1.44$ to $6.50) \times 10^{-3}$ mol^{-2}·l^2·s^{-1}	48	–179	327)

Synthesis and Investigation of Collagen Model Peptides

Eckhart Heidemann and Werner Roth*

Technische Hochschule Darmstadt, Fachgebiet Makromolekulare Chemie, Abteilung Eiweiß und Leder, Darmstadt, Federal Republic of Germany

So far, model peptides have been synthesized mainly by the liquid- or solid-phase technique in which tripeptides with negligible degree of racemization are coupled in a stepwise manner with polypeptides with defined chain lengths. Through cross bridging of such chains, triple helical models can be obtained containing 15–30 tripeptides which fold and unfold quickly and reversibly. This paper reviews the present state of knowledge in this area and discusses some consequences for the collagen structure based on experimental findings.

Introduction	145
1 Synthesis of Sequence Polymers	148
1.1 Polycondensation of Oligopeptides	148
1.2 Fractionation of the Polymers	159
1.3 The Solid-Phase Method	160
2 The Structure of Collagen-Model Peptides	161
2.1 X-Ray Analysis	161
2.2 Circular Dichroism	162
2.2.1 Dependence of the CD Signal on Chain Length	164
2.3 Influence of Solvent on Folding Characteristics	168
2.4 Folding Characteristics of Alternating Polyheterotripeptides	171
2.5 Structure of Block Copolymers	172
2.6 Influence of Charged Tripeptides on Collagen Folding	172
2.7 Structural Characteristics of Covalently Bridged Oligopeptides	174
2.8 Collagen-Model Peptides with Antiparallel Structure	177
3 Kinetic Aspects of Triple-Helix Formation of Peptide Models Compared with those of Collagen Peptides	179
3.1 General Considerations	179
3.2 The Role of *cis-trans* Isomerism	182

---

* We thank Mrs. L. Jourdan and Mr. Dr. H. G. Neiss for the assistance during the preparation of this paper

4 Thermodynamic Aspects of the Triple Helix-Coil Transition 186
  4.1 Cooperative Processes 186
  4.2 The All-or-None Model (AON) I 186
  4.3 The Staggering-Zipper Model (SZ) II 188
  4.4 Estimation of the Nucleation Parameters 190
  4.5 Influence of Proline and Hydroxyproline 194

5 Conclusions and Outlook on Future Developments 198

6 References 200

Introduction

Collagen is a widely occurring connective tissue protein accounting for about 30% of the protein content of vertebrates. It has mainly a supporting function and contributes to the construction of the organic matrix. It is found in the skin, cartilage, tendons, bones, and teeth. Collagen, isolated under native conditions from connective tissue, shows under the electron microscope thread-like structures, so-called fibrils. The fibrils consist of collagen molecules which are displaced with respect to each other. These collagen molecules are, in turn, built up of three polypeptide chains of the general sequence Gly-X-Y. Approximately three times 340 tripeptides give rise to a twisted rod-like structure of the length of 2800 Å (Fig. 1).

Each polytripeptide chain is twisted around a threefold screw axis and exists in a secondary structure, analogous to the left-handed polyproline II-helix, i.e. with *trans*-position of the peptide bond (pitch: 8.4 Å, 3 amino acids) (Figs. 2,3).

All the three chains together build up a right-handed triple helix (pitch 85.5 Å, 30 amino acids), which is stabilized through intermolecular hydrogen bonds. The structure proposed by Piez and Traub[1] contains side chains of every third amino acid directing toward the triple-helix axis. This means that, for steric reasons, these positions could only be taken by glycine. Hydrogen bridges result between the amino group of glycine and the carbonyl group of the amino acid in the neighboring chain. This is the so-called "one-

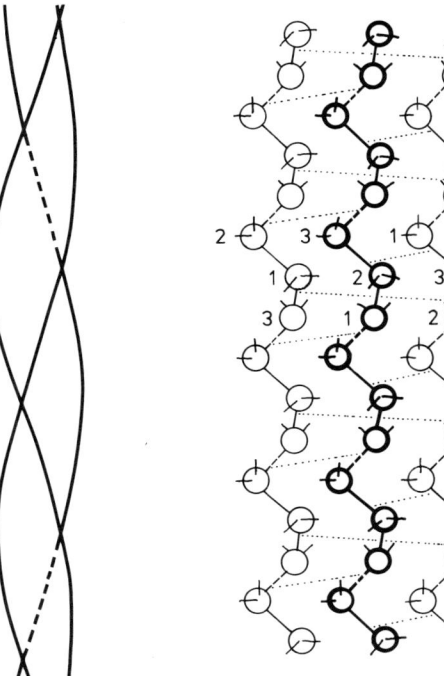

Fig. 1. Schematic representation of the chain alignment of a triple helix. Circles represent α-carbons, that of glycine is denoted number 1. *Heavy circles* indicate the chain in front, the N-terminal is at the *bottom*. The intrachain hydrogen bonds are designated by *broken lines*

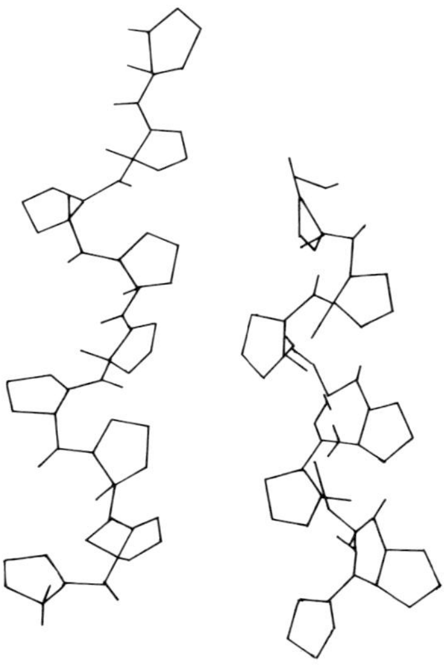

Fig. 2. Wire models of polyproline II (*left*) and I (*right*)

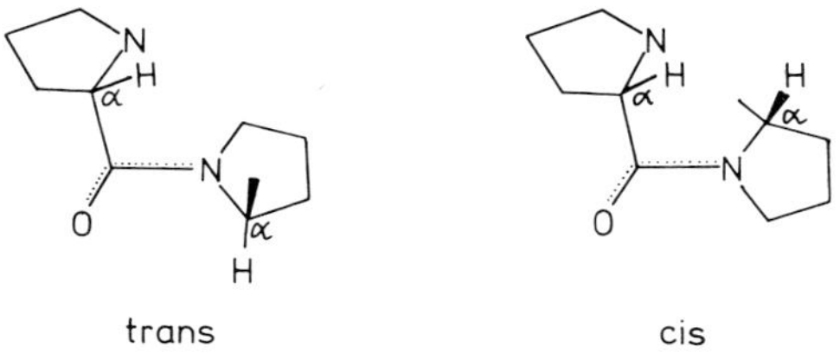

Fig. 3. *Cis* and *trans* forms of the X-L-Pro peptide bond

bonded-structure" because the amino acids in position Y remain free from intermolecular bonds. It is this system of intermolecular hydrogen bridges that imparts the triple helix its remarkable stability and tenacity. The imino residues of proline and hydroxyproline play a key role in the stabilization of the triple-helix in the collagens of all the different animals and organs[2, 3] and will be discussed later. The fixation of the dihedral angle between N and C_α is due to the pyrrolidine ring. Furthermore, the dihedral angle between C_α and the carbon atom of the carbonyl group is fixed by steric hindrance, due to the close neighbourhood of the pyrrolidine ring exactly in that position which leads to the threefold screw axis.

The detection of the collagen-like threefold symmetric polypeptides, polyglycine[4] and polyproline[5], was the first help to elucidate the collagen structure using a synthetic peptide model.

The amino acid sequence of the collagen type I (bone, skin, tendon) is nearly completely known[6]. The sequence of the different tripeptides in the α_1-chain shows a more or less statistic distribution. The content of the tripeptides in the α_1-chain of type I collagen, however, is quite different (Table 1).

A repeat of hexapeptides is found to be frequent, whereas the repeat of a nonapeptide is infrequent. The more frequently a single tripeptide occurs, the more frequent is its presence in such larger repeat hexa-, nona-, and higher peptides.

Table 1. Frequency of some tripeptides in the α_1-chain of type I collagen

Tripeptide	No.	Hydroxyproline containing tripeptides	No.	Proline containing tripeptides	No.
Gly-Ala-Asp	5	Gly-Gly-Hyp	1	Gly-Pro-Ala	31
Gly-Lys-Asp	3	Gly-Ala-Hyp	20	Gly-Pro-Val	2
Gly-Asp-Ala	5	Gly-Val-Hyp	2	Gly-Pro-Ile	4
Gly-Asp-Arg	4	Gly-Leu-Hyp	11	Gly-Pro-Pro	1
Gly-Glu-Ala	7	Gly-Phe-Hyp	7	Gly-Pro-Hyp	39
Gly-Glu-Thr	3	Gly-Pro-Hyp	39	Gly-Pro-Ser	10
Gly-Glu-Gln	3	Gly-Hyp-Hyp	1	Gly-Pro-Thr	2
Gly-Glu-Arg	10	Gly-Ser-Hyp	10	Gly-Pro-Met	3
Gly-Ala-Lys	9	Gly-Met-Hyp	1	Gly-Pro-Asp	1
Gly-Ala-Arg	9	Gly-Glu-Hyp	11	Gly-Pro-Gln	7
Gly-Asp-Arg	4	Gly-Asn-Hyp	1	Gly-Pro-Lys	7
Gly-Gln-Arg	3	Gly-Gln-Hyp	2	Gly-Pro-Arg	8
Gly-Lys-Arg	3	Gly-Lys-Hyp	1	Gly-Pro-Hyl	1
Gly-Lys-Asp	3	Gly-Arg-Hyp	5		

In the α_1-chain of the type I-collagen, the content[5] of the different categories of tripeptides is as follows:
a) Gly-Pro-X   25%
b) Gly-X-Hyp   25%
c) Gly-X-Y   50%
In a) and b) is included the most prominent tripeptide d) Gly-Pro-Hyp(Gly-Pro-Pro) 10%.

The part of non-imino acid-containing tripeptides amounts to nearly 50%. It is astonishing that at some places in the collagen chain, containing 1044 amino acid units, such sequences are accumulated, forming imino acid-free ranges of remarkable length.

Collagen undergoes a cooperative triple helix – coil – transition when warmed in aqueous solution. This means that the elementary reactions taking place at any part of the molecule are influenced not only by changes in the external conditions but by the constitution of the surrounding segments. The thermodynamic characterization of this process is, however, not possible, due to its irreversible course in the case of the native molecules. Also in the case of reversible transformations, an unambiguous correlation between the stability of the tertiary and primary structure would not be possible, because of the involvement of many different amino acids. The synthesis of model peptides with uniform sequences and chain lengths becomes, therefore, a prerequisite to the interpretation of equilibrium measurements on the basis of model concepts. This review reports the results of the research on collagen model polytripeptides of different sequences with proline or hydroxyproline, or without imino acids. Combinations of proline-rich with proline-free tripeptides in blocks or alternately in the form of polyhexapeptides were investigated. Most important are molecules of a defined chain length and high optical purity. The influence of different chain lengths on the folding behavior is a very fruitful approach. Now, in many laboratories the solid-phase technique is applied and highly sophisticated crosslinked model peptides could be built up. With the aid of these models the collagen triple-helix fold could be studied in much more detail.

1 Synthesis of Sequence Polymers

1.1 Polycondensation of Oligopeptides

Many polypeptides have been synthesized as models for the study of the natural collagen triple-helix structure (see reviews[12, 18, 28, 30]). In the early fifties, polyglycine and polyproline were synthesized via the NCAs of the respective amino acids[4, 5]. Polyhydroxyproline was also prepared and investigated by Sasisekharan[7]. The oligomers of $(Pro)_n$ and $(Hyp)_n$ were synthesized by Kobayashi, Isemura, and Sakakibara[8] and by Rothe[9], using classical peptide syntheses. The Merrifield technique used by Rothe et al.[9a] led to proline oligomers.

In the late fifties, the high concentration of the tripeptide sequence Gly-Pro-Hyp after hydrolysis of collagen with collagenase was detected[10]. Thus, the first step was to build up this sequence peptide as a model. In 1961, Andreeva et al.[11] reported that the synthesis of $(Gly-Pro-Hyp)_n$ yielded a crude product containing a collagen-like substance with a molecular weight of 25 000 (later corrected). The polycondensation of the unprotected tripeptide was carried out by activation with tetraethyl pyrophosphite. These polymers often contained a phosphite group on the N-terminal chain end. In the earlier sixties, collagen model polytripeptides were made by a Russian group[68], by the Israeli group around Katchalski, Berger, Traub et al.[13], by the group of these authors[14–17], and by the group around Blout (for a review see[18]). The polymers were formed by polycondensation of the respective active tripeptide esters.

A general step ahead in polycondensation was achieved by the application of the active ester method by DeTar et al.[19] and Kovacs et al.[29] Very soon, the nitrophenyl ester, the pentachlorophenyl ester, or the hydroxysuccinimido ester were used exclusively. The esters of the protected tripeptides could be purified by crystallization, then the N-protecting group was split off and the free peptide esters were purified again. Addition of base starts the polycondensation, resulting quickly in the formation of a viscous solution at low temperature.

The extensive purification of the tripeptide or hexapeptide activated esters seemed to be very important for achieving a high molecular weight[26]. Therefore, this technique was preferred to the one-pot method using simultaneously the free tripeptide, the hydroxy compound, and dicyclohexylcarbodiimide. N-hydroxysuccinimide[16], 3-hydroxy-4-oxo-3,4-dihydro-1,2,3-benzotriazine[16], 1-hydroxybenzotriazol[16] and catechol[25] have been used. The pentachlorophenyl active esters are more stable than the nitrophenyl esters. Because of the low solubility of all the by-products, the degree of purification was not so high as in the case of the nitrophenyl ester. The decision for one or the other ester depends on the sequence of the peptides and its solubility in different solvents. Minute amounts of free amines had to be removed. The use of succinimido esters, even after careful purification, yielded very different results[17, 26] which seemed to be due to the instability of these esters. Rapid purification and immediate use of the resulting products gave good results.

Now, it is widely known that proline at the N-terminal position causes problems of steric hindrance by using active ester couplings in the polycondensation step as well as in the synthesis of the tri- or hexapeptides. This is often a stringent restriction also if proline or glycine are intended to be in the C-terminal position.

Table 2. Survey of synthesized collagen models. Average molecular weights (M)

Polypeptide	Reference	Molecular weight or number of sequences	Method of preparation	Applied phys. or chem. methods
(Gly-Pro-Gly)$_n$	Shibnev et al. (1968)[40]	5000–9000	-OTCp	Sedimentation studies, IR
	Shibnev et al. (1969)[42]	9000	-OTCp	
	Shibnev et al. (1970)[41]			
	Shibnev et al. (1968)[40]	5000	-ONp	
	Shibnev et al. (1970)[41]			
	Bloom et al. (1966)[43]	6000		
	Shibnev et al. (1969)[44]	11 000	-ONSu	Sedimentation studies, optical rotation, melting properties in aqueous solution
	Shibnev et al. (1970)[41]		-OPfp	
	Shibnev et al. (1970)[41]		-OQu	
	Shibnev et al. (1965)[45]	3 200	S.P.	
	Rothe et al. (1970)[131]	n = 1–8		NMR, IR, optical rotation dispersion
(Pro-Gly-Gly)$_n$	Stewart (1965)[46]	6000	-ONp	X-ray diffraction studies
	Traub et al. (1966)[47]	3 500	TEPP	
	Traub (1969)[48]	1 400		
	Wolman et al. (1969)[49]		Hydrazide, NBS oxidation	End group assay
(Gly-Hyp-Gly)$_n$	Shibnev et al. (1967)[132]	4000	-OPy	
	Sakakibara et al. (1968)[31]	n = 10, 15, 20		Melting properties
	Kikuchi et al. (1969)[128]	n = 3, 5, 10, 15, 20		Substrate for protocollagen hydroxylase
	Kobayashi et al. (1970)[55]	n = 10, 15, 20		Melting properties, Optical rotation, Sedimentation studies, Titration curves, Effect of pH on T_m
(Pro-Pro-Gly)$_n$	Berg et al. (1970)[84]		S.P.	
	Olsen et al. (1971)[85]	n = 10, 20		Electr. microscopy of segments
	Kobayashi et al. (1973)[129]	n = 1, 2, 3, 4, 5, 10, 15		NMR
	Sakakibara et al. (1972)[130]			
	Okuyama et al. (1972)[86]	n = 10		X-ray diffraction studies

Table 2 (continued)

Polypeptide	Reference	Molecular weight or number of sequences	Method of preparation	Applied phys. or chem. methods
(Pro-Pro-Gly)$_n$	Shaw et al. (1975)[89]	n = 7, 8	S.P.	Optical rotation dispersion
				Melting properties in solution
				Detailed thermodynamic analysis
				Kinetic studies
	Engel et al. (1977)[92]	n = 3–7		Thermodynamic studies
	Bruckner et al. (1975)[33]		Classical synthesis	Sedimentation studies
				CD Measurements
	Sutoh et al. (1974)[35]	n = 10, 12, 14, 15		Sedimentation studies
				Thermal melting curves
				Thermodynamic analysis
				Hydrodynam. properties
(Gly-Pro-Pro)$_n$	Shibnev et al. (1966)[52]	100 000	– OCp	X-ray diffraction studies
	Shibnev et al. (1971)[53]			
(Gly-Pro-Pro)$_n$	Shibnev et al. (1969)[50]	11 700		Different M_n by end group assay and sedimentation studies
(Pro-Gly-Pro)$_n$	Engel et al. (1965)[54]	5000	TEPP	Melting properties in aqueous solution
	Engel et al. (1965)[13]	1000–12 000		
	Traub et al. (1966)[47]			X-ray diffraction studies
	Yonath et al. (1969)[56]			
	Shibnev et al. (1970)[41]	25 000	– ONSu	
	Shibnev et al. (1966)[57]		TEPP	
	(1969)[50]			
(Gly-Hyp-Hyp)$_n$	Andreeva et al. (1967)[58]	157 800	– OPCp	
	Andreeva et al. (1967)[59]	16 000	– OTCp	
	Shibnev et al. (1968)[60]			
	(1970)[41]			

Synthesis and Investigation of Collagen Model Peptides

Peptide	Reference	Method	MW / n	Studies
(Pro-Hyp-Gly)$_n$	DeTar et al. (1972)[61]	–ONp	45 000	Molecular weight by Archibald end group assay; Sedimentation studies
	DeTar (1966)[62]	S.P.	40 000	Sedimentation studies
	Sakakibara et al. (1976)[36]	S.P.	n = 5, 10	Melting properties in solution
(Pro-aHyp-Gly)$_n$	Inouye et al. (1976)[63]	S.P.	n = 10	Melting properties; CD Measurements
(Gly-Pro-Hyp)$_n$	Shibnev et al. (1970)[41]	–OPfp		
	Shibnev et al. (1970)[41]	–ONSu		
	Shibnev et al. (1970)[41] (1969)[44]	–OTCp	20 000	
	Shibnev et al. (1970)[41]	–OPCp	100 000	
	Poroshin et al. (1969)[65]	–OPCp	100 000	
	Huggins et al. (1966)[64]	–ODnp		
	Poroshin et al. (1969)[65]		36 000	
	Andreeva et al. (1961)[66]			
	Shibnev et al. (1964)[67]			
	Shibnev et al. (1969)[50]	TEPP		
	Shibnev et al. (1965)[68]			
	Rogulenkova et al. (1964)[69]			
	Huggins et al. (1966)[64]			
(Pro-Gly-Hyp(Ac))$_n$	Traub et al. (1966)[70]	TEPP	2 700–3 500	X-Ray diffraction studies
(Pro-Hyp-Gly)$_n$	Shibnev et al. (1970)[41]			
	Shibnev et al. (1969)[42]		n = 10	Sedimentation studies; CD measurements
	Weber et al. (1978)[34]			
(Pro-Hyp(Ac)-(Gly)$_n$	Weber et al. (1978)[34]	S.P.	n = 10	Melting properties
(Pro-Pro-Gly)$_n$				
(Pro-DL-Pro-Gly)$_n$	Frey et al. (1976)[71]		n = 10	

Table 2 (continued)

Polypeptide	Reference	Molecular weight or number of sequences	Method of preparation	Applied phys. or chem. methods
(Ala-Gly-Pro)$_n$		9000		Fractionation
				M_n by gel chromatography
(D-Ala-Gly-D-Pro)$_n$	Fairweather et al. (1972)[26]	12700	-OPCp	
(Ala-Gly-D-Pro)$_n$		3700		
(D-Ala-Gly-Pro)$_n$		4300		
(Ala-Gly-Pro)$_n$				
	Shibnev et al. (1968)[40]			
(Ala-Gly-Pro)$_n$	Shibnev et al. (1969)[50]			
(Pro-Ala-Gly)$_n$	Heidemann et al. (1968)[72]	15000	-ONSu	Vapor pressure
				Osmometry
		5000–15000	TEPP	Gel chromatography
	(1968)[73]			ORD
			-ONp	Melting properties
	Brown et al. (1972)[74]	12400	-ONp	M_n by Archibald method
				Gel chromatography
				CD measurements
(Pro-Ala-Gly)$_n$	Brown et al. (1972)[75]	>6900	-ONp	
	Heidemann et al. (1973)[16]		DCC/HOBn	Gel chromatography
	Heymer et al. (1976)[21]	n = 4, 5, 6, 7, 8, 12, 13, >31	DCC/HOBn	Thermodynamic analysis
	Jones (1969)[76]	12000		
	Bloom et al. (1966)[77]	15000	-ONp	M_n by Archibald method
	Oriel et al. (1966)[78]	14000		
(Gly-Pro-Ala)$_n$	Brown et al. (1969)[79]	14000		
	Shibnev et al. (1969)[44]	20000	-ONSu	Low-temperature
	Shibnev et al. (1968)[40]	4300	-OTCp	CD measurements
	Shibnev et al. (1969)[50]	2000		
(Gly-Ala-Pro)$_n$	Shibnev et al. (1969)[53]	10000	-TEPP	

Synthesis and Investigation of Collagen Model Peptides

Peptide	Reference	MW	Method	Studies
(Ala-Pro-Gly)$_n$	Segal et al. (1969)[80]	5 000–14 000	–ONSu	Equilibrium sedimentation studies; H-D exchange; Melting properties
	Blout et al. (1971)[81]	5 000–14 000	–ONp	M_n by Archibald method; End group assay
	DeTar et al. (1972)[61]	2 500		
	DeTar (1966)[62]	<500		
	Blout et al. (1971)[82]		–ONSu	IR studies; X-ray diffraction studies; CD measurements in different solvents
	Huggins et al. (1966)[64]		TEPP	
(Pro-Gly-Ala)$_n$	Heidemann et al. (1967)[83]	2 100–2 500		IR, H-D exchange; ORD; CD melting properties
(Gly-Ala-Pro)$_n$	(1968)[73]	2 500–11 000		
(Pro-Ala-Hyp)$_n$	Shibnev et al. (1969)[50]		–OPCp	X-Ray diffraction
	Andreeva (1967)[59]	13 500		
	Shibnev et al. (1970)[41]			
(Ala-Ala-Gly-Pro-Pro-Gly)$_n$	Segal (1969)[87]	2 000–12 000	–ONSu	Gel chromatography; Sedimentation studies; H-D exchange; Melting properties
(Pro-Ala-Gly-Pro-Pro-Gly)$_n$		2 000–12 000	–OSU	
(Ala-Pro-Gly-Pro-Ala-Gly)$_n$	Segal et al. (1969)[88]	2 000–12 000	–OSU	X-ray diffraction studies
(Pro-Gly-Val)$_n$	Rapaka et al. (1975)[90]	2 000–6 000	–OPCp	Gel chromatography
(Gly-Val-Pro)$_n$		2 500–6 000	–ODnp	
(Val-Pro-Gly)$_n$	Huggins et al. (1966)[64]	109 000	Stepwise classical approach	CD measurements in different solvents; IR; Viscosity; Melting properties
(NVa-Gly-Pro)$_n$	Bonara et al. (1974)[91]	10 000; n = 2–8		

Table 2 (continued)

Polypeptide	Reference	Molecular weight or number of sequences	Method of preparation	Applied phys. or chem. methods
(Leu-Gly-Pro)$_n$	Shibnev et al. (1969)[50]	8 700	TEPP	Gel chromatography
(Pro-Ser-Gly)$_n$	Brown et al. (1972)[74]	16 000	–ONp	M_n by Archibald method
				End-group assay
	(1972)[75]	18 000		CD measurements in different solvents
				Calculation of thermodynamic data
(Pro-Ser-Gly)$_n$		$M_{max} = 32 000$	TEPP	Gel chromatography
(Pro-Gly-Ser)$_n$		10 000	–ONSu	Molecular weight determination by end-group titration
				Vapor pressure osmometry
(Ser-Gly-Pro)$_n$	Heidemann et al. (1969)[15]	25 000	TEPP	Melting studies in water shows randomly coiled structure for all fractions
(Gly-Ser-Pro)$_n$		12 500		
(Gly-Pro-Ser)$_n$		17 000		
(Ser-Pro-Gly)$_n$		25 000		
(Pro-Ser-Gly)$_n$	Shibnev et al. (1970)[95]	4 000	POCp	Viscosity
(Ser-Gly-Pro)$_n$	Bell et al. (1975)[103]	6 000–10 000	POCp	
(Gly-Ser-Pro)$_n$	Shibnev et al. (1970)[96]	2 500	–OPCp	
(Ser-Pro-Gly)$_n$	DeTar et al. (1972)[61]	11 000	–ONp	Archibald end-group assay
		10 000		Equilibrium sedimentation
	(1966)[62]	10 000		
		10 000		
(Gly-Ser-Hyp)$_n$	Shibnev et al. (1970)[96]	6 080	–OPCp	
(Gly-Gly-Lys(Tos))$_n$	Poroshin et al. (1968)[99]	10 000	–ONp	

Peptide	Reference	Activation	MW	Methods
(Gly-Pro-Lys)	Shibnev et al. (1969)[50]	TEPP		
(Gly-Pro-Lys(Tos))$_n$	Shibnev et al. (1966)[100]	−ONp	7000–10000	
	(1967)[101]	−ONp	7000–10000	
	(1967)[101]	−OTCp		
(Gly-Pro-Lys(Tos))$_n$	Poroshin et al. (1970)[102]	−ONp		
(Lys-Gly-Pro)$_n$	Bell et al. (1975)[103]	−OPCp	6000–10000	Viscosity
	Heidemann et al. (1977)[22, 24]	−OSu	700– >13000	CD measurements
				Melting properties
				Gel chromatography
(Pro-Lys-Gly)$_n$	Heidemann et al. (1977)[22, 24]	−OPCp	700– >13000	
(Pro-Pro-Lys)$_n$		−OPCp	800– > 1800	
		−OPCp	290– > 2900	
		−OSu	400– 2900	
(Lys-Gly-Ala)$_n$		−OPCp	800– 3300	
(Orn-Gly-Pro)$_n$	Bell et al. (1975)[103]	−OPCp	7000	Gel chromatography
(Gly-Pro-Glu(OBzl))$_n$	Poroshin et al. (1969)[104]	−ONp	2500	CD measurements
(Pro-Glu-Gly)$_n$	Neiss et al. (1976)[20]	−OPCp	6000	Melting properties in different solvents
				Gel chromatography
(Ala-Glu(OET)-(Gly)$_n$	Anderson et al. (1970)[113]	−ONp	9000	X-ray diffraction studies, IR, light scattering, End-group assay
(Hyp-Glu-Gly)$_n$	Khalikov et al. (1968)[97]	−OTCp	8500	
(Hyp-Glu(OBzl)-(Gly)$_n$	Shibnev et al. (1970)[98]	−OTCp	8500	
(Glu-Gly-Ala)$_n$	Zegelmann et al. (1970)[105]	−ONp	8700	
		−OPCp	6780	
(Gly-OBzl)-(Gly-Ala)$_n$	Zegelmann et al. (1972)[106]	−ONp		

Table 2 (continued)

Polypeptide	Reference	Molecular weight or number of sequences	Method of preparation	Applied phys. or chem. methods
(Glu-Gly-Pro)$_n$	Beil et al. (1975)[103]	200–4500	–OPCp –ONSu	M_n by Archibald method X-ray diffraction studies CD measurements, IR
	Doyle et al. (1970)[107]			
(Ala-Ala-Gly)$_n$	Takahashi (1969)[108] (1970)[109] (1971)[110]	35 000	–ONp	
(Pro-Leu-Gly)$_n$	Kitaoka et al. (1958)[93]	6600 6600	TEPP DCC	
(Gly-Pro-Leu)$_n$	Kitaoka et al. (1953)[93]	6000	TEPP	
	Shibnev et al. (1969)[50]	6000		
(Gly-Pro-Leu)$_n$	Shibnev et al. (1969)[44]	4000–6000	–ONSu	
(Pro-Gly-Leu)$_n$	Rapaka et al. (1975)[90]	2000–4000	–OPCp	
(Pro-Gly-Phe)$_n$		2500–6000	–ONp	Gel chromatography
(Phe-Pro-Gly)$_n$	Tamburro et al. (1968)[94]	3500	–ONp –ONp TEPP	
(Pro-Phe-Gly)$_n$				
(Gly-Pro-Tyr)$_n$				
(Typ-Gly-Pro)$_n$	Bell et al. (1975)[103]	20 000	–OPCp	
(Pro-Ser-Gly)$_n$	Rapaka et al. (1975)[90]	2000–6000	–OPCp	Gel chromatography
(Ser-Pro-Gly)$_n$				
(Hyp-Ser-Gly)$_n$	Khalikov et al. (1968)[97] Shibnev et al. (1970)[98]	6130	–OTCp	
(Gly-Gly-Lys-Gly-Pro-Pro)$_n$	Khodadadeh et al. (1977)[22]	580–10000	–OPCp	Gel chromatography Melting properties in different solvents at different pH
	Erian (1978)[120]	n = 6, n = 18		
(Gly-Glu-Arg-Gly-Pro-Pro)$_n$	Erian (1978)[120]	n = 6		

Peptide	Reference	MW / n	Activation	Methods
(Gly-Pro-Lys-Gly-Pro-Pro)$_n$	Erian (1978)[120] Neiss (1977)[121] Khodadadeh (1976)[122]	13 000	−OPCp	Melting properties CD measurements in different solvents at different pH, fluorescence studies after labeling, melting properties
(Pro-Glu-Gly-Pro-Ala-Gly)$_n$	Neiss et al. (1975)[20] Neiss et al. (1975)[20] Neiss et al. (1975)[20]	3 000–6 000 3 000–7 000 3 000–7 000	EEDQ/DMF −OSu −OPCp	
(Gly-Pro-Glu-Gly-Pro-Pro)$_n$	Neiss et al. (1975)[20]		−OPCp	(Insoluble)
(Lys-Gly-Pro-Ala-Gly-Pro)$_n$	Heidemann et al. (1977)[24]	13 500 n = 8 n = 10	−OPCp	Gel chromatography Melting properties by CD measurements
	Khodadadeh (1976)[122]	12 000	−OSu	
(Ala-Ala-Gly)$_n$	Brack et al. (1972)[111]	14 000		Viscosity IR
(Glu-Glu(OBzl)-Lys(Tos))$_n$	Burichenko et al. (1970)[114]	3 200	−OPCp	X-ray diffraction studies CD measurements
(Glu-(OMe)-Gly-Lys(Tos))$_n$	Yusupov et al. (1969)[115]		−OTCp	
(Glu-(OBzl)-Gly-Lys(Tos))$_n$	Yusupov et al. (1969)[115]		−ONp	
	Yusupov et al. (1969)[115]		−OPCp	
(Glu-Ser-Gly)$_n$	DeTar et al. (1967)[116]	5 000–10 000	−ONp	
(Lys-Arg-Gly)$_n$	Burichenko et al. (1972)[117]	7 000	−OTCp	
(Lys-Ala-Gly)$_n$ (Ala-Gly-Lys(Z))$_n$	Johnson (1968)[118] Fairweather et al. (1972)[119]	13 100	−OPCp	
(Pro-Pro-Gly)$_n$(Ala-Pro-Gly)$_n$- (Pro-Pro-Gly)$_n$	Sutoh et al. (1974)[35]	n = 7 m = 1 n = 6 m = 3 n = 5 M = 5	S.P.	Melting properties CD measurements in different solvents

Table 2 (continued)

Polypeptide	Reference	Molecular weight or number of sequences	Method of preparation	Applied phys. or chem. methods
Boc(Gly-Pro-Pro-)$_5$(Gly-Pro-Leu)$_5$(Gly-Pro-Pro)$_n$	Roth et al. (1978)[37]		L.P.	Trimer crosslinked polymer
[(Gly-Pro-Pro-)$_n$Gly-Ala-Ala)$_3$Aha]$_3$Lys-Lys-Gly	Divanidis (1978)[161]	n = 4, 5, 6		CD measurements
PTC-[(Pro-Ala-Gly)$_{12}$]$_3$	Heidemann et al. (1977)[24]			Melting properties
				Kinetic measurements
[(Gly-Pro-Ala)$_{10}$Aha]$_3$-Lys-Lys-Gly	El Sheikh (1978)[123]			
PTC-[(Ala-Hyp-Gly)]$_3$	Roth et al. (1978)[37]	n = 5–10		
[(Ala-Gly-Pro)$_n$Aha]$_3$-Lys-Lys-Gly-OH		n = 5–15		
PTC-[(Gly-Pro-Lys)$_n$]$_3$	Erian (1978)[120]			
[(Gly-Pro-Lys)$_n$-Aha]$_3$-Lys-Lys-Gly-OME$_n$	Heppenheimer (1978)[124]	n = 7–14	LSP	CD measurements in different solvents and at different pH
				Melting properties
				Racermated, melting properties
[(Gly-Pro-Ala)$_n$-Aha]$_3$-Lys-Lys-Gly-OMe		n = 7–14		CD measurements in different solvents,
[(Gly-Pro-Pro)$_4$-Aha]$_3$-Lys-Lys-Gly-OMe		n = 5–14		Melting properties

Abbreviations of active esters or coupling methods.
S.P.: solid-phase technique; L.P.: liquid-phase technique; OPTcp: pentachlorophenyl ester; OPTcp: trichlorophenyl ester; TEEP: tetraethyl pyrophosphit; ONp: p-nitrophenyl ester; ONSu: N-hydroxysuccinimido ester; OPFp: pentafluorophenyl ester; OQu: 8-hydroxyquinyl ester; OPy: 3-hydroxypyridyl ester; ODnp: 2,4-dinitrophenyl ester; DCC: dicyclohexylcarbodiimide; HOBn: 3-hydroxy-4-oxo-3,4-dihydro-1,2,3-benzotriazin; Opi: N-hydroxypiperidine; EEDQ: 2-ethoxy-1-ethoxycarbonyl-1,2-dihydroquinoline; Tos: p-toluenesulfonyl; PTC: propanetricarboxylic acid; OBu^t: tert-butyl ester; Nva: norvaline; Aha: aminohexanoic acid; Orn: ornithine

Active esters are preferably synthesized by either the mixed anhydride method, e.g.[22, 26], or by the dicyclohexylcarbodiimide method[9, 32].

Table 2 shows a list of collagen model peptides which have been prepared. Many efforts have been made to prevent racemization. The polycondensation reaction seemed to be more sensitive to racemization than the coupling steps preparing the monomeric tripeptide. Therefore, the sequence of the monomer was selected with Gly or Pro at the C-terminal chain end, because racemization is mostly favored at the carboxy-activated amino acid, and these amino acids cannot racemize.

1.2 Fractionation of the Polymers

The next approach was to reduce the polydispersity of the chain length by fractionation. In the earliest preparation, fractionation was performed by precipitation or by dialysis, but the results were not very satisfying.

Gel filtration on Sephadex G 25, G 50 or G 75 beads was just available. After standardization with peptides and proteins of known length, the molecular weight of the respective fraction could be determined. Also cyanogen bromide peptides of collagen chains were available in the later sixties, thus leading to a more consistent standardization[15–17].

The separation on gel beads was more and more refined. Oligomer series of many polyhexapeptides from $n = 1$ to $n = 9$ could be completely separated[16, 20–24] (see Fig. 4). The hexapeptide oligomers could be separated better than the respective tripeptides.

The use of Sephadex G 50 superfine, Biogel P 150, and Biogel P 10, P 6, or P 4 gave the sharpest separations. With increasing degree of oligomerization the sharpness of the

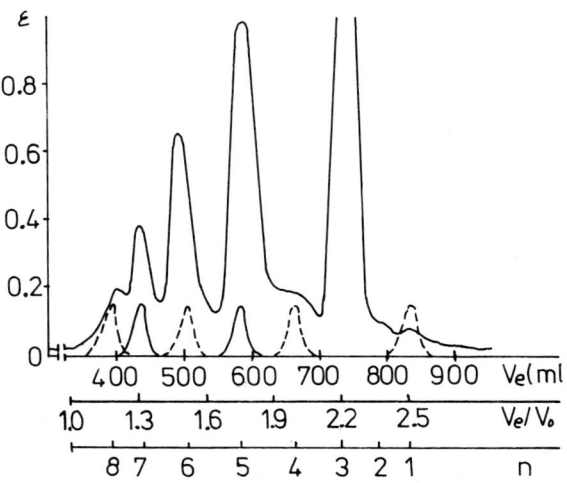

Fig. 4. Elution of the crude polyhexapeptide (Pro-Ala-Gly-Pro-Ala-Gly)$_n$ from a Biogel P 4 column. (34 × 1400 mm, 200–400 mesh, 0.05 M acetic acid, 50 ml/h, 20 °C). The upper curve was obtained by the first chromatography, the curves below were obtained by the fifth rechromatrography. ε = absorbance at 230 nm, V_e = elution volume, V_o = 310 ml, n = number of hexapeptide units

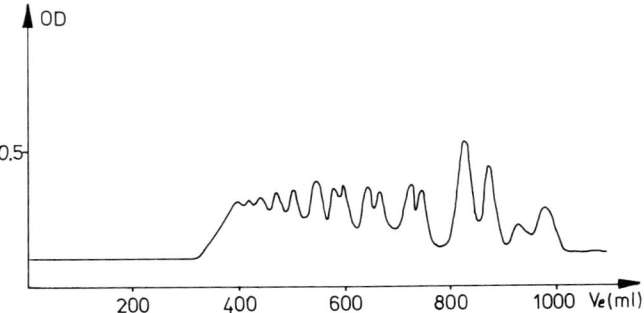

Fig. 5. Elution of a crude polytripeptide (Pro-Ala-Gly)$_n$ from a column of Biogel P4 (200–400 mesh, 20 °C, 0.05 M acetic acid, column 1700 × 35 mm, 40 ml/h); 100 mg applied. OD = optical density at 230 nm; V_e elution volume

separation decreased. The limit seems to be the hexapeptide oligomer with n = 10. Even in special cases is a separation of oligotripeptides possible, using very long columns of Sephadex G 50 superfine or Biogel P 4 (Fig. 5, Table 2).

1.3 The Solid-Phase Method

Difficulties due to side reactions (cyclization) and a broad molecular weight distribution accompanying the polycondensation of active esters led to the application of methods wherein the polymers are built up stepwise. In 1968, Sakakibara et al.[31] introduced the solid-phase technique using Merrifield's resin. By stepwise addition of *tert*-pentyloxycarbonyl tripeptides, they have synthesized (Pro-Pro-Gly)$_n$ with n = 5, 10, 15 and 20.

The peptide is removed from the polystyrene resin by means of hydrogen fluoride. The couplings were nearly complete by using a threefold excess of the N-protected tripeptide. A fractionation of the resulting oligotripeptide, however, has been performed. The folding behavior in water was much more pronounced than in the case of the polymers obtained by the "old" TEPP method[13].

Independently, Rothe et al.[131] have synthesized the oligomers of (Gly-Pro-Gly)$_n$, n = 1–8, using the solid-phase method. The sequences Z(Gly-Pro-Pro)$_n$O-(t-But), n = 3–7, have been synthesized by Weber and Nitschmann[34], (Pro-Hyp-Gly)$_n$, n = 5–10, by Sakakibara et al.[36], (Pro-Pro-Gly)$_n$, n = 10, 12, 14, 15 by Sutoh and Noda[35].

A new modification, which promises much more application, is the liquid solid-phase method applied by W. Roth et al.[37] to the preparation of highly sophisticated collagen model peptides. Due to the great difficulties and low coupling yields in the preparation of (Pro-Ala-Gly)$_n$ using the Merrifield technique, a generally applicable method was looked for. The principle is based on a proposal made by Bayer and Mutter[39] employing a soluble poly(ethylene glycol) resin which yields better coupling results. Activation is performed preferably by the symmetric anhydride method. After coupling, the liberated protected amino acid is very easily recovered and used again in the next step. A twofold excess of a tripeptide active ester gives a nearly complete coupling in many cases. Examination of the degree of coupling and repeating the coupling steps again, if necessary, can easily be performed.

The peptide is removed from the resin by treating the peptide-containing resin with triethylamine in methanol for a longer time. Extensive column chromatography purification, however, is necessary in each case.

2 The Structure of Collagen-Model Peptides

2.1 X-Ray Analysis

The tertiary structure of proteins can be determined directly only in the crystalline state making use of X-ray analysis. The molecular conformation of the collagen molecule has been derived mainly from its wide-angle X-ray diffraction pattern. Triple-helix structure fiber photographs were interpreted by Ramachandran and Kartha in 1955 as well as by Rich and Crick (for a review see[1]). The X-ray patterns of polyproline I and II and of polyglycine II[4], with indications for the threefold screw axis, show a resemblance to the collagen pattern. Thus, the first investigation on the collagen-like polytripeptide (Gly-Pro-Hyp)$_n$ was made by Andreeva et al.[11] and by Rogulenkowa et al.[69] on oriented film preparations which show the main features of collagen. The prominent reflection is an equatorial reflection at 11.9–12.5 Å and a meridional reflection at 2.82 Å, and further two near meridional reflections on the third and on the seventh layer lines are observed. Since that time, many polytripeptides built up from tripeptide sequences found in collagen have been investigated by X-ray analysis on oriented specimens, namely by the group of Traub et al.[47, 48] at the Weizmann Institute, and by the Russian group around Shibnev[57], indicating the similarity to collagen. The meridional reflection from 2.75–2.92 Å obviously corresponds to the translation of an amino acid residue in the direction of the fiber axis. The near meridional reflection is due to the side-chain distance. All the X-ray photographs show a diffuse halo which is interpreted as a pattern resulting from the geometry of the triple-helix backbone.

Besides these general features, synthetic polypeptides have a much more regular sequence and could give X-ray diffraction patterns with more reflections, depending on their crystallinity. Crystallization and formation of the helix structure depend strongly on the solvent and its water content. In the case of collagen models with only few differences in the primary structure, unit cell parameters can be compared and a similarity to the collagen triple-helix structure can be observed. Although in many cases the polymers show a threefold screw axis and a translation in fiber direction as well as a rotational angle of 108°, the molecular packing of the chains is not always triple-helical (Segal and Traub[80], 1969, with (Ala-Pro-Gly)$_n$). Yonath and Traub[56] (1969) elucidated the structure of the Gly-Pro-Pro sequence. Only a one-bonded structure is possible in this case. A two-bonded structure model (two hydrogen bonds on one tripeptide), however, can be built up with polymers of the sequence Gly-amino acid-imino acid. It has, however, been shown that the sequence polymers (Gly-Gly-Pro)$_n$ and (Gly-Ala-Pro)$_n$ do not behave as two-bonded structures. Studies of polyhexapeptides by Segal[87], in which a strong helix-forming tripeptide and a weak helix-forming tripeptide are combined to give the monomeric hexapeptide, have revealed that the fold is forced to a collagen-like triple helix, e.g. with (Gly-Ala-Pro-Gly-Pro-Pro)$_n$ and (Gly-Ala-Pro-Gly-Pro-Ala)$_n$. This shows the positive influence of a strong helix-forming sequence on a weak helix-forming sequence.

Since these investigations could be carried out only in the crystalline state, the question of the dynamics of the triple-helix formation and of the correlation of its stability with the amino acid sequence could be answered only with the help of other methods working in solution.

The shape of the collagen molecule is a characteristic rod: It is very long, thin and rigid. Because of the dramatic steric changes of collagen in the helix-coil (and vice versa) transition, many of the physical methods (e.g. ultracentrifugation, light scattering, viscosity, birefringence, electron microscopy and others) can be used for the determination of the transition. Model peptides, however, are much shorter and produce much smaller effects because of the small differences in the length-to-thickness ratio. Moreover, the rod-like or extended molecule character need not necessary be caused by a triple-helical structure.

2.2 Circular Dichroism

Chirooptical properties give more subtle information on the conformational behavior of biopolymers and peptides in solution. In early experiments, optical rotation and optical rotatory dispersion (ORD) have been recognized as valuable techniques, followed more recently by significant progress and refinements in the equipment which have resulted in the routine measurements of applied circular dichroism (CD).

Native collagen exhibits a large levorotation in the visible and near ultraviolet regions and a sudden decrease to less negative values upon denaturation. The near-UV visible ORD obeys the one-term Drude equation. In the far ultraviolet, a large negative (intrinsic) Cotton effect centered around 205 nm is observed, similar to that of polyproline II, which is, however, markedly displaced to longer wavelengths. Subsequent curve fitting analysis in the far ORD of polyproline II revealed at least the presence of two optically active transitions at 207 and 221 nm implying a hidden positive Cotton effect with weak rotational strength at 221 nm[18] (Figs. 6 and 7).

These optically active absorption bands are more easily resolved in the corresponding CD spectra (Figs. 8 and 9).

Neutral salt-soluble collagen as well as acid-soluble collagen show a CD spectrum (Fig. 8) having bands at 198 nm, $\Theta = -53\,000$ (deg $\cdot$ cm^2 $\cdot$ dmol^{-1}), and at 223 nm, $\Theta = 7500$ (deg $\cdot$ cm^2 $\cdot$ dmol^{-1}), the ratio between both being 7:1. Pysh has calculated the CD spectra of α-helix, β-structure, polyproline I and II[127].

He has confirmed the correct correlation between sign and order of magnitude of the observed circular dichroism bands on the one hand and the geometry of the polypeptide structures on the other. He found (loc. cit.) "the agreement between the calculated spectra and experimental data to be better than qualitative for the four polypeptide configurations. Two of these conformations, the α-helix and the poly(L-proline) I helix, can be considered as favorable cases in the sense that one term in the complete expression for rotational strength dominates, resulting in two large π-π^* CD bands of opposite sign. The effect of other contributions is then merely to modify these bands by some relatively smaller amount. For the antiparallel pleated sheet and the poly(L-proline) II helix, it seems that no single term dominates, but it nevertheless appears that evaluation

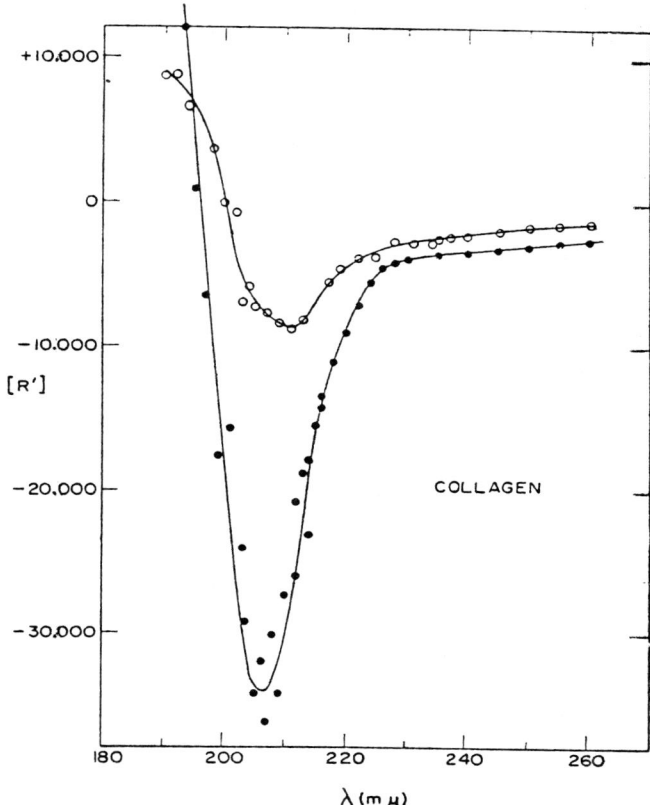

Fig. 6. The far ultraviolet rotatory dispersion of native calf skin collagen ●–●–● in 0.01 molar acetic solution. The ultraviolet rotatory dispersion of the same preparation of calf skin collagen heated at 50 °C for 30 min, cooled to 25 °C and measured immediately, O–O–O. Concentrations were between 0.0076 and 0.076%. Data from Blout et al.[18a]

of all terms in a consistent manner reproduces the sign of the dominant π-π^* CD-band observed: negative for poly(L-proline) II and positive for the antiparallel pleated sheet. The sign of the n-π^* rotational strength is correctly predicted in all cases."

So far, the CD of poly(L-proline) II has been interpreted without ambiguity for a rather strong left-handed helix.

Arguing in a similar way, one can assume that a significant degree of regularity also in the case of polypeptides with repetitive Gly-X-Y-sequences indicates a left-handed helical structure. Figure 10 illustrates the dependence of possible assemblies of the polypeptides on different proline contents. If glycine occurs in every third position, the formation of a triple helix is undoubtedly preferred, but hexagonal or sheet-like aggregation cannot be excluded principally, provided that the aggregation pattern fits the angle of about 120° to enable the formation of hydrogen bonds[47, 56]. As will be described later, these assumptions are valid. For example, the CD of collagen and that of the segment α1-CB2 containing 36 amino acid residues are very close to each other[1a]. Furthermore, certain

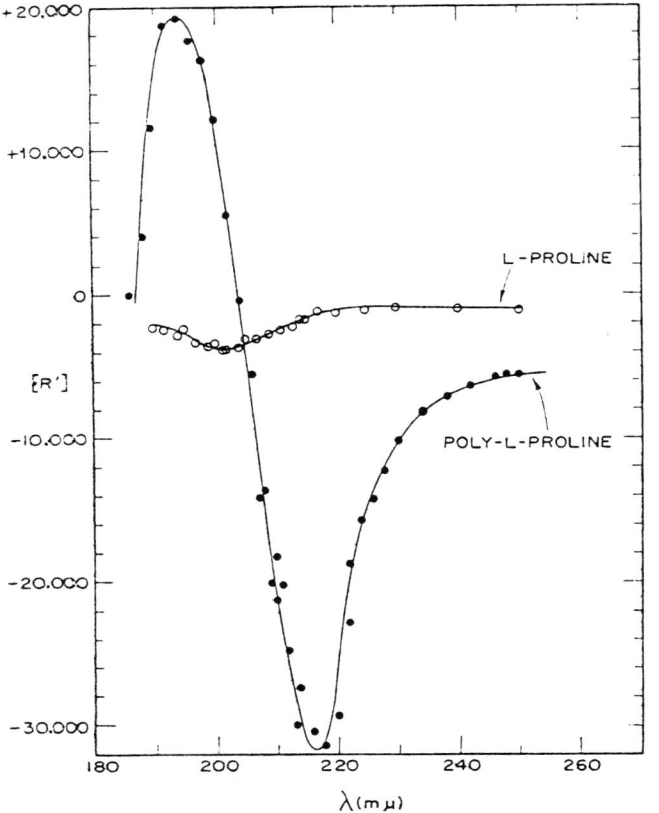

Fig. 7. The far ultraviolet rotatory dispersions of poly-L-proline II ●–●–● and L-proline ○–○–○ in aqueous olution. Concentrations were between 0.015 and 0.147% in a 1-mm cell. Reproduced with permission from Blout et al.

model peptides exhibit the denaturation characteristics of collagen and its segment α1-CB 2. By heating the solution, the positive band vanishes, resulting a small negative plateau at about 225 and the negative band increases to less negative values. This transition exhibits a sigmoidal dependence on temperature. The mechanism of the triple-helix folding or unfolding will be discussed later.

2.2.1 Dependence of the CD Signal on Chain Length

Investigations of the oligomers of the hexapeptide (Pro-Ala-Gly-Pro-Ala-Gly)$_n$ are representative for studies on many other oligomers. At low temperature, these oligomers show an increase of the negative band around 198 nm and a change of the curve-position at 223 nm from negative values to less negative values and finally to a more or less distinct maximum, depending on chain length.

Synthesis and Investigation of Collagen Model Peptides

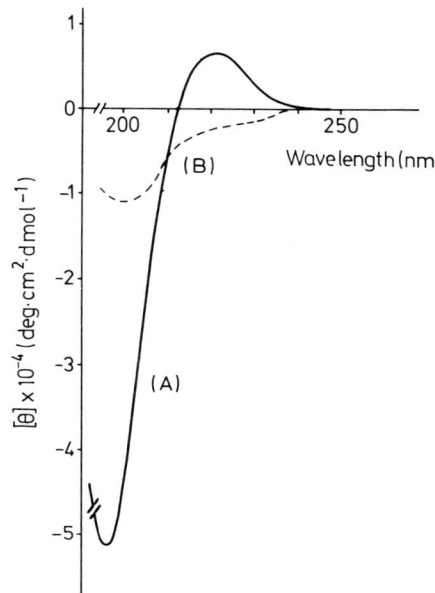

Fig. 8. CD spectra of collagen in 0.01 M acetic acid. (A): At 24 °C in the native state, (B): at 44 °C in a randomly coiled state. Data from Piez et al.[1a]

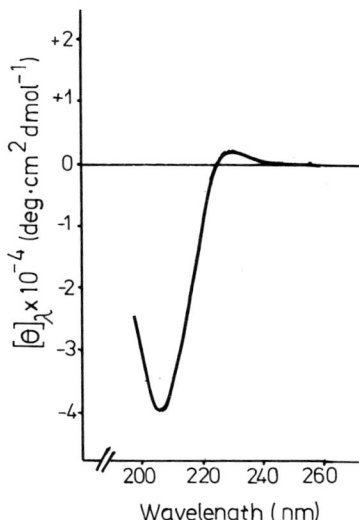

Fig. 9. CD for poly(L-proline) II helix in water. Data from Carver et al.[18]

These curves were obtained after 14–21 days' incubation at 1 °C. By heating, the 198 nm band shifts to less negative and the band at 223 nm to more negative values (Fig. 11).

The time needed for the coil-to-helix transition is relatively long, whereas it is much shorter for the inverse reaction. As the time needed for reaching the equilibrium in the

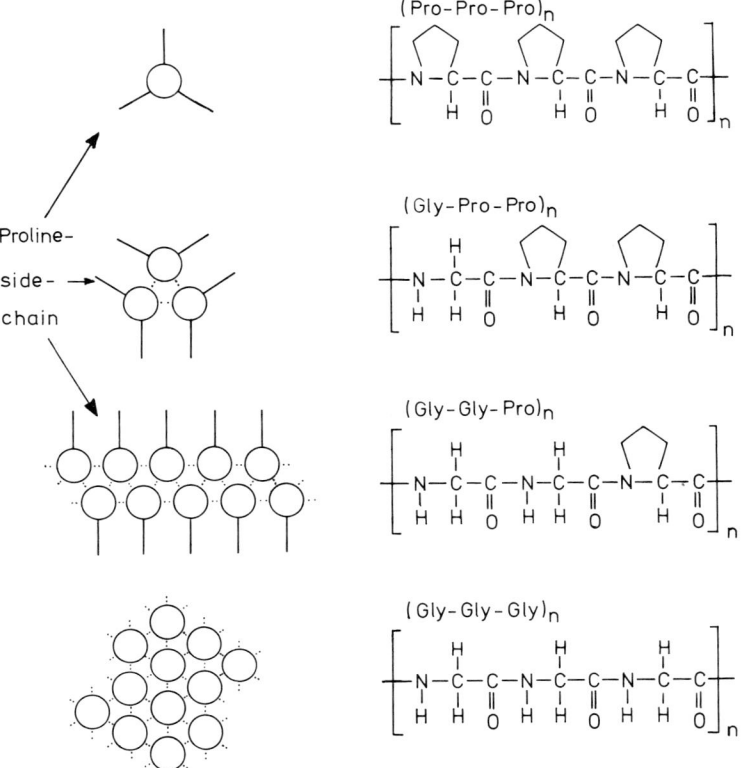

Fig. 10. Possible collagen-linke structure depending on the hydrogen-bonding capacity of certain polypeptides based on their proline content

first mentioned case is very long, one can either wait until the equilibrium is established or measure immediatly. We call these cases measuring under helix or coil conditions, respectively.

Under random coil conditions, the solution is heated to 50–70 °C for 10 min, then quickly cooled to the respective temperatures between 0 and 70 °C and measured instantly.

Figure 12 shows the linear dependence of the ellipticity on temperature (measured at 198 nm).

Figure 13 shows that the analogous dependence on temperature after longer incubation time at 1 °C becomes nonlinear with higher oligomers. Helix formation starts with n = 6. The ellipticity, depending on the chain lengths at random coil conditions, is shown in Fig. 14.

The magnitude of the CD signal increases with increasing chain length. The amplitude, however, approximates a constant value at long chain lengths, minimizing the chain N-end and C-end effects. Heymer[21] calculated from these data that the influence of each chain end affects a range of 7 hexapeptide residues. Different chain segment mobilities can be the reason for this dependence of the CD signals in the oligomer series.

Synthesis and Investigation of Collagen Model Peptides

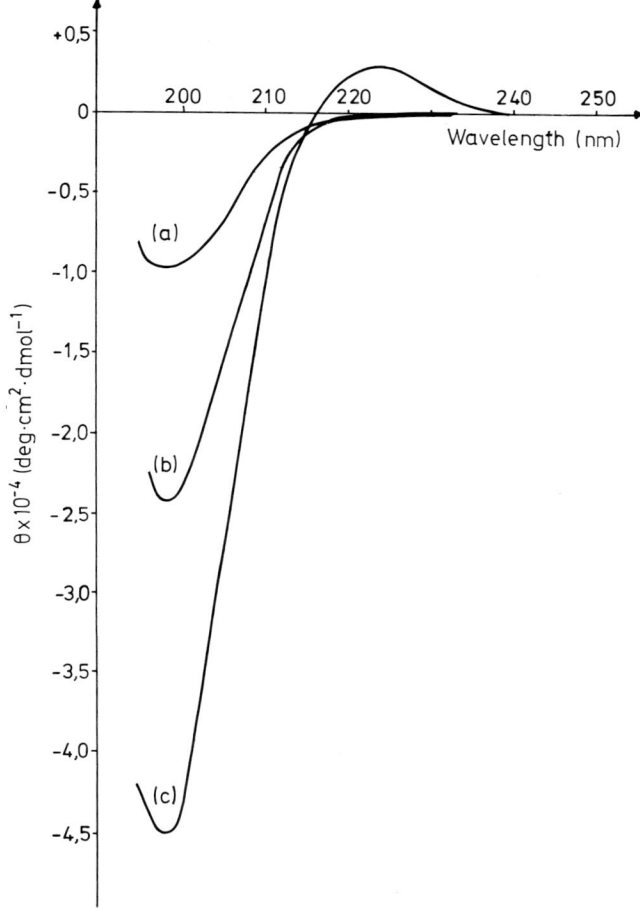

Fig. 11. Ellipticity of (Pro-Ala-Gly-Pro-Ala-Gly)$_n$ n = 12–13. (a): At 1 °C after 1 day; (b): at 1 °C after 14 days; (c): of (Pro-Ala-Gly)$_{31}$ after 14 days at 3 °C in water

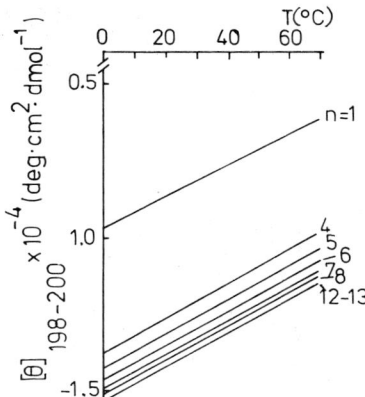

Fig. 12. Ellipticity of (Pro-Ala-Gly-Pro-Ala-Gly)$_n$ with n = 1–13 in water, depending on the temperature under random coil conditions

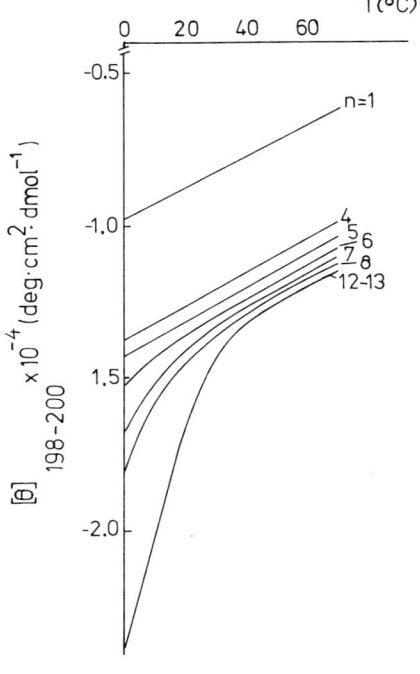

Fig. 13. Ellipticity of (Pro-Ala-Gly-Pro-Ala-Gly)$_n$ with n = 1–13 in water, depending on the temperature under helix conditions

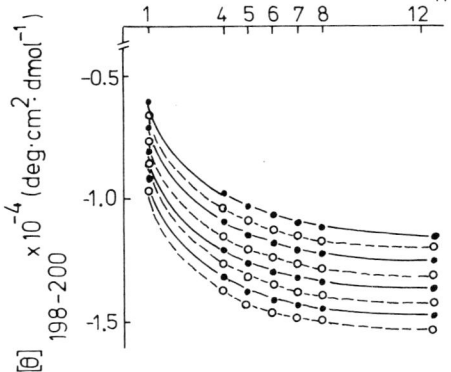

Fig. 14. Ellipticity of (Pro-Ala-Gly-Pro-Ala-Gly)$_n$ with n = 1–13 in water in the presence of 0.15 M NaF at pH 5 at different temperatures, depending on the chain lenghts under random coil conditons. The curves are measured (from below) at 1 °C (○) and 10 °C (●), 20 °C (○) and 30 °C (●), 40 °C (○) and 50 °C (●), 60 °C (○) and 70 °C (●)

2.3 Influence of Solvent on Folding Characteristics

Figures 15 and 16 demonstrate folding in the 1,1,1,3,3,3-hexafluoro-2-propanol/ethylene glycol (HFP/EG) mixture (1:2) and in 1,3-propandiol in comparison to Fig. 13, which describes helix formation in water. The structure formation is much more pronounced. This is indicated by the more negative signals of the CD spectrum at 198 nm. The negative values of Θ for the octamer increase from -1.8×10^{-4} deg · cm^2 · dmol^{-1} in

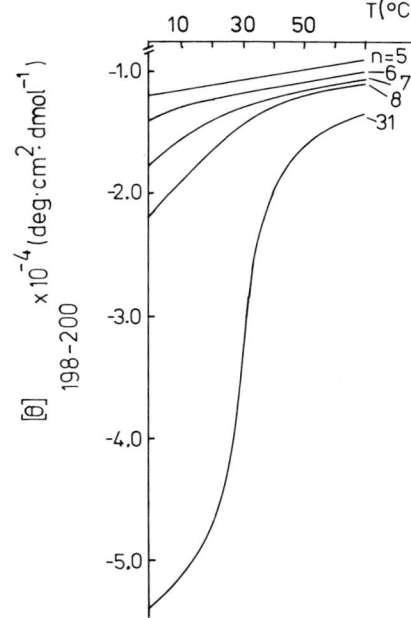

Fig. 15. Ellipticity of (Pro-Ala-Gly-Pro-Ala-Gly)$_n$ (n = 5–8) in 1,1,1,3,3,3-hexafluoro-2-propanol/ethylene glycol (1:2), depending on the temperature under helix conditions

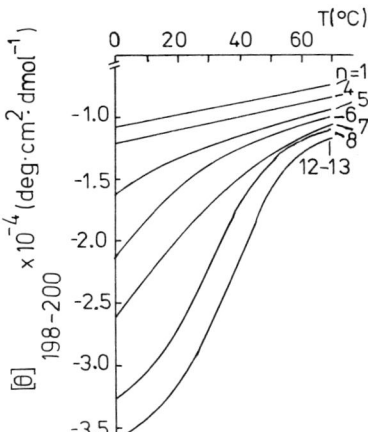

Fig. 16. Ellipticity of (Pro-Ala-Gly-Pro-Ala-Gly)$_n$ (n = 1–13) in 1,3-propandiol, depending on the temperature under helix conditions

water to -2.2×10^{-4} deg · cm^2 · dmol^{-1} in HFP/EG to -3.3×10^{-4} deg · cm^2 · dmol^{-1} in 1,3-propandiol at 0 °C. Sigmoidally shaped curves arise, indicating a cooperative transition. 1,3-propandiol enhances the velocity of folding, shown in Fig. 17.

Even under random coil conditions at rapid cooling and instant measuring, no linear temperature function can be observed. The deviation from linearity is the higher, the more the temperature decreases and the chain length increases.

The enhancement of the fold in structure-forming solvents leads to a strong increase of the values near 223 cm^{-1} to a positive peak (see Fig. 18).

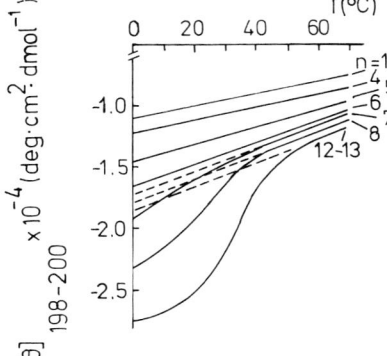

Fig. 17. Ellipticity of (Pro-Ala-Gly-Pro-Ala-Gly)$_n$ (n = 1–13) in 1,3-propandiol, depending on the temperature under random coil conditions

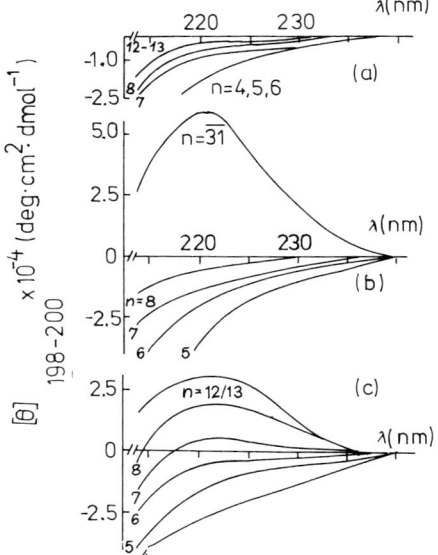

Fig. 18 a–c. Ellipticity or (Pro-Ala-Gly-Pro-Ala-Gly)$_n$ (n = 1–13). (**a**) In water; (**b**) in 1,1,1,3,3,3-hexafluoro-2-propanol/ethylene glycol (1:2); (**c**) in 1,3-propanediol, depending on the chain length under helix conditions

In some cases, these organic solvents cause no stronger folding but adversely (Lys-Gly-Pro)$_n$ folds to a lower extent in 1,1,1,3,3,3-hexafluoro-2-propanol/ethylene glycol or in 1,3-propandiol than in water (Table 3).

The sequence Gly-Glu-Arg... folds better in water than in methanol as shown below. Many folding investigations have shown clearly different influences of the solvents. Even in many cases, folding can be enhanced in methanol, or methanol/water mixtures, trifluoroethanol, buffer solutions, or higher concentrated salt solutions.

Table 3. CD parameters of some polyhexapeptides

Polypeptide	n	Solvent[a]	$[\Theta]$198 nm[b]	$[\Theta]$225 nm[b]
(Pro-Ala-Gly-Pro-Ala-Gly)$_n$	12/13	Water	−20 000	−300
	31	Water	−45 000	+2430
	31	HFiP/EG	−52 000	+5700
(Pro-Ala-Gly-Pro-Ala-Gly)$_n$, tBu	12/13	1,3-Propandiol	−35 000	+2950
(Pro-Glu-Gly-Pro-Ala-Gly)$_n$, OH	10	Water	−20 000	+360
	10	HFiP/EG	−37 000	+1100
(Pro-Glu-Gly-Pro-Ala-Gly)$_n$	>10	HFiP/EG	−47 000	+4700
(Gly-Pro-Glu-Gly-Pro-Pro)$_n$, tBu tBu	>5	HFiP/EG	−27 000	+2700
(Pro-Glu-Gly-Pro-Glu-Gly)$_n$	=10	HFiP/EG	−16 250	−300
(Lys-Gly-Pro-Lys-Gly-Pro)$_n$	>10	Water	−15 800	−100
	>10	HFiP/EG	−12 800	
	>10	1,3-Propandiol	−5 200	
	>10	MeOH/Water	−17 900	
(Lys-Gly-Pro-Ala-Gly-Pro)$_n$	>10	Water	−15 000	−200
	>10	HFiP/EG	−12 700	
	>10	1,3-Propandiol	−10 000	
(Gly-Pro-Lys-Gly-Pro-Pro)$_n$	>10	Water	−35 000	+2200
	>10	HFiP/EG	−12 000	−500
(Gly-Pro-Lys-Gly-Pro-Pro)$_n$, OH	>10	1,3-Propandiol	−19 000	+1000
(Gly-Glu-Lys-Gly-Pro-Pro)$_n$	>10	Water (pH 5.5)	−20 000	+400

[a] HFiP-1,1,1,3,3,3-hexafluoro-2-propanol; EG = ethylene glycol
[b] In deg · cm^2 · dmol^{-1}

2.4 Folding Characteristics of Alternating Polyheterotripeptides

Segal[87], using Gly-Pro-Pro, was the first who studied the influence of two tripeptides combined in an alternating sequence on the polymer model peptide. For the evaluation of the mutual influences, the magnitude of the CD bonds has been utilized. Polyheterotripeptides clearly show an average in fold strength of the single sequences if they are investigated in the form of polytripeptides. This is proved by using combinations of Gly-Pro-Pro and Gly-Pro-Ala with Gly-Ala-Pro and Gly-Ala-Ala. The latter two show a poor or no fold at all in the form of a polytripeptide.

Table 3 indicates, however, that (Pro-Glu-Gly-Pro-Ala-Gly)$_n$ exhibits a more pronounced folding than (Pro-Ala-Gly)$_n$ or (Pro-Glu-Gly)$_n$ alone. Moreover, (Gly-Pro-Lys-Gly-Pro-Pro)$_n$ displays a very strong folding like (Gly-Pro-Pro)$_n$, although Gly-Pro-Lys has a much poorer folding strength. This shows that combinations between different tripeptides do not lead necessarily to an average in the folding strength of both compenents. We guess that more positive neighbouring effects exist than expected before.

2.5 Structure of Block Copolymers

The concept of block copolymers as introduced by Sutoh and Noda[35] in the field of collagen-model polytripeptides was a valuable and interesting alternative to the heteropolytripeptides with two alternating tripeptides. By solid-phase technique, they have synthesized oligomers bearing on both the head and the tail of the chain a block of (Pro-Pro-Gly)$_n$ with n = 5, 6, or 7. Between these strong helix-forming blocks, a block of poor helix former, (Ala-Pro-Gly)$_m$ with m = 5, 3, 1, respectively, was inserted, resulting in total chain lengths of tripeptides with 2n + m = 15. The block copolymers have been compared with polymers (Pro-Pro-Gly)$_n$ (n = 10, 12, 14) representing the situation with the middle block with m = 0. Of course, the middle block reduces the transition temperature, but only by a few Kelvin (Table 4).

Table 4. Transition temperatures of the block copolymers (Pro-Pro-Gly)$_n$-(Ala-Pro-Gly)$_m$-(Pro-Pro-Gly)$_n$

		Transition temperature (°C) for –(Ala-Pro-Gly)$_m$	
		m = 0	m = 5, 3, 1
(Pro-Pro-Gly)$_n$	2n = 10	26	23
	2n = 12	39	35
	2n = 14	50	46

This experiment shows that folding of the non-helix or poor helix-forming sequence is encouraged in the triple helix too, even if this part of the helix contributes to a decrease of the folding strength. Sedimentation and hydrodynamic measurements clearly show the formation of a trimer aggregation and of an extended shape.

Roth and Heppenheimer[37] found that the non-helix forming[93] sequence (Gly-Pro-Leu) inserted between two blocks of (Gly-Pro-Pro) 5 is folded to a triple helix in methanol/water (volume ratio 9:1); (Gly-Pro-Leu)$_n$ alone does not fold (Fig. 19).

2.6 Influence of Charged Tripeptides on Collagen Folding

Charged groups on side chains could lead to repulsive forces. Thus, the aggregation of three chains to a triple helix should be hindered or disturbed. As A. Schmidt[125] has found, the charged groups in collagen are concentrated in 56 clusters along the rod-like molecule. It is astonishing that only 14 of these clusters are charged in the same sense. These groups could lead to repulsion. 42 clusters are built up simultaneously from positively and negatively charged amino acids. These clusters often comprise some tripeptides bearing charged side-chain groups.

In the α1-chain of hide collagen, there are even many tripeptides containing positive as well as negative charges. This is shown by the occurrence of the following sequences: Gly-Glu-Arg = 10 times, Gly-Glu-Lys = once, Gly-Asp-Arg = 4 times, Gly-Asp-Lys =

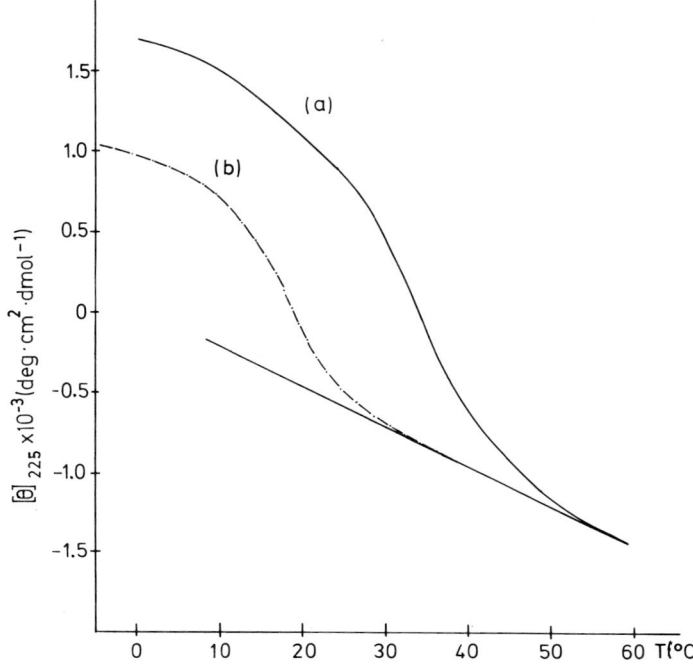

Fig. 19. Transition curve of (Gla-Pro-Pro)$_5$(Gly-Pro-Leu)$_5$(Gly-Pro-Pro)$_5$ in methanol/water. (a): Volume ratio 9:1; conc.: 0.86 mmol · L^{-1} = 301.2 K; (b): Volume ratio 3:1; conc.: 0.76 mmol · L^{-1} = 289.2 K; λ = 225 nm

once, Gly-Lys-Asp = 3 times, Gly-Arg-Asp = once, Gly-Arg-Glu = once, Gly-Lys-Glu = once. There are 22 tripeptides which have positive and negative charges simultaneously. In this case, we can assume the positive and negative charges to attract each other. Erian[120] has studied the folding of two of these sequences in water and in methanol under acidic, neutral, and alkaline conditions. The sigmoidal shape of the

Table 5. Formation of the negatively and positive sequences in water and methanol, indicated by the collagen-like CD spectra and by the shape of the temperature transition curve

Sequences	pH	Structure in water	in methanol
(Gly-Glu-Arg-Gly-Pro-Pro)$_6$	7.2	random coil	random coil
	2.5	triple helix	
	9.5	triple helix	
(Gly-Glu-Lys-Gly-Pro-Pro)$_6$	7.2	random	
	2.5	coil	
	9.5		
(Gly-Glu-Lys-Gly-Pro-Pro)$_{18}$	7.2	random	triple helix
	2.5	coil	
	9.5	triple helix	

temperature-transition curve as well as the formation of positive CD signals at 223 nm were the criteria for the triple helix.

The polyhexapeptide from Gly-Glu-Arg combined with Gly-Pro-Pro forms a triple helix in water, but only in the charged state. In methanol, no structure formation has been observed. The polyhexapeptide containing Gly-Glu-Lys forms in water a relatively poor structure only in a negatively charged state. In methanol, however, the folding strength increases strongly. However, the differences in folding at various pH decrease. The charges seem to be shielded by the organic solvent. We guess that in the collagen sequence in the natural state either other groups or absorbed counterions surrounding the charged groups can shield the charge and depress repulsion. Model-peptide chains with arginine and glutamic acid, however, show a better folding in a state either minus or plus charged, obviously due to the more stretched form.

A very similar behavior has been found by Goren et al.[126] in their investigation of the polymer (Lys-Ala-Glu)$_n$. The higher the water content in mixtures with methanol and the more the pH shifts up or down relative to the neutral state, the charge increases and the CD signals assume more the form of the characteristic β-structure, thus losing the form of an α-helix.

2.7 Structural Characteristics of Covalently Bridged Oligopeptides

Covalent bridging of biopolymers is one of the widely occurring principles in nature for increasing the stability of the tertiary structure, for example the disulfide bridges in keratin and ribonuclease.

It is well known that native collagen contains tripeptide sequences, which alone are not capable of building up a triple helix (e.g. Gly-Pro-Leu, Gly-Pro-Ser) when they exist as homopolypeptides. The synthesis of threefold covalently bridged peptide chains opens up the possibility of investigating the folding properties of such "weak" helix formers, because the bridging reduces the entropy loss during triple-helix formation and thereby increases the thermodynamic stability of the tertiary structure. Therefore, we have

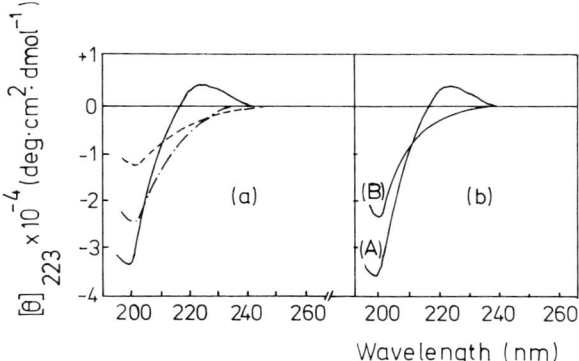

Fig. 20 a, b. CD Spectra of (**a**) monomer (I) (---), dimer II (...), trimer III (——) of crosslinked (Pro-Ala-Gly)$_n$ (n = 12) at 8 °C; (**b**) III at 8 °C (A), at 40 °C (B). Solvent: water; concentration: 2 mg/ml

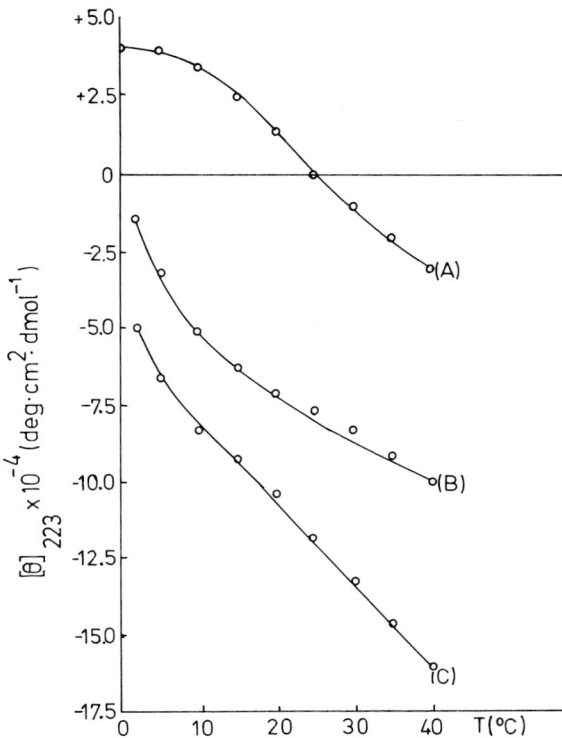

Fig. 21. Thermal transition curves of (A): the monomer, (B): the dimer, and (C): the trimer of crosslinked (Pro-Ala-Gly)$_n$ (n = 12). Solvent: water concentration: 2 mg/ml; λ = 223 nm

synthesized two bridging reagents, namely propane tricarboxylate tris(pentachlorophenyl)1,2,3-(PTC) and N-tris(6-amino hexanoyl)-lysyl-lysine (Lys-Lys)[24].

PTC was coupled with (Pro-Ala-Gly)$_n$, n = 12. The covalently bridged three-chain compound could be separated by gel chromatography from the respective two chains and the single chain in low yield, PTC-[(Pro-Ala-Gly-)$_{12}$]$_3$ shows a strong fold (Figs. 20, 21) in comparison to the dimer or the monomer.

The maximum at 223 nm is remarkably high, though the chain is relatively short. Also remarkable is the higher folding velocity in comparison to the use of single-chain collagen-peptide models as shown in Fig. 22. The folding is also completely reversible.

In the same manner, oligomers having a sequence (Ala-Hyp-Gly)$_n$ (n = 7, 8, 9, 10) were bridged by covalent bonds to PTC and investigated in solvents of various polarities (dilute acetic acid, pH = 3, and methanol/water (1:1)). Starting with n = 8, cooperative transformations could well have been observed but the measured effects were too small for thermodynamic evaluation possible. The major temperature-dependence in the helix region leads to the conclusion that the peptide chains exhibit disordered regions in the helix state (Fig. 23). This was also confirmed by recent investigations from Lazarev et al.[38] on model peptides with the sequence (Gly-Ala-Pro)$_n$ and (Gly-Ala-Hyp)$_n$ using IR spectroscopy and deuterium exchange. According to these authors, the sequence (Gly-Ala-Hyp)$_n$ fails to build up a triple-helix conformation even in the crystalline state but

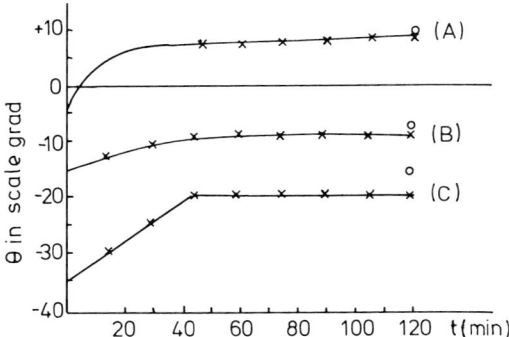

Fig. 22. Relaxation time curves after temperature jump from 35 to 37 °C for trimer (A), dimer (B) and monomer (C) of crosslinked (Pro-Ala-Gly)$_n$ (n = 12) (scale grand). Points (○) denote the Θ values after 2 weeks at 5 °C. Solvent: water; concentration: 2 mg/ml

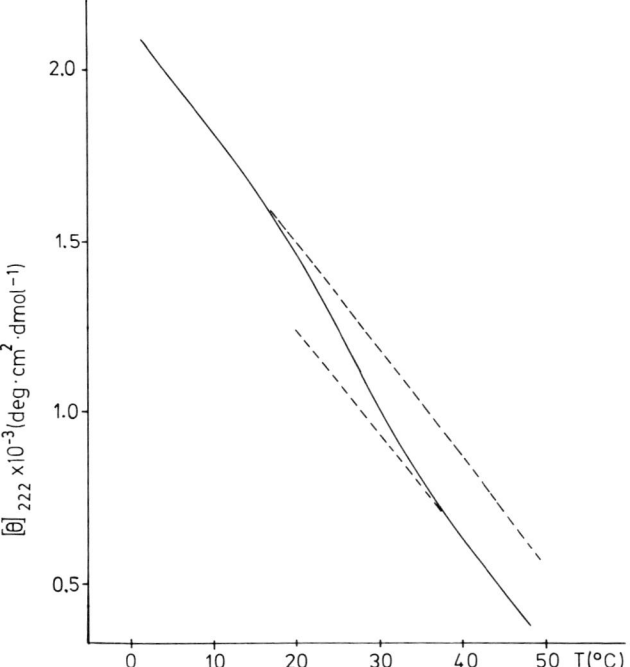

Fig. 23. Temperature transition curve of PTC [(Ala-Hyp-Gly)$_{10}$]$_3$ in water/methanol; volume ratio 1:1

rather tends to build up polymorphous structures, as it was already proven by Doyle et al[160] in the case of (Gly-Ala-Pro)$_n$.

In the case of (Ala-Gly-Pro)$_n$ with n = 5–15, the tripeptide chains were synthesized by the liquid-solid phase technique. As mentioned above, the coupling of longer preformed peptide chains was difficult and the yield of the trimer was low. Therefore, a liquid-solid phase technique was applied in which a trimer was grown in a stepwise manner, beginning from a trifunctional crosslinked base **A**.

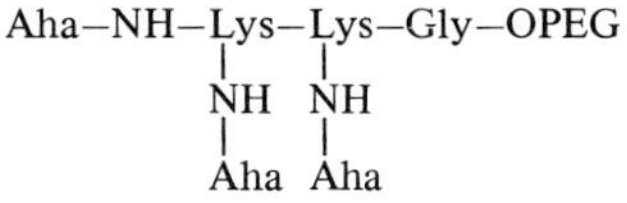

Aha–NH–Lys–Lys–Gly–OPEG
           | |
          NH   NH
           | |
          Aha Aha

Aha = 6-aminohexanoyl   **A**

The completely synthesized and covalently bridged chains where separated from poly(ethylene glycol) (PEG) by saponification with triethylamine in methanol.

2.8 Collagen-Model Peptides with Antiparallel Structure

In 1967 McBride et al.[134] found that the one-chain ascaris collagen forms a triple helix by reverse folding. The hydrogen bonds formed are intrachenar. Rat-skin collagen, however, exhibits reverse folding, also[135] from its very low concentration in water. This intrachenar reaction is preferred to the three-chain interchain combination. Engel[13] found that the molecular weight ratio of the renatured model peptide $(Pro-Pro-Gly)_n$ does not exactly correspond to the molecular weight ratio derived from the process 3 coiled chain → 1 triple helix. He concluded that the model peptide $(Pro-Pro-Gly)_n$ might form an antiparallel structure. In 1968 Ramachandran, Doyle and Blout[136] described a space-filling model of $(Gly-Pro-Pro)_n$ with antiparallel structure. In 1970, Andries and Walton[137] obtained spherulites by crystallization of $(Gly-Pro-Pro)_n$ with different chain length. They found a correlation between the thickness of the crystals and the chain length. It was remarkable that the length of the folded molecule was just one third of the chain length, obviously due to reverse folding. Berg, Olson and Prockop[84] studied folded and unfolded $(Gly-Pro-Pro)_n$ (n = 10) chains by titration with NaOH. They pointed out that the amino group in the helix state has 0.5 pH units lower at the pK. This should be an indication of a parallel structure. For an antiparallel structure, the pK should increase by interaction of the carboxy group with the amino group. In 1976, Seba[138] studied the folding kinetics of the α1-CB 4 peptide and observing a first-order reaction at low concentrations due to reverse folding. It should be necessary to have a strong indication for the backfolded antiparallel structure. Covalent bridged three-chain models with parallel and antiparallel alignment in $(Pro-Ala-Gly)_n$ were synthesized (Fig. 24). The covalently bridged peptide models have been constructed by fixation of the chains on a trivalent bridge compound. The parallel bridged model was built in the same way as earlier by El Sheikh[123]. We used tris-(1,3,5-trinitrophenyl) 1,2,3-propane tricarboxylate for coupling

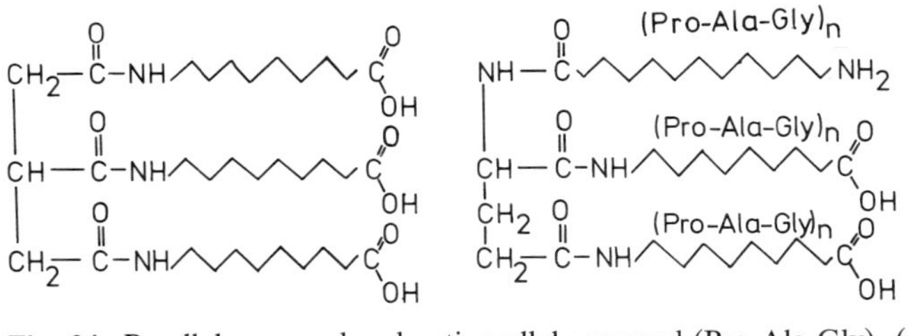

Fig. 24. Parallel arranged and antiparallel arranged $(Pro-Ala-Gly)_n$ (n = 8, 12, 16). In accordance with the peptide nomenclature in which the amino terminal end, letters with mirror image were chosen in this figure (see top right). The parallel arrangement could be depicted only in this way

the chains (Pro-Ala-Gly)$_n$ (n = 8, 12, 16). The antiparallel bridged model was obtained by coupling of N-α-Z-Glu-α,γ-dinitrophenyl ester at first with two chains. After splitting off the N-α-Z-group, coupling of (Pro-Ala-Gly)$_n$-nitrophenyl ester was carried out. After each step, extensive purification by gel chromatography on Sephadex G 50 superfine was performed. The substances used for the investigation are completely free from single or double chains. The three-chain models dissolved in water at different concentrations were incubated at 4 °C for 3 days. The CD spectra show that the parallel as well as the

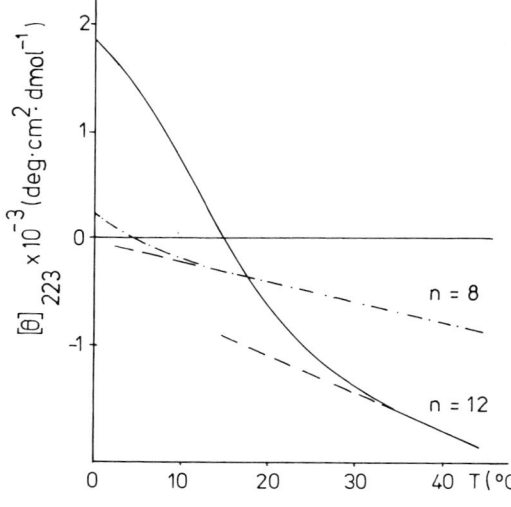

Fig. 25. Transition curves of the parallel covalently bridged (Pro-Ala-Gly)$_n$ (n = 8, 12)

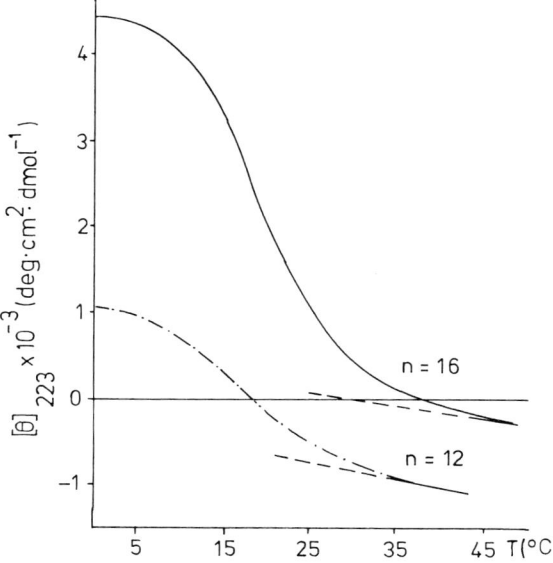

Fig. 26. Transition curves of the antiparallel crosslinked trimer (Pro-Ala-Gly)$_n$ (n = 12, 16)

antiparallel model peptides fold in a very similar manner. The longer the chain, the more pronounced is the typical maximum at 223 nm, indicating the similarity to the collagen triple-helix structure[25]. The molar ellipticity depends on the temperature as shown in Figs. 25 and 26.

The reaction enthalpy and reaction entropy were derived from the curves comparing with data for the "all-or-none" model[140] using the computer program of Rosenbrock[139] (Table 6).

Table 6. Thermodynamic parameters of (Pro-Ala-Gly)$_n$ (n = 12, 16)

	Parallel covalently bridged		Antiparallel covalently bridged	
	ΔH/(kJ · mol^{-1})	ΔS/(J · mol^{-1} · K^{-1})	ΔH/(kJ · mol^{-1})	ΔS/(J · mol^{-1} · K^{-1})
n = 12	−129.7	−454	−123.6	−429.
n = 16	−	−	−158.1	−542.

As Table 6 shows the values of the parallel structure are not so much different from those of the antiparallel structure with the same chain length. The structure of the stars with n = 8 was too poor, and the parallel model with n = 16 insoluble. Reverse folding of such short chains seems to be impossible[21]. The structure observed is conclusively due to an antiparallel arrangement. By a space-filling model it can be shown that no great steric problems arise in the formation of the hydrogen bridges which are necessary for stabilizing the structure.

3 Kinetic Aspects of Triple-Helix Formation of Peptide Models Compared with those of Collagen Peptides

3.1 General Considerations

The in vivo biosynthesis of collagen proceeds up to the triple-helix within 10 min. The velocity of renaturation in vitro is much slower and the native state of longer chains cannot be reached even after days and weeks[149]. Kühn, Engel et al.[141] succeeded in proving by ultracentrifugation measurements that, besides the trimers, also oligomers of higher molecular weights are formed in such temperature-quenched collagen solutions. These results have been interpreted as the inability of the chains to reunite perfectly aligned but to form staggered products. The non-helical chain segments are able to associate with other unbonded chains and thus to form high molecular weight products.

Harrington studied the kinetics of renaturation in dilute solutions of α_1-chains[135] and reported that α_1-chains assumed a triple-helical structure also by back-folding of single chains. This monomolecular reaction in dilute solutions proceeds more rapidly than the trimerization of three α_1-chains. Besides, a negative apparent enthalpy of activation has been measured. Figure 27 shows schematically the possible ways of renaturation proposed by Harrington.

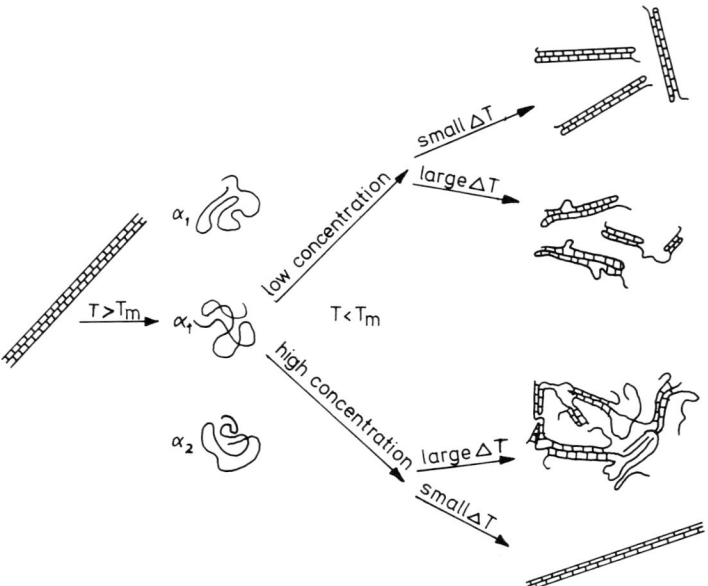

Fig. 27. The possible pathways of collagen renaturation proposed by Harrington et al.[135]

It is evident that with growing length of the peptide chains the probability of incorrect nucleations between staggered chains increases. The rate of dissociation of such a staggered trimer/oligomer is the lower, the larger the length of the helical sequence. Due to this fact products which may form long helical sequences can, indeed, no longer dissociate. Nevertheless, at fixed pressure, temperature and concentration, the most thermodynamic stable conformation is not formed yet. This leads to a distribution of oligomer chains which does not correspond to the real equilibrium but rather to a distribution depending on the history of the process. Thus, the transition curves of denaturation and renaturation form hystereses which were measured with long peptides resulting from cyanogen-bromide cleavage of the α_1-chain by Saygin and Heidemann[3]. Therefore, the value obtained by kinetic experiments on native collagen is of limited applicabily.

Weidner and Engel[142] used the relatively short collagen peptide $\alpha 1$-CB_2 as a model peptide for kinetic measurements. They observed that the rate at the beginning of the helix formation, starting from purely coiled chains, obeys the following equation:

$$\left.\frac{d\Theta}{dt}\right|_{t=0} = k \cdot C_g^\alpha$$

where C_g is the total concentration of peptide single chains, $\alpha = 2.9 \pm 0.2$, and Θ the degree of conversion.

All relaxation curves exhibited more than one phase at various degrees of conversion and at different temperatures. This clearly rules out the all-or-none mechanism (AON) although the AON model is able to fit easily to the measured equilibrium transition curve. However, a mechanism has been proposed which allows the existence of side

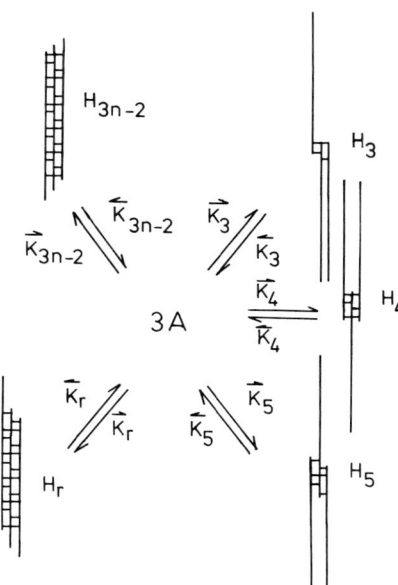

Fig. 28. The "star"-model proposed bei Weidner and Engel[142]

products (star model, Fig. 28). This model finally describes the dependence of the mean relaxation time (found experimentally from the rate of conversion at the beginning) on the degree of conversion Θ.

Comparing the experimental with the computed data Weidner obtained parameters for the nucleation and propagation steps and for the apparent enthalpy of activation. The length of the nucleus H_x could be estimated from the apparent enthalpy of activation, $E_a = -54.4$ kJ/mol triple helix, and the enthalpy of reaction, $\Delta H^0 = -11.7$ kJ/mol tripeptide for an elementary step of the pre-equilibria. The length of the nucleus was found to be six tripeptide units. Considering the nucleation as one single step including the pre-equilibria, the following Eq. (1) can be formulated:

$$3\,C \underset{}{\overset{\sigma \cdot s^x}{\rightleftharpoons}} H_x \underset{}{\overset{S}{\rightleftharpoons}} H_{x+1} \underset{}{\overset{S}{\rightleftharpoons}} \ldots \underset{}{\overset{S}{\rightleftharpoons}} H_{3n-2} \tag{1}$$

for which C = coil, H_x = nucleus containing x H-bonds, $3n - 2$ = maximum number of possible H-bonds, σ = nucleation parameter, S = equilibrium constant of the propagation step.

The constant of equilibrium of the whole reaction may be formulated as product of the constants of elementary steps, because the same heat and entropy of formation is expected for every single step.

$$K_c = \sigma \cdot S^{3n-2} = \frac{[H_{3n-2}]}{[C]^3} \,;\, S = \exp(-\Delta H_s^0/RT) \cdot \exp(\Delta S_s^0/R) \tag{2}$$

Since the enthalpy of activation of an elementary step cannot be negative, the measured negative apparent enthalpy of activation is explained by a fast pre-equilibrium preceding the rate-determining step (formation of H_x in Eq. (1)). Thus, the sum of the negative reaction enthalpies of the fast pre-equilibrium and the real positive enthalpy of activation of the rate-determining step are measured experimentally. This can be expressed the other way round, if instable intermediates are formed in a fast exothermic pre-equilibrium, their concentration decreasing with increasing temperature. This means that the rate of formation of the resulting nucleus H_x decreases with rising temperature.

3.2 The Role of *cis-trans* Isomerism

In comparison to the constant of propagation of the α-helix formation ($k_p \simeq 10^{10} \, s^{-1}$) and the double-helix formation ($k_p \simeq 10^7 \, s^{-1}$), a comparatively small parameter concerning the formation of triple helix has been found ($k_p \simeq 8 \times 10^{-3} \, s^{-1}$). A higher entropy of activation is assumed as the main cause of this occurrence which means a lower frequence factor in the Arrhenius equation.

These results show unambiguously that the formation of the triple helix in the in vivo collagen biosynthesis is governed by a mechanism which differs from that of in vitro renaturation. Later measurements made with the amino terminal segment Col 1–3 of type III procollagen led to the following results[143, 144]. The peptide Col 1–3 shows an unusual fast triple-helix formation being concentration-independent and reversible. It seems that three neighboured interchenar disulfide bridges are located at the carboxylic end of the triple-helical part, thus enableing the formation of an effective nucleus without the necessity of fast pre-equilibria. Thus, the difficulty of nucleus formation is reduced in a way that it is not any longer rate-determining. The observed rate constant of $8 \times 10^{-3} \, s^{-1}$ at 20 °C thus corresponds to the propagation of a triple helix, generated from an existing nucleus. The observed positive apparent enthalpy of activation of +97 kJ/mol per triple helix in 4 M guanidine hydrochloride leads to the conclusion that the rate of folding is governed by the *cis-trans*-isomerism of the peptide bonds in direct neighborhood to the helical part. At 37 °C, it takes minutes to complete the folding and thus it is comparable with the velocity of the biosynthesis in vivo.

This gives an important hint at what kind of model peptides are synthesized to obtain detailed information about the thermodynamics and kinetics of the collagen triple-helix formation. A first success was already achieved by synthesizing peptides of the following general structure[37]:

```
H–(X–Gly–Pro)ₙ⌇⌇⌇⌇─────────┐
                            │
H–(X–Gly–Pro)ₙ⌇⌇⌇⌇──Lys–Lys–Gly–OMe
                            │
H–(X–Gly–Pro)ₙ⌇⌇⌇⌇─────────┘
```

In the case of X = Ala, cooperative equilibrium transition curves have been found starting from n = 8 tripeptide units. The reasons for this extraordinary behavior have been discussed concerning the peptide Col 1–3. CD measurements on peptides having chain lengths of 6 and 7 tripeptide units, seem to point out that these peptides are able to form a small amount of triple helix in water near 0 °C. Thus, the sequence (Ala-Gly-

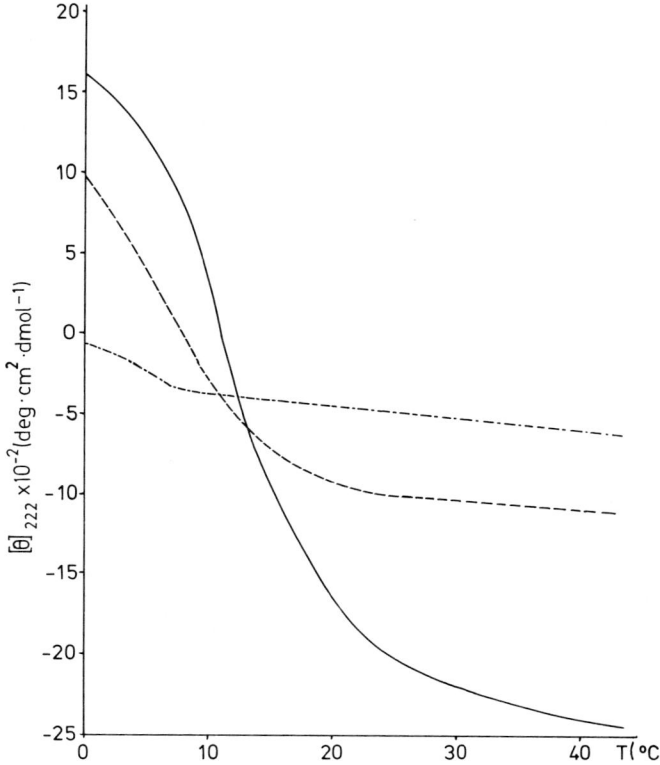

Fig. 29. Equilibrium melting transition curves of [(Ala-Gly-Pro)$_n$]$_3$ in water /HAc (99:1); n = 6 (---); n = 7 (-·-·-); n = 8 (——)

Pro)$_n$ must have the length of 8 tripeptide units, thus being at least able to built up a stable triple helix in water at a temperature of 1 °C (Fig. 29).

The transition took a course independent of concentration, reversible and extraordinarily fast. Transition temperature and kinetics were independent of pH and of the concentration of the salt. Measurements have been made either in acetic acid, (pH 3.0) or in 50 mM phosphate buffer at pH 7.5.

In methanol (100%) starting from a chain length of n = 5 tripeptide units, complete transition curves could be measured (Fig. 30).

Temperature-jump experiments showed an evident increase of the rate of transition by using methanol as solvent instead of water. According to Fig. 31, this is mainly caused by the increase of the fast kinetic phase at the expense of the following slow phase.

In this case, we point to the fact that a fast ($\tau < 5$ s) and a slow phase have been observed in temperature-jump experiments also with the peptide Col 1–3. The slow phase – as already mentioned – has been associated with the *cis-trans* isomerism of peptide bonds in the direct neighborhood of the helical part. Only peptide bonds to which proline or hydroxyproline contribute their secondary nitrogen are able to assume a *cis*-configuration at equilibrium (*cis* to *trans* ratios of 1:40 to 1:1)[145]. Therefore, the fast

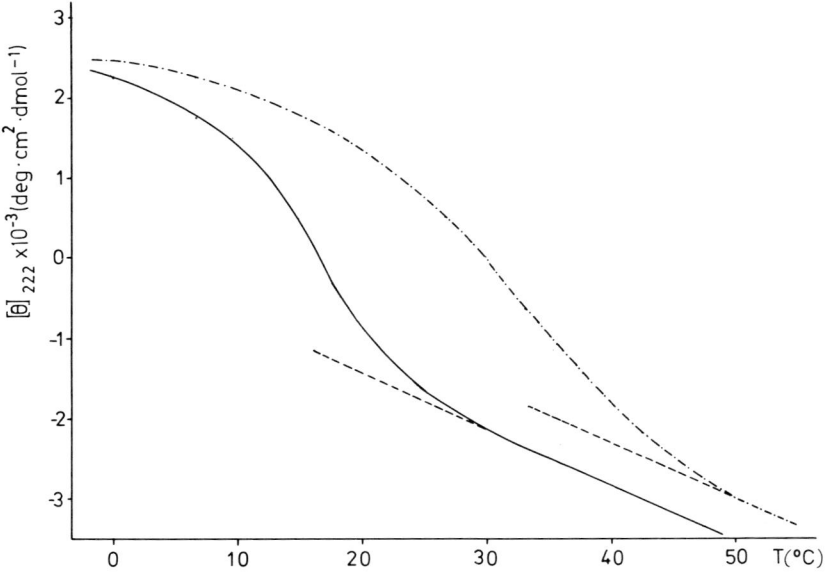

Fig. 30. Equilibrium melting transition curves of [(H-(Ala-Gly-Pro)$_n$]$_3$ in methanol; n = 5 (——); n = 7 (–·–·–)

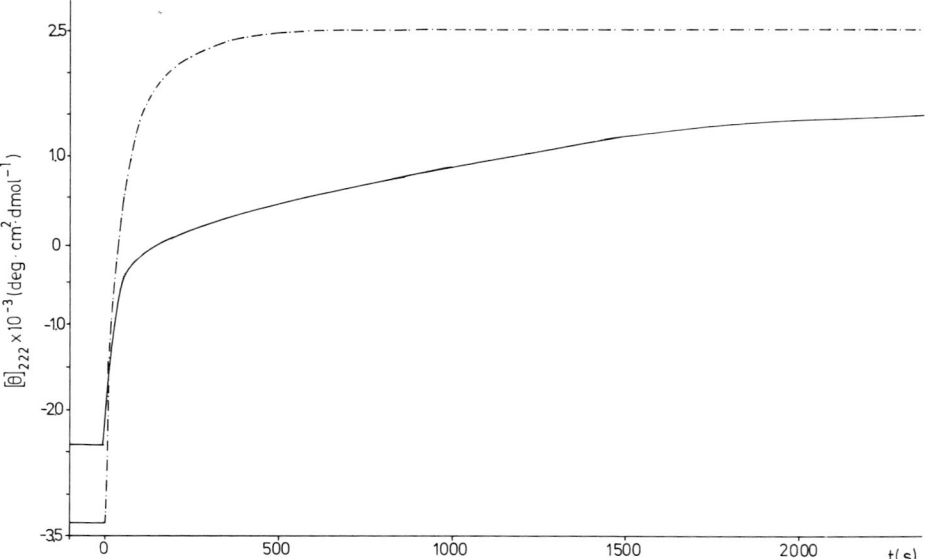

Fig. 31. T-jump experiments of [H-(Ala-Gly-Pro)$_n$]$_3$ in water/HAc (99:1) (n = 8, T = 40 °C, T_2 = 0.5 °C (——)) and in methanol (n = 7, T_1 = 55.7 °C, T_2 = 0.5 °C (–·–·–)) correspond to the helical and coiled state, respectively

phase corresponds to the insertion of amino acids into the triple helix, which exists in the *trans*-form at the moment of folding. The magnitude of the fast kinetic phase thus gives information on the percentage of *trans*-peptide bonds in the coiled state.

Thus, we conclude that, concerning the peptide $[\text{H}(\text{Ala-Gly-Pro})_n]_3$ in methanol, the percentage of *trans*-peptide bonds is higher than in water.

To answer the question whether the *cis*-transisomerization of the bridged polypeptides with a Ala-Gly-Pro sequence represents the rate-determining step, the following experiment was carried out: The polypeptide with a chain length n = 8 was denaturated in a rapid reaction with a temperature jump from 9.2 to 30 °C and subjected to renaturation at 9.2 °C after an incubation time of 25 s. In a second and a third experiment, the incubation in the coiled state was prolonged respectively to 75 and 125 s. It could be observed that the amplitude of the rapid phase depends on the time that lapses between the denaturation and renaturation (Fig. 32).

One may conclude that the rate-determining step of the renaturation is at least partly influenced by the *cis-trans* isomerization of the peptide bond the secondary nitrogen atom of which arises from proline. Otherwise, only the entropy-controlled slow nucleation should be observed kinetically. The covalent bridging through Lys-Lys, therefore, gives rise not only to thermodynamic stabilization of the triple helix but also to kinetic properties which have hitherto been observed in the case of type III procollagen[146] and its aminoterminal fragment Col 1–3[144].

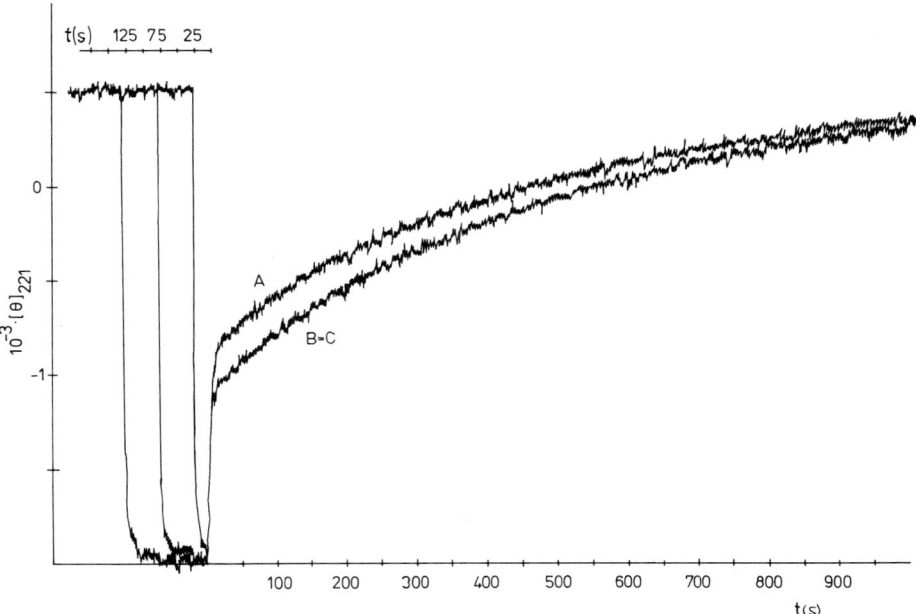

Fig. 32. Double-jump experiments of unfolding and refolding. The peptide $[(\text{Ala-Gly-Pro})_8]_3$ in 50 ml phosphate buffer (pH 7.5) was incubated at 9.2 °C and quickly unfolded by a first temperature jump from 9.2 to 30 °C. This process took 25 s, the time needed to reach the final temperature. In a first experiment (*curve A*), the second jump back from 30 to 9.2 °C followed immediately after complete unfolding of the peptide, i.e. 25 s after the first jump. In a second and a third experiment (*curve B, C*), the time lapse between the first and the second jump was 75 and 125 s, respectively

4 Thermodynamic Aspects of the Triple Helix-Coil Transition

4.1 Cooperative Processes

In the case of cooperative processes, the formation of a nucleus, already discussed from the kinetical point of view, plays a crucial role. The steady state described by Eq. (1) depicts the formation of a triple helix as the simplest model by the formation of a nucleus H_x through fast pre-equilibria and subsequent propagation steps, H_x in this case is a triple-helical intermediate with x tripeptide units (that means x hydrogen bonds) in the helical state. The final product H_{3n-2} possesses two hydrogen bonds less than tripeptide units because the three single chains are staggered at one amino acid residue each.

For this kind of cooperative processes, it is characteristic that the formation of the nucleus is thermodynamically more difficult than for further propagation steps (positive cooperativity). This implies that the elementary transition step of an individual chain segment (tripeptide unit) is influenced by the state of adjacent segments through intramolecular interactions.

σ is a measure of the difficulty of forming a helix nucleus in comparison with that of propagation. Zimm and Bragg[147] related the chain segments c, being in a coiled state, with the statistical weight 1. Every helical segment within a helical sequence is associated with the statistical weight s and every helical segment h not following a helical segment is associated with the statistical weight $\sigma < 1$.

In the case of $0 \leq \sigma < 1$, we talk of positive cooperativity which leads inevitably to the following characteristic qualities. The chain segments prefer the same state as their immediate neighbors. Then, comparable small variations of exterior parameters (temperature, pH, solvent etc.) cause practically complete transitions. Besides, one can see an increasing sharpness of transition with growing chain length because in this case, the number of propagation steps rises in comparison with the nucleation steps; as a result, the number of segments, being cooperatively converted within a macromolecule, increases.

Several methods have been developed for the quantitative description of such systems. The partition function of the polymer is computed with the help of statistical thermodynamics which finally permits the computation of the degree of conversion Θ. In the simplest case, it corresponds to the linear Ising model according to which only the nearest segments interact cooperatively[149]. The second possibility is to start from already known equilibrium relations and thus to compute the relevant degree of conversion Θ.

4.2 The All-or-None Model (AON) I

All-or-none transitions occur if the chain length is relatively short (n ≤ 15 tripeptide units) and if the cooperativity is high ($\sigma \ll 1$) since in this special case, the concentration of intermediates is negligibly low. Besides, in the case of short chains we may conclude that back folding and oligomerization are negligibly small because of the shortness of the chain ends beyond the helical part. A further simplification is the assumption that only "one helical sequence" exists, which excludes the formation of loops within a helical part, because of reasons of stability. Under these circumstances, only two different products exist in a measurable concentration at equilibrium.

$$3\,C \underset{}{\overset{K_c}{\rightleftharpoons}} H_{3n-2}$$

According to the mass-action law, the equilibrium is described by Eq. (3)

$$K_c = \frac{[H_{3n-2}]}{[C]^3} = \frac{\Theta}{3\,C_g^2(1-\Theta)^3} \tag{3}$$

$$\Theta = \frac{3[H_{3n-2}]}{C_g} = \text{degree of conversion}$$

C_g = total concentration of single chains

The solution of Eq. (4) gives for Θ a third-power polynomial of K_c, its real root being computed with the Cardan formula[148] resulting in the expression:

$$\Theta = 1 - \sqrt{\frac{1}{2K_x} + \sqrt{\left(\frac{1}{2K_x}\right)^2 + \left(\frac{1}{3K_x}\right)^3}} - \sqrt{\frac{1}{2K_x} - \sqrt{\left(\frac{1}{2K_x}\right)^2 + \left(\frac{1}{3K_x}\right)^3}}$$

$$K_x = 3\,C_g^2 \cdot K_c = 3\,C_g^2 \cdot \exp(-\Delta H^\circ/RT + \Delta S^\circ/R)$$

A fit of this relation to the experimentally found equilibrium transition curve, $\Theta = F(T)$, yields the parameters ΔH° and ΔS°.

In a first approximation, it is possible to obtain ΔH° from the slope in the middle of the equilibrium transition curve. At $T = T_m$ and $\Theta = 0.5$, according to van't Hoff ($K_c = 4/3 \cdot C_g^2$), the following expressions result:

$$\Delta H^\circ = 8\,R T_m^2 \left(\frac{d\Theta}{d\tau}\right)_{\Theta = 0.5} \tag{4}$$

$$\Delta S^\circ = \frac{\Delta H^\circ}{T_m} - R \ln(3\,C_g^2/4) \tag{5}$$

Equation (5) reveals that T_m depends on the concentration and can be computed from it.

This model does not say anything about the mechanism of triple-helix formation, because even in the case of an AON mechanism, nucleation may take place at many positions of the chains and may lead to products the chains of which are staggered. The AON model is based on the assumption that these products are too instable to exist in measurable concentration. As already mentioned, Weidner and Engel[142] succeeded in proving by relaxation measurements of $\alpha 1$ CB 2 that the kinetics of in vitro triple-helix formation is governed by more than one relaxation time. This rules out an AON mechanism, but the fitting to the experimentally found equilibrium transition curves nevertheless showed good accommodation and ΔH° computed from these curves could be confirmed by calorimetric measurement.

4.3 The Staggering-Zipper Model (SZ) II

A further model can be derived from the principle of the "staggering-zipper". It takes into account the existence of measurable concentrations of helical intermediates, the chains of which are staggered, during the transitions. If, at a given staggering, the highest possibility of hydrogen bonds r_{max} is realized, the individual equilibrium constants $\sigma \times S^{r_{max}}$ of the corresponding products must be multiplied with the statistical weights, because there are several possibilities of staggering which depend on chain length n. These statistical weights are expressed by the following equation according to Weidner and Engel[142]:

$$g_r = \begin{cases} n - v, & \mu = 0, 2 \\ 0, & \mu = 1 \end{cases} \quad \text{for } r = 3v + \mu \quad \begin{cases} v = 1, \ldots n - 1 \\ \mu = 0, 1, 2 \end{cases}$$

In the case in which a helical product with a certain orientation of the single chains forms not only the maximum number, r_{max}, of hydrogen bonds – which means that the ends of the helical parts are coiled – the statistical weights are computed according to[142] (Fig. 33):

$$g_r = (n - v + 1)^{2-\mu} \cdot (n - v)^{1+\mu}$$

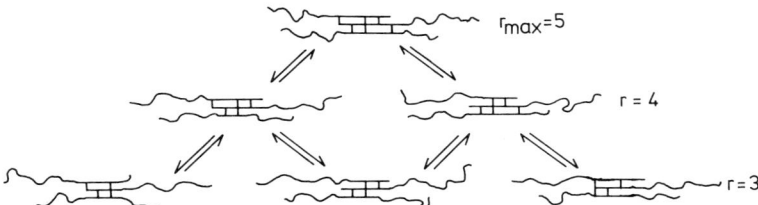

Fig. 33. Unzippering of a staggered helical state according to model II[148]

For both cases, the assumption is valid that only one helical sequence exists and that products with the same number of hydrogen bonds have the same stability. Considering the statistical weights of the possible intermediates, the whole measurable degree of conversion, Θ_{total}, is computed by the mass-action law and can be derived from Eq. (6)[149].

$$\Theta_{total} = F(n, C_g, \sigma, \Delta H^\circ, \Delta S^\circ) \tag{6}$$

$$\Theta_{total} = \frac{3 C_g^2 \cdot \sigma \cdot f_k^3}{3n - 1} \sum_{v=1}^{n-1} \sum_{\mu=0}^{2} (3v - \mu) \cdot (n - v)^3 \left(\frac{n - v + 1}{n - v}\right)^{2-\mu} \cdot S^{3v+\mu}$$

$$f_k = \frac{C_{coil}}{C_g}$$

Besides the concentration and chain length n, the nucleation parameter is given, too. Merely, ΔH° and ΔS° are computed by fitting of Eq. (6) to the experimental found equilibrium transition curve. The parameters ΔH_s° and ΔS_s° respectively, refer to one propagation step, assuming that the only temperature-dependent parameter is the equilibrium constant s of the propagation step (Eq. (2)).

The difficulty of nucleation, and thus σ, might be caused entropically in case of triple-helix formation ($\Delta H_\sigma^\circ \simeq 0$). The molecular reason for this is that ΔH_s° of an $-NH \cdots O = C$-hydrogen bond is gained during one step of propagation, while the energy $T \cdot \Delta S_s^\circ$ is needed. In this case, ΔS_s° is the loss of entropy of the freezing of the $NH-C_\alpha$ and $C_\alpha-CO$ bonds, freely rotating in the coiled state. Compared with this, a four fold loss of entropy takes place during the nucleation[92]. This is due to
1) the loss of entropy caused by the trimerization of three single chains, and
2) the partial fixation of the tripeptides adjacent to the nucleus, occuring only once in the formation of the nucleus and caused by short contacts of adjacent chains.

Since σ can be separated from the propagation step only because it is independent of temperature, additional influences which may exist remain undiscovered as far as they are connected with a gain or a loss of enthalpy (forces between the single chains and interactions with solvent). A possibly appearing effect of enthalpy, which occurs only during nucleation is then distributed among the propagation steps.

Figure 34 shows two equilibrium transition curves, one of them being computed with the help of the AON model (I), the other by means of the SZ model (II) for arbitrarily chosen parameters.

At higher temperatures ($T > T_m$), model II predicts larger values of Θ, because this increases the number of possibilities of forming a helix, in contrast to the AON model (I). On the other hand, at lower temperatures, model I gives more possibilities of the realization of the coiled state and thus the transition curve near $\Theta = 1$ is flat in comparison with the AON case.

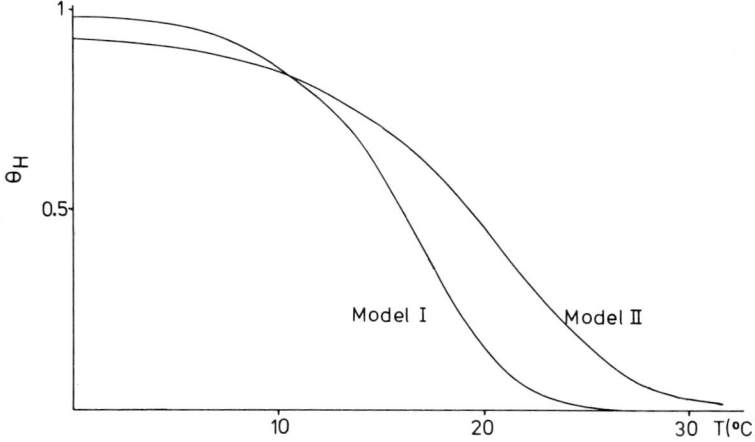

Fig. 34. Computed equilibrium melting transition curves according to modell I and II. The parameters are $\Delta H_s^\circ = -10.4$ kJ/mol tripeptide, $\Delta S_s^\circ = -7.7$ j(K · mol) tripeptide, $\sigma = 5 \times 10^{-4}$, $C_g = 1.24 \times 10^{-3}$ mol/l, n = 16

4.4 Estimation of the Nucleation Parameters

According to Eq. (1), the triple-helix formation can be imagined as a process which may be divided into $3n - 2$ steps of equilibrium with the same values of ΔH_s° and ΔS_s° each. This means that ΔG° is a linear function of $3n - 2$ for the whole process.

$$\begin{aligned}\Delta G^\circ &= \Delta G_\sigma^\circ + (3n - 2)\Delta G_s^\circ \\ \Delta H^\circ &= (3n - 2)\Delta H_s^\circ ; \quad \Delta H_\sigma^\circ = 0 \\ \Delta S^\circ &= \Delta S_\sigma^\circ + (3n - 2)\Delta S_s^\circ\end{aligned} \quad (7)$$

The parameters indexed with σ are connected with the nucleation step or other effects occurring only once per triple helix. Parameters denoted by s are related with the equilibrium constants of the propagation steps and are ordered to be independent of the position of the reacting chain segment. This implies that end effects are neglected. Since the same dependences are valid for ΔH° and ΔS°, with the help of their chain length dependence we can determine ΔG_σ° by extrapolation up to $3n - 2 = 0$, and thus, σ can be estimated it depends neither on temperature nor on the chain length.

$$R \ln \sigma = \Delta S_\sigma^\circ; \quad \Delta H_\sigma^\circ = 0 \quad (8)$$

Engel et al.[92] have estimated the nucleation parameter for H(Pro-Pro-Gly)–OH by computing ΔG° and ΔS° with Eq. (5) and reported melting temperatures T_m and ΔH° values determined calorimetrically. Utilizing the chain length dependence, they obtained the following parameters (in diluted acetic acid at 25 °C; n = 5, 10, 14, 15):

$\Delta G_\sigma^\circ = +20.9$ kJ/mol triple helix
$\Delta G_s^\circ = -2.05$ kJ/mol tripeptide
$\Delta H_\sigma^\circ = 0$ kJ/mol triple helix
$\Delta H_s^\circ = -7.74$ kJ/mol tripeptide
$\Delta S_\sigma^\circ = -71.1$ J/kmol triple helix
$\Delta S_s^\circ = -18.8$ J/kmol tripeptide
$\sigma_{app} = 2 \times 10^{-4}$ l^{-2} mol^{-2}

They interpreted their measurements pointing out that the experimentally found ΔG_σ° and σ at first should be regarded as apparent nucleation parameters. ΔG_σ° also seems to include end effects because in the AON case, ΔG° for nucleation and ΔG° for end effects are additive. The occurrence of end effects is revealed by the dependence of nucleation parameters on the solvent found by Engel et al.[92]. This effect is explained by the existing repulsion of equally charged end groups, which is stronger in water than in organic solvents in which ΔG° has a larger negative value.

The nucleation parameters have to be determined with the help of the chain length dependence of T_m, since no calorimetric data are available. Using Eqs. (2, 7, 8) we obtain

$$\frac{1}{T_m} = \frac{1}{3n - 2} \cdot \frac{R}{\Delta H_s^\circ} \cdot \ln\left(\frac{3 \cdot C_g^2 \cdot \sigma_{app}}{4} + \frac{\Delta S_s^\circ}{\Delta H_s^\circ}\right) \quad (9)$$

Go and Suezaki[149] thus determined for H-(Pro-Pro-Gly)$_n$ with n = 10, 15, 20

ΔH_s° = −8.2 kJ/mol tripeptide
ΔS_s° = −19.2 J/(kmol) tripeptide
σ_{app} = 10^{-2}–10^{-6} $1^{-2} \cdot mol^{-2}$

For monomolecular transitions Eq. (9) may be simplified to following term

$$\frac{1}{T_m} = \frac{1}{3n-2} \cdot \frac{R}{\Delta H_s^\circ} \cdot \ln\left(\sigma_{app} + \frac{\Delta S_s^\circ}{\Delta H_s^\circ}\right) \tag{9a}$$

From the chain length dependence of T_m, the expression $\Delta S_s^\circ / \Delta H_s^\circ = T_{m,\infty}^{-1}$ is obtained where $T_{m,\infty}$ cannot be the real transition temperature for very long chains because the AON model is not valid for the latter.

Some results of the chain length dependence of the monomolecular transition of the peptide [H(Ala-Gly-Pro)$_n$]$_3$ (Table 7) have already been reported[151].

In 1% aqueous acetic acid, the peptides of the sequence (Ala-Gly-Pro)$_n$, bridged with Lys-Lys and beginning with n = 8 show a cooperative transition, which was interpreted as a triple helix-coil transition (see Figs. 35, 36).

The reversibility of the reaction guarantees that each measurement represents an equilibrium state. Furthermore, the position of the equilibrium turned out to be independent of the initial peptide concentration. Therefore, the thermodynamic parameters

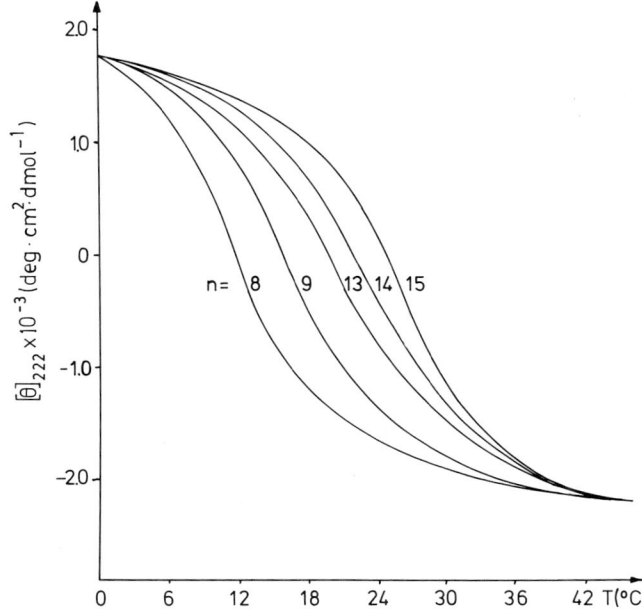

Fig. 35. Equilibrium melting transition curves of [H-(Ala-Gla-Pro)$_n$]$_3$ in water/HAc (volume ratio 99:1); n = 8, 9, 13, 14, 15

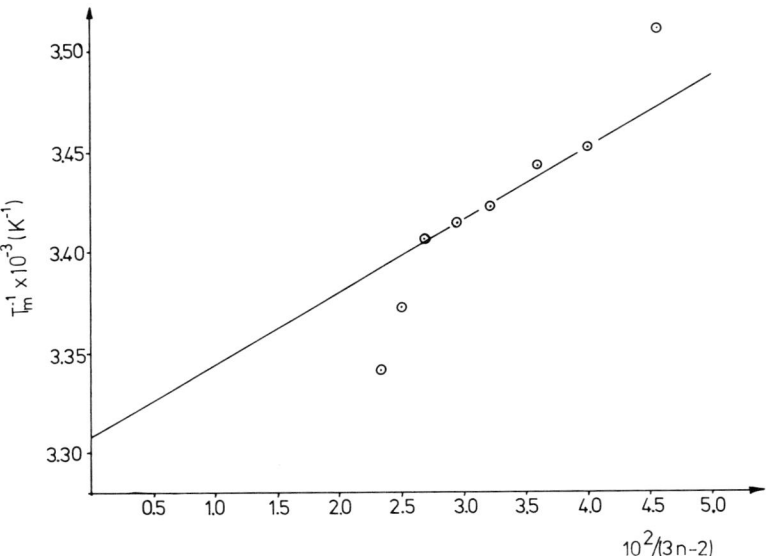

Fig. 36. Chain lenght dependence of T_m of [H-(Ala-Gly-Pro)$_n$]$_3$ (n = 8–15)

Table 7. Thermodynamic parameters of [H(Ala-Gly-Pro)$_n$]$_3$-Lys-Lys in water/HAc (99:1)

n	ΔH° (kJ · mol^{-1})	ΔS° (J · mol^{-1} · K^{-1})	T_m (K)
8	–	–	284.9
9	–183.0	–637.5	289.7
10	–187.3	–644.0	290.5
11	–206.2	–713.0	292.2
12	–247.5	–839.8	292.9
13	–210.5	–710.4	293.5
14	–212.4	–720.0	296.8
15	–230.5	–774.0	299.4

could be determined by applying the mass-action law to the corresponding monomolecular reaction. This was done by adapting the theoretical transition curves according to the "AON" model to the experimental data. Thus, the enthalpy and entropy change relating to one propagation step could be derived.

ΔH_s° = –7.0 kJ/mol tripeptide (from Table 7, n = 9–13)
ΔS_s° = –23.1 J/K · mol tripeptide (from $\Delta H_s^\circ / T_{m,\infty} = \Delta S_s^\circ$)

The melting point T_m and kinetics are independent of pH and of the salt concentration. This was found by studies in 1% aqueous acetic acid, pH 3.0 as well as in 50 mM phosphate buffer, pH 7.5). Recently, Greiche and Heidemann[23] described the synthesis of oligopeptides covalently bridged at the N-terminus by 1,2,3-propanetricarbocylic acid (PTC) (see Sect. 2.8). The investigation of the triple helix-coil transition revealed smaller enthalpy changes than when the bridge link Lys-Lys was used (Table 8). Accordingly, the different thermal stability of the tertiary structure possessing the same chain length could solely be attributed to the steric relations at the bridge heads.

Table 8. Thermodynamic parameters of the coil-to-helix transition of collagen-model peptides, covalently linked with 1,2,3-propanetricarboxylic acid (PTC) and Lys-Lys, respectively. Solvent: 1% aqueous acetic acid (pH 3.0)

Peptide	$\Delta H°$ (kJ · mol^{-1})	$\Delta S°$ (J · mol^{-1} · K^{-1})	T_m (K)
[(Ala-Gly-Pro)$_8$]$_3$Lys-Lys	–	–	284.9
PTC[(Pro-Ala-Gly)$_{12}$]$_3$	−129.7	−454.0	285.2
[(Ala-Gly-Pro)$_{12}$]$_3$Lys-Lys	−247.7	−845.2	292.9

The use of 6-aminohexanoyl residues as spacers obviously allows an effective extension of the lysyl-lysine side chains, thus facilitating a tension-free alignment of tripeptides in the triple helix. With the use of PTC (1,2,3-propanetricarboxylic acid) as a bridging link, on the other hand, the peptide chains are forced to undergo looping at the N-terminus during triple-helix formation. This also reasonably results from space-fitting models. Hence, the diminished $\Delta H°$ values in the case of PTC-bridged polypeptides are probably due to the loop formation at the N-terminus, i.e. the resulting reduction in the number of hydrogen bonds.

As already mentioned, the enthalpy change $\Delta H_s°$ involved in an elementary propagation step corresponds to the equilibrium constant S. The parameter σ, however, is purely entropically influenced mainly due to the steric restrictions during the formation of a helical nucleus. The determination of σ, since it is related to the same power $(3n-2)$ of s, requires the consideration of the dependence of the thermodynamic parameters on the chain length (Eq. (9a)).

Since the parameter σ can be separated from the propagation step only because it is independent of temperature, additional influences which may exist remain undiscovered as far as they are connected with a gain or a loss of enthalpy (forces between the single chains and interactions with solvent). A possibly occurring effect of enthalpy, which is observed only during nucleation, is then distributed among the propagation steps.

Figure 35 shows how the equilibrium transition curves of [(Ala-Gly-Pro)$_n$]$_3$Lys-Lys shift to higher temperatures with increasing chain length. The continuous curves represent the numerical fit to the experimental transition data.

As shown in Fig. 36, in 1% aqueous acetic acid (pH = 3.0), a linear rise in T_m from the chain length n = 9 to n = 13 is observed. The apparent nucleation parameter for n = 9–13 thus obtained according to Eq. (9a) is

$\sigma_{app} \simeq 5 \times 10^{-2}$

The covalent bridging of the three polypeptide chains with the sequence Ala-Gly-Pro has facilitated, as expected, the nucleation step for triple-helix formation. This is also manifested by the marked increase in the renaturation rate as compared to that of the individual chains. This could be explained through entropy considerations; in the case ofe the covalently bridged polypeptide chains, the formation of a trimer molecular contact arising from three single (individual) chains via a diffusion-controlled pre-equilibrium is no more essential. The bridged carboxylic end should, therefore, be considered as the starting point for nucleation which leads to helix formation without causing a disorder in the peptide chains.

4.5 Influence of Proline and Hydroxyproline

In the ideal case, the well-known relationsship $\Delta G° = -RT \ln K_c = \Delta H° - T\Delta S°$ is valid for the triple helix-coil transition. Accordingly, for $T = T_m$ and $\Theta = 1/2$, we get Eq. (5).

$$T_m = \frac{\Delta H°}{\Delta S° + R \cdot \ln(0.75 \cdot C_g^2)} \tag{5}$$

It has been proved by many experiments that in the case of an increasing percentage of imino acids, T_m increases in the case of native collagen as well as in the case of model peptides (Fig. 37 and 39). This can be attributed to the following causes (seee Eq. (5)):
– Increase of enthalpy gain ($\Delta H°$ changes to higher negative values)
– and/or decrease of the entropy losses ($\Delta S°$ changes to higher positive values)
– enthalpy overcompensation of increasing entropy losses
– entropy overcompensation of increasing enthalpy losses

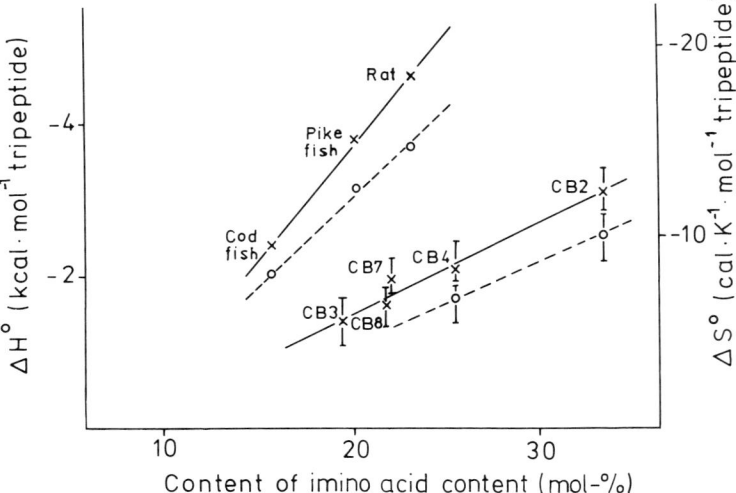

Fig. 37. Dependence of the thermodynamic parameters ΔH and ΔS of triple-helix formation on the imino acid content of the peptides (obtained by cleavage of calf skin-type I collagen with cyanogene and subequent isolation by column chromatrography)[3] and of the native neutral salt-soluble skin collagene of various animals. The entropy values are denoted by dotted lines

According to the collagen structure proposed by Rich and Crick[155] (3 polypeptide chains of the sequence Gly-X-Y joined by one hydrogen bond per tripeptide), $\Delta H°$ must remain constant with increasing percentage of imino acids, if the stabilization of the triple helix is effected only by hydrogen bonds. Besides, the entropy losses, caused by the rigidity of the pyrrolidine bonds in peptides with a high percentage of imino acids, should be raised.

Sutoh and Noda[154] succeeded in proving, by synthesizing block copolymers of the structure $(Gly-Pro-Pro)_n(Gly-Ala-Pro)_m-(Gly-Pro-Pro)_n$, that with increasing imino acid content, $\Delta S°$ changes to higher positive values. They do, however, not relate this to lower entropy losses of conformation but to hydrophobic interactions of the proline residues in the helical state.

This interpretation is in accordance with the observation that $\Delta H°$ rises with insertion of additional proline residues. Since the thermostability also increases, obviously an increase in the entropy of the whole solvent-polymer system is to be expected, causing an overcompensation of the enthalpy losses. Thus, besides the hydrogen bonds, special solvent effects have to be taken into account. Since the interaction between pure solvent molecules is expected to be different from the solvent-protein interactions, the state of the solvent molecule in immediate neighbourhood of the protein surface differs from the state of the pure solvent. Since different conformations mean different extents of contact between protein and solvent, the corresponding conformation is stabilized, depending on the composition of the solvent. We know that the transport of short-chain hydrocarbons from the gas phase or an apolar solvent into water is connected with high entropy losses[155]. This is also the reason for the low solubility of hydrocarbons in water. Proteins with many hydrophobic side chains, therefore, prefer such conformations by means of which they may build up regions without any water. The mentioned increase of entropy in the formation of such hydrophobic bonds points to a breakdown of an adjacent regular water structure.

In contrast to the results of Sutoh and Noda[154] in the case of collagen and CB peptides resulting from it, a strong increase of $\Delta H°$ and $\Delta S°$ with rising imino acid content and increasing thermostability[3, 151, 152] is obtained. It proves the existence of an enthalpy overcompensation of the entropy losses, which cannot be explained in detail at the present state of research. It seems to be sure that besides the number of hydrogen bonds, also other factors determine the thermostability of collagen and its model peptides.

Privalov and Tiktopulo[152] could prove by calorimetric measurements that, when heating a solution of tropocollagen in water, not only the triple helix itself melts cooperatively but also larger layers of water, adjacent to the regular collagen structure. This effect could be detected only for native collagen and might be one reason for either $\Delta H°_s$ or $\Delta S°_s$ (per mole tripeptide) changing to higher positive values in the following order at the same imino acid content: tropocollagen > CB-peptides > synthetic model peptides. (The role of specific interactions between side chains will be discussed later.)

Menashi et al.[153] could confirm the results of Privalov and Tiktopulo[152] and interprete the described effects as follows: In the case of native tropocollagen, the pyrrolidine residues are probably directed away from the fibrillar axis and are mostly coated by water which is structured in the immediate neighbourhood to the pyrrolidine residues. During the denaturation these pyrrolidine residues form hydrophobic bonds with each other or with other apolar residues within the same chain (endothermic interaction) while the structure of water breaks down (increase of entropy).

Because of these observations, comparative experiments with peptides of different proline content in a solvent less polar than water, are recommended. (Pro-Pro-Gly)$_n$ and (Pro-Ala-Gly)$_n$, in methanol/acetic acid (volume ratio 9:1) show a temperature-induced triple helix-coil transition which is characterized by the following parameters[92, 150]:
(Pro-Pro-Gly)$_n$: $\Delta H_s^\circ = -1.9$ kJ/mol tripeptide; $\Delta S^\circ = -5.4$ J $\cdot$ mol^{-1} $\cdot$ K^{-1}
(Pro-Ala-Gly)$_n$: $\Delta H_s^\circ = -0.9$ kJ/mol tripeptide; $\Delta S^\circ = -3.8$ J $\cdot$ mol^{-1} $\cdot$ K^{-1}

According to these data, in methanol, increasing entropy losses are overcompensated with respect to enthalpy when replacing alanine by proline. Methanol is not able to build up a three-dimensional layer in the immediate vicinity of the triple helix. Nevertheless, the thermodynamic data also reveal here, that an interaction between methanol and the triple helix occurs, the extent of which is caused by the percentage of proline. Comparing peptides of the same sequence in different solvents, one can observe the following tendency. Using solvents of different polarity, ΔH° changes to higher negative values with increasing polarity whereas T_m decreases because ΔS° also assumes higher negative values[92].

The interpretation of these results is, however, problematic since no data on the absolute enthalpy and entropy of the respective triple helix and coiled state are available. Though it may be taken as an established fact that the entropy of conformation of a (Pro-Pro-Gly)$_n$ coil is lower than in the case of a (Pro-Ala-Gly)$_n$ coil, we are not sure whether the entropy of the triple helix depends on the imino acid content.

Calorimetric data seem to confirm that tropocollagen has regions which melt separatively and cooperatively, because in this case, alternating regions with different imino acid percentage are present, i.e. different degrees of order. Concerning the model peptides dealt with, we conclude that (Pro-Ala-Gly)$_n$ forms a triple helix which is more flexible than the (Pro-Pro-Gly)$_n$ helix and thus not all of the possible hydrogen bonds of (Pro-Ala-Gly)$_n$ are enclosed in the triple-helical state.

All results, except those of Sutoh and Noda, seem to confirm that the higher thermostability with increasing imino acid content is caused by an enthalpy overcompensation of the entropy losses. ΔH° and also ΔS° are obviously caused by different effects. This is shown by Eq. (10) and Fig. 38

$$\Delta G^\circ = \Delta H^\circ_{H_2O} + \Delta H^\circ_{hydrophob.} + \Delta H^\circ_{solvent} - T\Delta S^\circ_{solvent} - T\Delta S^\circ_{conf.} \qquad (10)$$

Since 1973, several authors have proved that there is a relationship between thermostability of collagen and the extent of hydroxylation of the proline residues[31, 34]. Equilibrium measurements of the peptides α1-CB 2 of rat tail and rat skin revealed a higher $T_{m,\infty}$ for α1-CB 2 (rat skin)[157]. The sequence of both peptides is identical except that in the peptide obtained from rat skin, the hydroxylation of the proline residues in position 3 has occurred to a higher extent than in the case of α1-CB 2 (rat tail). Thus, a mere difference of 1.8 hydroxy residues per chain causes a $\Delta T_{m,\infty}$ of 26 K. Obviously, there are different stabilizing interactions in the triple-helical state, that means α1-CB 2 (rat skin) forms more exothermic bonds than α1-CB 2 (rat tail) in the coil triple-helix transition. This leads to an additional gain of enthalpy which overcompensates the meanwhile occurring losses of entropy.

Ward and Mason[157] explain these effects by direct hydrogen bonds between the hydroxy group of hydroxyproline in position 3 and the carboxylic side chain of glutamic acid in the adjacent chains. From the sequence we know that in the case of parallel and

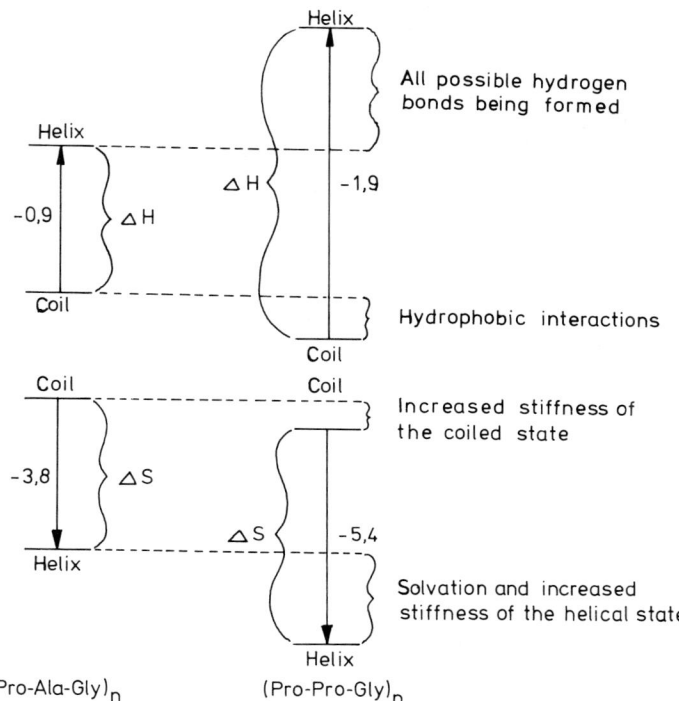

Fig. 38. $\Delta H°/\Delta S°$ diagram of the coil $\rightleftharpoons$ helix transition of (Pro-Ala-Gly)$_n$ and (Pro-Pro-Gly)$_n$, respectively. ($\uparrow$: H changes to higher negative values, ($\downarrow$: S changes to higher positive values). $\Delta H°$ is expressed in kJ · mol^{-1} tripeptide, $\Delta S°$ is expressed in J · mol^{-1}K^{-1} tripeptide

perfectly aligned chains, four of the six hydroxyproline residues are adjacent to glutamic acid residues. This indication of special interactions between adjacent amino acid residues has been proved by comparing $\Delta H°$ of CB-peptides with $\Delta H°$ of synthetic peptides[142]. The comparison of $\Delta H°$ values of (Gly-Pro-Pro)$_{10}$ and (Gly-Pro-Hyp)$_{10}$ points out that in the hydroxylation of one proline residue, $\Delta H°_{OH}$ is increased to the amount of -1.2 kJ/mol tripeptide. If we substract this amount from $\Delta H°$ of the CB-peptides corresponding to the respective percentage of proline, we can regard these CB-peptides as peptides being formed by interchanging hydroxyproline with proline (in the case of simple additivity of $\Delta H°$) (Fig. 39).

The $\Delta H°$ values of the synthetic polypeptides being much too small and corresponding to the degree of imino acids, seem to suggest that despite the dominant role of the imino acids, a further contribution of a kind already discussed must exist. Besides the possibility of direct hydrogen bonding of the hydroxy group of hydroxyproline to a carbonyl group of an adjacent chain[158], other authors propose models in which water forms an interchenar and/or intrachenar bridge between hydroxyproline and an adjacent carbonyl group[159, 160].

Recently, Engel et al.[92] critically discussed these models and concluded by equilibrium measurements in different solvents that bridging of the hydroxy group of hydroxyproline by means of a water molecule must be an unspecific reaction and can be caused as

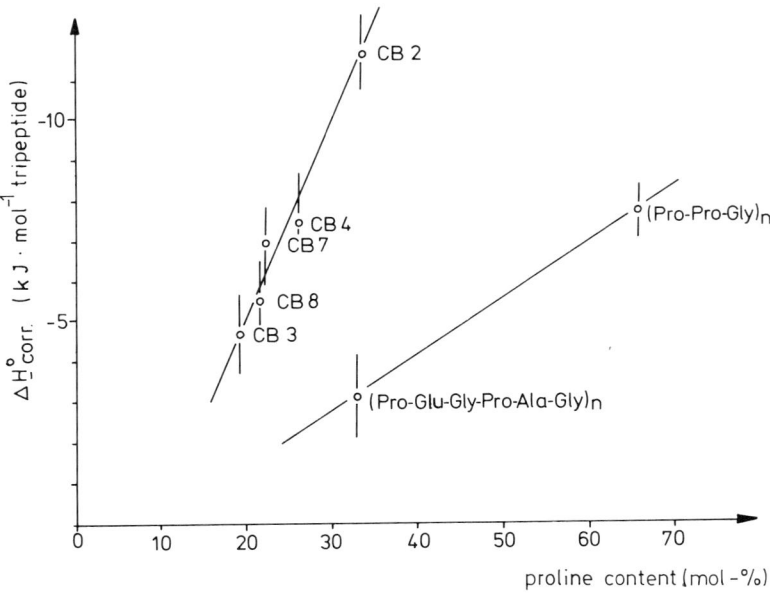

Fig. 39. Dependence of $\Delta H°$ of triple-helix formation on the proline content of the CNBr peptides and synthetic model peptides, respectively

well (if occurring at all) by 1,2-propane-diol, methanol, or acetic acid. A further stabilizing effect could be the increased solvation of the triple helix caused by the incorporation of hydroxyproline.

5 Conclusions and Outlook on Future Developments

Because of the large number of amino acids involved in the triple helix – coil transition of collagen, it is reasonable to synthesize simple model peptides in order to establish unambiguous relations. The thermodynamic characterization of these model peptides supplies information about their thermostability as well as about the enthalpy and entropy changes associated with the formation of the tertiary structure. The aim of these investigations is ultimately to establish a correlation between the primary structure and the stability of the tertiary structure of the highly sophisticated three-chain macromolecular structure of collagen. The dependence of the thermodynamic parameters on the chain lengths provides further information about the nucleation step, the molecular interpretation of which is the key to the understanding of the cooperative processes.

The model peptides should contain a well-defined sequence, should be pure and possess a uniform chain length. These are the prerequisites to the interpretation of the results of the thermodynamic measurements on the basis of the model concepts (e.g. "AON model").

The amino acids proline and hydroxyproline exert a stabilizing influence on the triple helix as described in detail in Sect. 4.5. By examining the CB peptides of collagen, a structural stability which is directly proportional to the imino acid content may thus be found. It has, however, not been possible to synthesize model peptides displaying structural stability comparable to that of the native peptides having corresponding amino acid contents.

It is known that native collagen contains tripeptide sequences which, because of being homopolypeptides, are not able to give rise to triple-helical tertiary structures (e.g. Gly-Pro-Leu, Gly-Pro-Ser). The reason for this and for the above-mentioned low thermostability of the synthetic homopolypeptides is presumably to be found in the fact that in the case of the model peptides with their monotonously repeated tripeptide sequences, special interactions between the side chains of the different amino acid residues as postulated by Ward and Mason are no more possible[157].

Model peptides that could build up quarternary fibrillar structures are not yet known. Though complete explanation of the interdependence between the primary structure and the stability of the quarternary structure has not yet been possible, i.e. the role of the different amino acids in collagen could be understood completely only in correlation with the fibril formation (formation of polar and hydrophobic "clusters").

Through the synthesis of block copolymers of the type $(GPP)_n(GPX)_m(GPP)_n$ it is possible to investigate the folding properties of even such sequences with the residue X which, as homopolypeptides, cannot give rise to a triple helix. This is especially true if the peptide chains are threefold covalently bridged. The entropy losses accompanying triple-helix formation should thereby by reduced and consequently the thermodynamic stability of the tertiary structure be increased. This effect is of great importance in connection with the desired "all-or-none" behavior, since it helps to keep the model peptide chains short; furthermore, the thermodynamic and kinetic models are simplified. Ultimately, the bridged peptides represent a model of the amino terminal segment of Col. 1–3 in type III-procollagen. This peptide shows a comparatively rapid ($k_{exp} = 8 \times 10^{-3} s^{-1}$) reversible and concentration-independent coil-triple-helix transition, the rate-determining step of which being the *cis-trans* isomerization of the peptide bond with its secondary nitrogen atom from proline[144].

It was, therefore, inferred that the disulfide bridges occurring between the chains of the appendix peptides of procollagen play a key role during the triple-helix formation in vivo. The simulation of the kinetic properties of Col. 1–3 was made possible for the first time through the synthesis of the threefold covalently bridged model peptides of the sequence $(Ala-Gly-Pro)_n$, $n = 8$. For a further refinement of the model peptides, it is planned to bridge the Lys-Lys covalently bridged peptide from both sides by coupling two cysteine residues per individual chain. One may expect a further increase in the thermodynamic stability of the tertiary structure as well as the exclusion of end effects. These end effects are undesirable in the determination of the nucleation parameter, because their influence on the experimentally determined nucleation parameter cannot be separated from the nucleation process below the cooperative length.

6 References

1. Traub, W., Piez, K. A.: Adv. Protein Chem. *25*, 243 (1971)
1a. Piez, K. A., Sherman, M. R.: Biochem. *9*, 4129 (1970)
2. Privalov, P. L., Tiktopulo, E. J.: Biopolymers *9*, 127 (1970)
3. Saygin, Oe., Heidemann, E.: Biopolymers *17*, 511 (1978)
4. Crick, F. H. C., Rich, A.: Nature *176*, 780 (1955)
5. Berger, A., Kurtz, J. Katchalski, E.: J. Am. Chem. Soc. *76*, 5552 (1954)
6. Fietzek, P. P., Kühn, K.: Int. Rev. Connect. Tissue Res. *7*, 1 (1976)
7. Sasisekharan, V.: Acta Crystallogr. *12*, 903 (1959)
8. Okabayashi, H., Isemura, T., Sakakibara, S.: Biopolymers *6*, 323 (1968)
9. Rothe, M., Theyson, R., Steffen, K. D.: Tetrahedron Lett. 4063 (1970)
9a. Rothe, M., Mazáne, K. J.: Tetrahedron Lett. *1972*, 3795; Rothe M., Rott, H., Mazáne, K. J.: Proc. 14th Eur. Peptide Symp., Wepion, p. 309, 1976
10. Schrohenloher, R. E., Ogle, J. D., Logan, M. A.: J. Biol. Chem. *234*, 58 (1959); Graßmann, W., Nordwig, A., Hörmann, H.: Z. Phys. Chem. *323*, 48 (1961); see also the review article: Nordwig, A., Hannig, K.: Das Leder *14*, 281 (1963)
11. Andreeva, N. S., Millionova, M. J., Chirgadze, Y. N.: Aspects of Protein Structure, Academic Press, New York 1963; Biofizika *6*, 244 (1961)
12. Goren, H. J.: CRC Crit. Rev. Biochem. *2*, 197 (1974)
13. Engel, J. et al.: J. Mol. Biol. *17*, 255 (1965)
14. Jäger, G.: doctoral thesis, TH Darmstadt 1966
    Bernhardt, H. W.: doctoral thesis, TH Darmstadt 1968
15. Heidemann, E., Nill, H. W.: Z. Naturforsch., *B 24*, 837, 843 (1969)
16. Heidemann, E., Meisel, H. D.: Makromol. Chem. *166*, 1 (1973)
17. Heidemann, E., Harrap, B. S., Schiele, H. D.: Biochemistry *12*, 2958 (1973)
18. Carver, J. P., Blout, E. R., in: Treatise on Collagen I. Ramachandran, G. N. (ed.), p. 441 ff. Academic Press, New York 1967
19. DeTar, D. F. et al: J. Am. Chem. Soc. *78*, 947 (1963)
20. Neiss, H. G., Heidemann, E. R.: Makromol. Chem. *177*, 701 (1976)
21. Heymer, G., Heidemann, E. R.: Makromol. Chem. *177*, 3299 (1976)
22. Khodadadeh, K., Heppenheimer, K., Heidemann, E. R.: Makromol. Chem. *178*, 1897 (1977)
23. Greiche, Y., Heidemann, E. R.: Biopolymers *18*, 2359 (1979)
24. Heidemann, E. et al.: Polymer *18*, 420 (1977)
25. Cowell, R. D., Jones, J. H.: J. Chem. Soc. C *1971*, 1082
26. Fairweather, R., Jones, J. H.: J. Chem. Soc., Perkin I *1972*, 1908
27. Lorenzi, G. P., Doyle, B. B., Blout, E. R.: Biochemistry *10*, 3046 (1971)
28. Johnson, B. J.: J. Pharm. Sci. *63*, 313 (1974)
29. Kovacs, J., Ballina, R., Rodin, R. L.: J. Am. Chem. Soc. *89*, 119 (1965)
30. Kobayashi, Y., Isemura, T.: Progr. Polym. Sci., Japan *3*, 315 (1972)
31. Sakakibara, S. et al.: Bull. Chem. Soc. Japan *41*, 1273 (1968)
32. Rothe, M., Kunitz, F. W.: Liebigs Ann. Chem. *609*, 88 (1957)
33. Bruckner, P. et al.: Helv. Chim. Acta *58*, 1276 (1975)
34. Weber, R. W., Nitschmann, H.: Helv. Chim. Acta *61*, 701 (1978)
35. Sutoh, K., Noda, H.: Biopolymers *13*, 2385 (1974)
36. Sakakibara, S. et al.: Biochim. Biophys. Acta *303*, 198 (1973)
37. Roth, W., Heppenheimer, K., Heidemann, E.: Makromol. Chem. *180*, 905 (1979)
38. Lazarev, Y. A. et al.: Biopolymers *17*, 1197 (1978)
39. Bayer, E., Mutter, M.: Chem. Ber. *107*, 1344 (1974)
40. Shibnev, V. A. et al.: Izv. Akad. Nauk. SSSR, Ser. Khim. *1968*, 2564
41. Shibnev, V. A., Chuvaeva, T. P., Poroshin, K. T.: Izv. Akad. Nauk. SSSR, Ser. Khim. *1970*, 121
42. Shibnev, V. A. et al.: Izv. Akad. Nauk. SSSR, Ser. Khim. *1969*, 637
43. Bloom, S. M.: J. Am. Chem. Soc. *88*, 2035 (1966)
44. Shibnev, V. A. et al.: Izv. Akad. Nauk. SSSR, Ser. Khim. *1969*, 2532
45. Shibnev, V. A. et al.: Izv. Akad. Nauk. SSSR, Ser. Khim. *1968*, 144

46. Stewart, F. H. C.: Aust. J. Chem. *18*, 887 (1965)
47. Traub, W., Yonath, A.: J. Mol. Biol. *16*, 404 (1966)
48. Traub, W.: J. Mol. Biol. *43*, 479 (1969)
49. Wolmann, Y., Gallop, P., Patchornik, A.: J. Am. Chem. Soc. *83*, 1263 (1961)
50. Shibnev, V. A., Lazareva, A. V.: Izv. Akad. Nauk. SSSR, Ser. Khim. *1969*, 398
51. Shibnev, V. A. et al.: Izv. Akad. Nauk. SSSR, Ser. Khim. *1967*, 1634
52. Shibnev, V. A., Lisovenko, A. V.: Izv. Akad. Nauk. SSSR, Ser. Khim. *1966*, 1287
53. Shibnev, V. A. et al.: Dokl. Akad. Nauk. SSSR *1971*, 198, 862
54. Engel, J. et al., in: Structure and Function of Connective and Skeletal Tissue. Jackson, S. F. (ed.), p. 241. Butterworth, London 1965
55. Kobayashi, Y. et al.: Biopolymers *9*, 415 (1970)
56. Yonath, A., Traub, W.: J. Mol. Biol. *43*, 461 (1969)
57. Shibnev, V. A., Lisovenko, A. V., Rogulenkova, V. N., Millionova, I., Esipova, N. G., Chirgadze, Y. N.: Biofizika *11*, 1067 (1966)
58. Andreeva, N. S., Esipova, N. G., Millonova, M. I., Rogulenkova, V. N. Shibnev, V. A.: Mol. Biol. *1*, 657 (1967)
59. Andreeva, N. S. et al. in: Conformation of Biopolymers. Ramachandran, G. N. (ed.), p. 469. Academic Press, New York 1967
60. Shibnev, V. A., Chuvaeva, T. P., Poroshin, K. T.: Izv. Akad. Nauk. SSSR, Ser. Khim. *1968*, 225
61. DeTar, D. F., Albers, R. F., Gilmore, F.: J. Org. Chem. *37*, 4377 (1972)
62. DeTar, D. F.: Peptides, Proc. 8th Eur. Pept. Symp. 1966
63. Inouye, K., Sakakibara, S., Prockop, D. S.: Biochim. Biophys. Acta *420*, 133 (1976)
64. Huggins, M. L., Ohtsuka, K., Morimoto, S.: J. Polym. Sci., Part C, *23*, 343 (1968)
65. Poroshin, K. T., Chuvaeva, T. P., Shibnev, V. A.: Dokl. Akad. Nauk. Tadzik. SSR *12*, 21 (1969)
66. Andreeva, N. S. et al.: Biofizika *6*, 244 (1961)
67. Shibnev, V. A., Debabov, V. G.: Izv. Akad. Nauk. SSSR, Ser. Khim. *1964*, 1043
68. Shibnev, V. A., Rogulenkova, V. N., Andreeva, N. S.: Biofizica *10*, 164 (1965)
69. Rogulenkova, V. N., Millionova, M. I., Andreeva, N. S.: J. Mol. Biol. *9*, 253 (1964)
70. Traub, W., Yonath, A.: J. Mol. Biol. *16*, 404 (1966)
71. Frey, P., Nitschmann, H.: Helv. Chim. Acta *59*, 1401 (1976)
72. Heidemann, E., Bernhardt, H. W.: Das Leder *19*, 281 (1968)
73. Heidemann, E., Bernhardt, H. W.: Nature *220*, 1326 (1968)
74. Brown, F. R. et al.: J. Mol. Biol. *63*, 85 (1972)
75. Brown, F. R., Hopfinger, A. J., Blout, E. R.: J. Mol. Biol. *63*, 101 (1972)
76. Jones, J. H.: J. Chem. Soc. D. *1969* 1436
77. Bloom, S. M. et al.: J. Am. Chem. Soc. *88*, 2035 (1966)
78. Oriel, P. J., Blout, E. R.: J. Am. Chem. Soc. *88*, 2041 (1966)
79. Brown, F. R., Carver, J. P., Blout, E. R.: J. Mol. Biol. *39*, 307 (1969)
80. Segal, D. M., Traub, W.: J. Mol. Biol. *43*, 487 (1969)
81. Blout, E. R., Lorenzi, G. P., Doyle, B. B.: Biochemistry *10*, 3046 (1971)
82. Blout, E. R. et al.: Biochemistry *10*, 3052 (1971)
83. Heidemann, E., Bernhardt, H. W.: Nature *216*, 263 (1967)
84. Berg, R. A., Olsen, B. R., Prockop, D. J.: J. Biol. Chem. *245*, 5759 (1970)
85. Olsen, B. R. et al.: J. Mol. Biol. *57*, 589 (1971)
86. Okuyama, K. et al.: J. Mol. Biol. *72*, 571 (1972); ibid. *72*, 371 (1972)
87. Segal, D. M.: J. Mol. Biol. *43*, 497 (1969)
88. Segal, D. M., Traub, W., Yonath, A.: J. Mol. Biol. *43*, 519 (1969)
89. Shaw, B. S., Schurr, I. M.: Biopolymers *14*, 1951 (1975)
90. Bonara, G. N., Toniolo, C.: Biopolymers *13*, 1055 (1974); *13*, 1067 (1974)
92. Engel, J., Han-Ten Chen, Prockop, D. S.: Biopolymers *16*, 601 (1977)
93. Kitaoka, H., Sakakibara, S., Tani, H.: Bull. Chem. Soc., Japan *31*, 802 (1958)
94. Tamburro, A. M., Scatturin, A., Marchiori, F.: Gazz. Chim. Ital. *98*, 638 (1968)
95. Shibnev, V. A. et al.: Izv. Akad. Nauk. SSSR, Ser. Khim. *1970*, 2822
96. Shibnev, V. A. et al.: Izv. Akad. Nauk. SSSR, Ser. Khim. *1970*, 339
97. Khalikov, S. K., Poroshin, K. T., Shibnev, V. A.: Izv. Akad. nauk. Tadzh. SSR *11*, 28 (1968)

98. Shibnev, V. A. et al.: Izv. Akad. Nauk. SSSR, Ser. Khim. *1970*, 880
 99. Poroshin, K. T., Burichenko, V. K.: Dokl. Akad. Nauk. Tadzh. SSR *11*, 28 (1968)
100. Shibnev, V. A., Poroshin, K. T., Grechishko, V. S.: Izv. Akad. Nauk. SSSR, Ser. Khim. *1966*, 1493
101. Shibnev, V. A., Grechishko, V. S., Poroshin, K. T.: Izv. Akad. Nauk. SSSR, Ser. Khim. *1967*, 2327
102. Poroshin, D. T. et al.: Dokl. Akad. Nauk. Tadzh. SSR *13*, 19 (1970)
103. Bell, J. R., Jones, J. H., Webb. T. C.: Int. J. Pept. Protein Res. *7*, 235 (1975)
104. Poroshin, K. T, Chuvaeva, T. P., Shibnev, V. A.: Dokl. Akad. Nauk. Tadzh. SSR *12*, 15 (1969)
105. Zegelman, A. B., Yusupov, T. Y., Poroshin, K. T.: Dokl. Akad. Nauk. Tadzh. SSR *13*, 25 (1970)
106. Zegelman, A. B. et al.: Dokl. Akad. Nauk. Tadzh. SSR *15*, 33 (1972)
107. Doyle, B. B. et al.: J. Mol. Biol. *51*, 47 (1970)
108. Takahashi, S.: Bull. Chem. Soc., Japan *42*, 521 (1969)
109. Iio, T., Takahashi, S.: Bull. Chem. Soc., Jpn. *43*, 515 (1970)
110. Iio, T., Takahashi, S.: Bull. Inst. Chem. Res., Kyoto Univ. *49*, 80 (1971)
111. Brack, A., Spach, G.: Biopolymers *11*, 563 (1972)
112. Brack, A., Spach, G.: C. R. Acad. Sci., Paris, Ser. C *271*, 916 (1970)
113. Anderson, J. M., Rippon, W. B., Walton, A. G.: Biochem. Biophys. Res. Commun. *39*, 802 (1970)
114. Burichenko, V. K., Morozova, L. V., Poroshin, K. T.: Dokl. Akad. Nauk. Tadzh. SSR *13*, 26 (1970)
115. Yusupov, T. Y. et al.: Dokl. Akad. Nauk. Tadzh. SSR *12*, 32 (1969)
116. DeTar, D. F., Vajda, T.: J. Am. Chem. Soc. *89*, 998 (1967)
117. Burichenko, V. K. et al.: Izv. Akad. Nauk. SSSR, Ser. Khim. *1972*, 2597
118. Johnson, B. J.: J. Chem. Soc. C. *1968*, 3008
119. Feairweather, R., Jones, J. H.: J. Chem. Soc., Perkin Trans. I. *1972*, 2475
120. Erian Beshava, N.: Doctor-Thesis, University Hohenheim 1978
121. Neiss, H. G.: unpublished results
122. Khodadadeh, K.: doctoral thesis, TH Darmstadt 1976
123. El Sheikh, M.: doctoral thesis, TH Darmsatdt 1978
124. Heppenheimer, K.: unpublished results 1978
125. Schmidt, A.: doctor thesis, TH Darmstadt 1975
126. Goren, H. J., McMillin, C. R., Walton, A. G.: Biopolymers *16*, 1527 (1977)
127. Pysh, E. S.: J. Mol. Biol. *23*, 587 (1967);
     Pysh, E. S.: J. Chem. Phys. *52*, 4723 (1970)
128. Kikuchi, Y., Fuchimoto, D., Tamiya, N.: Biochem. J. *115*, 569 (1969)
129. Kobayashi, Y., Kyogoku, Y.: J. Mol. Biol. *81*, 337 (1973)
130. Sakakibara, S. et al.: J. Mol. Biol. *65*, 371 (1972)
131. Rothe, M., Schneider, H. J.: Angew. Chem. *82*, 557 (1970); Angew. Chem. Int. Ed. Engl. *9*, 535 (1970)
132. Shibnev, V. A. et al.: Izv. Akad. Nauk. SSSR, Ser. Khim. *1967*, 1634
133. Lazarev, Yu. et al.: Biopolymers *17*, 1215 (1978)
134. McBride, O. W., Harrington, W. F.: Biochemistry *6*, 1499 (1967)
135. Harrington, W. F., Rao, N. V.: Biochemistry *9*, 3714 (1970);
     Harrington, W. F., Karr, G. M.: Biochemistry *9*, 3725 (1970);
     Hauschka, P. V., Harrington, W. F.: Biochemistry *9*, 3725 (1970)
136. Ramachandran, G. N., Doyle, B., Blout, E. R.: Biopolymers *6*, 1771 (1968)
137. Andries, J. C., Walton, A. C.: J. Mol. Biol. *54*, 579 (1970)
138. Seba, J.: diploma thesis, TH Darmstadt 1976
139. Rosenbrock, H. H.: Comput. J. *3*, 175 (1960)
140. Pörschke, D., Eigen, M.: J. Mol. Biol. *62*, 361 (1971)
141. Kühn, K. et al.: Arch. Biochem. Biophys. *105*, 387 (1965)
142. Weidner, H., Engel, J., Fietzek, P.: Peptides, Polypeptides and Proteins, Part V., John Wiley and Sons, New York 1974
143. Nowack, H.: Eur. J. Biochem. *70*, 205 (1976)

144. Bächinger, H. P. et al.: Eur. J. Biochem. *90*, 605 (1978)
145. Grathwol, C., Wüthrich, K.: Biopolymers *15*, 2043 (1976)
146. Bächinger, H. P. et al.: Eur. J. Biochem. *106*, 619 (1980)
147. Zimm, B. H., Bragg, J. K.: J. Chem. Phys. *31*, 526 (1959)
148. Ising, E.: Z. Phys. *31*, 253 (1925)
149. Saygin, Ö.: doctoral thesis, TH Darmstadt 1976
150. Go, N., Suezaki, Y.: Biopolymers *12*, 1927 (1973)
151. Roth, W., Heidemann, E.: Biopolymers *19*, 1909 (1980)
152. Privalov, P. L., Tiktopulo, E. J.: Biopolymers *9*, 127 (1970)
153. Menashi, S. et al.: Biochim. Biophys. Acta *444*, 623 (1976)
154. Sutoh, K., Noda, H.: Biopolymers *13*, 2461 (1974)
155. Rich, A., Crick, F. H.: J. Mol. Biol. *3*, 483 (1961)
156. Nemethy, G., Scheraga, H. A.: J. Chem. Phys. *36*, 3387 (1962)
157. Ward, A. R., Mason, P.: J. Mol. Biol. *79*, 431 (1973)
158. Berg, R. A. et al.: Biochim. Biophys. Acta *328*, 553 (1973)
159. Ramachandran, G. N., Bansal, M., Bhatnagar, R. S.: Biochim. Biophys. Acta *322*, 166 (1973)
160. Traub, W.: Isr. J. Chem. *12*, 435 (1974)
161. S. Divanidis: Diploma thesis, Darmstadt 1978

Received October 30, 1979/April 13, 1981
W. Kern (editor)

Thermodynamics and Kinetics of Orientational Crystallization of Flexible-Chain Polymers

Galina K. Elyashevich

Institute of Macromolecular Compounds of the Academy Science of the USSR, Leningrad, USSR

Crystallization of flexible-chain polymers from isotropic solutions and melts usually leads to the formation of folded-chain crystals. Mechanical properties, particularly the tenacity of these samples, are limited by the small number of tie chains. Crystallization in deformed melts proceeds by a substantially different mechanism involving extension of chains and formation of extended-chain crystals: this process is to some extent similar to that of crystallization of rigid-chain polymers leading to polymers of high tenacities. This method which promotes the formation of extended-chain crystals during crystallization of flexible-chain polymers – the orientational crystallization – is suitable for the preparation of polymer films and fibers from flexible-chain polymers with high tenacity and high modulus. The paper presents a comparison of orientational crystallization with other methods for obtaining high-strength polymer materials. Technological prospects of this method are also considered.

1 Introduction: Fundamental Aspects of Orientational Processes in Crystallizable Polymers 207
   1.1 Anisotropy of Physical Properties as the Main Feature of the Oriented State of Polymers 208
   1.2 Polymer Flexibility as a Decisive Parameter of the Thermodynamic and Kinetic Features of Orientation Processes 208

2 Methods for Preparing Oriented Crystallizable Polymers 211

3 Main Features of Crystallization from Oriented Solutions and Melts 214

4 Thermodynamics of Crystallization of Flexible-Chain Polymers Under Conditions of Molecular Orientation 217
   4.1 Effect of Molecular Orientation on Crystallization of Flexible-Chain Polymers 217
   4.2 Topomorphism of Polymer Crystals 226
   4.3 Formation of Extended-Chain Crystals (ECC) Under Equilibrium Conditions 229
   4.4 Mechanism of the Formation of ECC Under Conditions of Molecular Orientation 230
   4.5 Melting of ECC 234

5 Structure and Properties of Systems Obtained Under Conditions of Molecular Orientation 237

6 Conclusions: Prospects of Processing of High-Tenacity Polymer Materials . . . 241

7 References 244

1 Introduction: Fundamental Aspects of Orientational Processes in Crystallizable Polymers

At present, the attention of the researchers in the field of the physical chemistry and technology of polymers is directed to the problems of the preparation of high-strength materials based on highly oriented polymers. These materials are now widely used. Highly oriented polymers are obtained by two methods: (1) pre-orientation of polymer systems in the viscous (or glassy) state with subsequent fixation of this orientation during the formation of the final material and (2) preparation of a weakly oriented "intermediate" and achievement of its high physico-mechanical properties by further treatment (drawing in various media, thermal treatment, etc.). This paper deals mainly with the former relatively new method which is of particular interest since it allows to prepare high-strength polymer materials satisfying all recent technical requirements for a single-stage process.

The problem of obtaining high-strength polymer materials is closely related to the development of completely oriented structures in these polymers. At present, the most widely used methods for obtaining highly oriented polymers are based on their crystallization from stirred solutions or oriented melts during flow. Mechanical treatment is a powerful factor affecting the thermodynamic state of the solution or of the melt undergoing deformation and can drastically change the thermodynamics and kinetics of crystallization and formation of the supramolecular structure in general. The structure and properties of such materials differ essentially from those of the materials obtained by the usual method (without deformation). The change in the Gibbs free energy of the melt or solution undergoing deformation is due to a change in the conformation of macromolecules. During the phase transition referred to as *orientational crystallization*, at least a portion of chains is crystallized in an extended conformation forming extended-chain crystals (ECC). The presence of crystals of this type leads to a change in all physico-mechanical characteristics of polymers and, in some cases, the corresponding properties observed are unique and cannot be attained by any other method.

Two approaches to the attainment of the oriented states of polymer solutions and melts can be distinguished. The first one consists in the orientational crystallization of flexible-chain polymers based on the fixation by subsequent crystallization of the chains obtained as a result of melt extension. This procedure ensures the formation of a highly oriented supramolecular structure in the crystallized material. The second approach is based on the use of solutions of rigid-chain polymers in which the transition to the liquid crystalline state occurs, due to a high anisometry of the macromolecules. This state is characterized by high one-dimensional chain orientation and, as a result, by the anisotropy of the main physical properties of the material. Only slight extensions are required to obtain highly oriented films and fibers from such solutions.

These two different approaches for attaining an oriented state in flexible-chain and rigid-chain polymers indicate that the fundamental property of macromolecules – their flexibility – is of great importance to the orientation processes. However, the mechanism of the transition into the oriented state and the properties of highly oriented systems exhibit many features characteristic of both rigid- and flexible-chain polymers.

1.1 Anisotropy of Physical Properties as the Main Feature of the Oriented State of Polymers

The processes of ordering in polymer systems consisting of linear polymers are related, at least on one level of supermolecular organization, to the development of a predominant localization of macromolecules (or their parts) along some directions: the orientation axes, i.e. to the transition of the system into the oriented state. The most simple and most widely spread type of polymer orientation is the uniaxial orientation, i.e. the one-dimensional orientation in the direction of the axes of macromolecules.

The ability of being converted into the oriented state and the related appearance of a sharp anisotropy of the physical properties of the resulting material is one of the main features of polymers distinguishing them from low-molecular weight simple substances. The possibility of this transition is inherent in their structure: their anisotropy due to their chain structure, i.e. the existence of the predominant direction of interatomic forces along the main chains of the macromolecules. When the polymer system undergoes orientation, the latent (or, more precisely, the local) anisotropy of the internal field becomes explicit as the macroscopic anisotropy of all properties. Since the energy of interaction between the molecules in polymeric substances is always much lower than that of the chemical bonds between the atoms in the chain, the macromolecule retains its individuality in any system and the polymer system as a whole exhibits the anisotropy of properties characteristic of a single macromolecule.

Usually, the transition of polymer systems into the oriented state occurs as a result of deformation e.g. upon exposure to external stress. When the polymers undergo deformation both the macromolecule as a whole and its parts (segments) can undergo orientation. The rates of these orientation processes are very different and, hence, the orienting forces affect first of all the orientation of chain segments and subsequently that of a chain molecule as a whole. However, by varying the extension velocity and the temperature, only the overall orientation process may predominate, thus extension of all chains occurs in a single act.

Any extended part of a linear polymer molecule exhibits a strong anisotropy of many properties since its atoms and atomic groups are oriented and the macromolecule is actually a "one-dimensional crystal". The parallel packing of these parts during the formation of a uniaxially oriented polymer substance imparts the anisotropie properties of a single molecule to the whole polymeric material.

1.2 Polymer Flexibility as a Decisive Parameter of the Thermodynamic and Kinetic Features of Orientation Processes

Usually, dilute polymer solutions are isotropic systems, i.e. macromolecular chains can exist in these solutions independently of each other with a random distribution of orientations of the long axes of coils. The solutions of flexible-chain polymers remain isotropic when the solution concentration increases whereas in concentrated solutions of macromolecules of limited flexibility the chains can no longer be oriented arbitrarily and some direction of preferential orientations of macromolecular axes appears, i.e. the mutual orientations of the axes of neighboring molecules are correlated. This means that the system changes from the state with an isotropic distribution of elements to that of a local parallel order of relatively long chain segments exhibiting a structure similar that of low molecular weight lyotropic liquid crystals.

As was shown by Flory[1], starting from a certain concentration, it is impossible in principle (for purely geometric reasons) to fill a larger fraction of the volume with these chains when they are randomly arranged. This critical concentration depends on chain flexibility which Flory characterized by the fraction f of folded (gauche) isomers in a polymer molecule

$$f = (Z-2)\exp\left(-\frac{\varepsilon}{kT}\right) / \left[1 + (Z-2)\exp\left(-\frac{\varepsilon}{kT}\right)\right] \tag{1}$$

where Z is the coordination number in the quasi-lattice model used by Flory, and ε the difference in the free energy of a gauche and trans conformation i.e. the effective energy of a single break in the rod-like (in the initial state) molecule. By minimizing the Gibbs free energy of the system, Flory determined the critical equilibrium value of the flexibility parameter f_{cr} for the stability limit of the isotropic phase. For a non-diluted polymer (the volume fraction of the polymer in solution equals unity); this value is determined by the limiting value of the inequality

$$f > 1 - \frac{1}{e}\left(\frac{ZN}{2e}\right)^{\frac{1}{N-2}} \tag{2}$$

For high degrees of polymerization N, this inequality becomes

$$f > 1 - \frac{1}{e} \cdots \approx 0.63, \tag{3}$$

i.e. for high N, the value of f_{cr} is independent of Z.

This condition means that for $f < 0.63$ the disordered arrangement of molecules is thermodynamically unstable and the system is spontaneously reorganized into an ordered liquid crystalline phase of a nematic type (Flory called this state "crystalline"). This result has been obtained only as a consequence of limited chain flexibility without taking into account intermolecular interactions.

Hence, Flory's theory offers an objective criterion for chain flexibility and makes possible to divide all the variety of macromolecules into flexible-chain ($f > 0.63$) and rigid-chain ($f < 0.63$) ones. In the absence of kinetic hindrance, all rigid-chain polymers must form a thermodynamically stable organized nematic phase at some polymer concentration in solution which increases with f. At $f > 0.63$, the macromolecules cannot spontaneously adopt a state of parallel order under any conditions.

Hence, the transition of a polymer system into the oriented state is a result of the competition of two fundamental properties of a polymer molecule: (1) its inherent anisotropy which is the main reason for the ability of polymer systems to form an oriented phase and (2) its flexibility which favours coiling of a long molecule. The result of this competition is determined by the chemical nature of the molecule; however, kinetic hindrance can prevent the transition into the oriented state.

For rigid-chain crystallizable polymers, spontaneous transition into the nematic phase is accompanied by crystallization: intermolecular interactions should lead to the formation of a three-dimensional ordered crystalline phase.

The ability of rigid-chain molecules to form a thermodynamically stable nematic phase accounts for the ability of these systems to undergo a transition into the fibrous state[2]. The technological process of the preparation of fibres from solutions of rigid-chain polymers is based on this phenomenon. The nematic domains formed play the part of stocks in the structure of the future fibre. Even at low draw ratios, the boundaries between the domains disappear and uniaxial parallel packing of the macromolecules occurs in the whole volume of the system. It is fixed by the removal of the solvent and crystallization accompanied by chain extension, since rigid chains can crystallize only in this manner. This structure, in which almost 100% of chains passes through the fibre cross-section and the only type of defects are the joints between the ends of macromolecules randomly distributed in the volume of the crystal, ensures high values of tenacity (of the order of magnitude of 2000 MPa) and of elastic modulus (not less than 10^5 MPa) (Fig. 1).

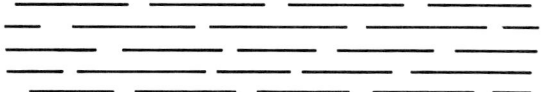

Fig. 1. Schematic representation of the structure of a crystallizable rigid-chain polymer; point defects are located at end joints

In contrast, for flexible-chain polymers, the transition into the ordered state is possible only if the flexibility can be decreased to values below f_{cr} (in the absence of external deformational fields, the crystallization of flexible-chain polymers occurs by the mechanism of chain folding).

This decrease in effective flexibility can be attained by extending the macromolecules in a mechanical or stationary hydrodynamic field[3]. The extending fields decrease the freedom of internal rotation of the chain units, and the new effective flexibility parameter now becomes:

$$f = (Z-2)\exp\left(-\frac{\varepsilon}{kT} - \frac{Fl}{kT}\right) / \left[1 + (Z-2)\exp\left(-\frac{\varepsilon}{kT} - \frac{Fl}{kT}\right)\right] \quad (4)$$

where F is the extending force applied to a segment of length l.

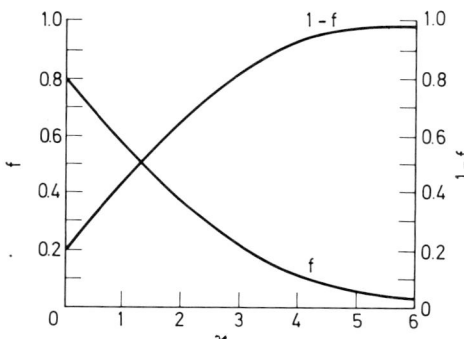

Fig. 2. Induced rigidity as a function of $\varkappa$ determined by the extending force

Figure 2 shows the increase in the rigidity (1 – f) of macromolecules induced by the field as a function of the parameter $\varkappa = \varepsilon/kT + Fl/kT$. As soon as the flexibility decreases to f < 0.63, a system of molecules flexible in the state of rest will undergo a "spontaneous" transition into a nematic oriented state upon the action of the stretching field, just as it occurs for rigid molecules at rest.

Hence, the main aim of the technological process in obtaining fibres from flexible-chain polymers is to extend flexible-chain molecules and to fix their oriented state by subsequent crystallization. The filaments obtained by this method exhibit a fibrillar structure and high tenacity, because the structure of the filament is similar to that of fibres prepared from rigid-chain polymers (for a detailed thermodynamic treatment of orientation processes in polymer solutions and the thermokinetic analysis of jet-fibre transition in longitudinal solution flow see monograph[3]).

2 Methods for Preparing Oriented Crystallizable Polymers

Usually, crystallization of flexible-chain polymers from undeformed solutions and melts involves chain folding. Spherulite structures without a preferred orientation are generally formed. The structure of the sample as a whole is isotropic; it is a system with a large number of folded-chain crystals distributed in an amorphous matrix and connected by a small number of tie chains (and an even smaller number of strained chains called "loaded" chains). In this case, the mechanical properties of polymer materials are determined by the small number of these ties and, hence, the tensile strength and elastic moduli of these polymers are not high.

One of the main methods for improving the mechanical properties of linear polymers is their drawing that can be uniaxial (fibres), biaxial (films), planar symmetrical (films-membranes) etc. As a result of polymer deformation, the system changes into the oriented state fixed by crystallization.

Uniaxial orientation is one of the main methods for the preparation of high-strength polymer materials. In the technology of artificial fibres and films the preparation consists in thermal and orientational drawing of systems that have already undergone crystallization. Hence, the structure and the mechanical properties of the fibre or film are to a great extent determined by the structure of the starting crystalline material which, in the case of flexible-chain polymers always exhibits many defects. This limits the possibilities of increasing the strength of polymer materials by orientational drawing (see below).

For crystallizable polymers, orientational drawing is usually carried out in the temperature range between the glass transition temperature of amorphous regions and the melting temperature of polymer crystals. Below the glass transition temperature, unoriented crystalline polymers undergo brittle fracture during deformation without changing into the oriented state. At low polymer deformation (relative elongation from a few percents to several tens of percents), the deformation of the sample and its structure are completely reversible; although relatively high degrees of orientation are attained and the shape of spherulites changes, the sample rapidly restores its initial shape and structure, after the load is removed and/or the polymer is heated.

A further increase in extension leads to irreversible changes which immediately precede the transition of the polymer into the oriented state. During this transition, the spherulites undergo considerable structural changes and are thus converted qualitatively into different structural elements i.e. macrofibrils[4]. After a certain critical elongation has been attained, the initial crystallites collapse and melt and a new oriented structure is formed in which the "c" axes of crystals are oriented in the direction of extension.

The transition into the oriented state is accompanied by the formation of a "neck", a sharp and abrupt local constriction of the sample, in which the extent of orientation and the degree of extension are much higher than in the rest of the polymer. After the neck has been formed, further orientation of the sample occurs by spreading of the neck to the entire length of the polymer. When the sample is extended after passing into the oriented state, it undergoes further deformation and at some critical extension it breaks.

Hence, the extension of an isotropic unoriented partially crystalline polymer leads to the formation of a highly organized material with a characteristic fibrillar structure. The anisotropy of the sample as a whole is expressed by a higher modulus, tenacity and optical anisotropy. It would seem that the increase in strength in the drawing direction suggests that the oriented samples consist of completely extended chains. However, while the strength of such perfect structure for polyethylene has been evaluated as 13 000 MPa[5], the observed values for an oriented sample are 50 to 30 MPa.

The reason is that the polymer in the oriented state retains its semicrystalline structure in which the amorphous regions represent structural defects even after orientational drawing. As has been shown by many authors (e.g., refs. 6, 7), and important feature of the structure formed as a result of orientational drawing is the fact that the crystallites in oriented polymers are almost entirely oriented along the drawing axis and the chain orientation in them is close to 1 at high drawing ratios[7], whereas in the amorphous regions the macromolecular segments are disoriented. Although the increase in the drawing ratio leads to increasing chain orientation in the amorphous regions, this rise is slight and approaches a limiting value already at low drawing rations.

According to Hosemann-Bonart's model[8], an oriented polymeric material consists of plate-like more or less curved folded lamellae extended mostly in the direction normal to that of the sample orientation so that the chain orientation in these crystalline formations coincides with the stretching direction. These lamellae are connected with each other by some amount of tie chains, but most chains emerge from the crystal bend and return to the same crystal-forming folds. If this model adequately describes the structure of oriented systems, the mechanical properties in the longitudinal direction are expected to be mainly determined by the number and properties of tie chains in the amorphous regions that are the "weak spots" of the oriented system (as compared to the crystallite)[9].

In fact, the chains in these regions are overloaded compared to those in the crystallites because their amount in the cross-section is smaller than in the crystallites: according to Flory's estimates[10] no more than 50% of chains can pass from an arbitrarily chosen folded crystallite into the neighboring crystallite. Moreover, chains in amorphous regions are of different lengths and are oriented differently and, hence, the strain in them is distributed non-uniformly. The chains of minimum length equal to the width of the amorphous region undergo the highest strain. Peterlin[11] called them "loaded" (or "carrying load"), i.e. strength determining. The amount of chains under load does not exceed 5% of the total numbers of chains passing through the crystallite cross-section normal to the crystallographic c-axis[11] and determines the numerical tenacity deficit in oriented flexible-chain polymers. Therefore, the strength is low as compared to the theoretical estimate. During thermal and orientational drawing the loaded chains are broken first. However, just these chains transfer the load across the sample so that, as drawing continues, the possibility of establishing local strains required for "unfolding" the folds of the crystallites and for obtaining a structure with extended chains (of the fringed micelle type) gradually decreases.

Hence, orientational drawing prevents a marked increase in the number of tie chains. Therefore, a rise in the strength of polymer articles exceeding 1000 MPa is not achieved by drawing, i.e. the strength deficiency is retained even after repeated and complex drawings. Therefore, various indirect methods have been suggested. The most efficient is the method used by Savitsky and Levin[12]. A crystalline fibre is subjected to short heating at a temperature far exceeding its melting temperature with simultaneous strong drawing. In this case, it is actually the melt that undergoes drawing. This method makes possible to obtain, at least partially, extended-chain crystals and to increase the sample tenacity up to 2000 MPa.

However, the use of this method on an industrial scale is cumbersome and the question arises whether it is reasonable to form the fibre and then melt it, in order to change its structure completely. Is it not better to form a structure with a great number of tie chains required for the attainment of high strength at once during crystallization of the melt?

It has been found that the strenght deficiency of flexible-chain polymers is determined by their ability to form macromolecular coils and folded-chain crystals and that the pronounced chain folding during fibre spinning is the necessary disadvantage in the formation of articles from these polymers. However, this strength deficiency can be strongly reduced and the mechanical properties can approach those of rigid-chain polymers if one tries to extend the chains in the melt and to make them crystallize by the extension technique rather than by the folding technique. Then the subsequent crystallization can yield directly extended-chain crystals similar to those obtained from rigid-chain polymers and, hence, having a high number of tie chains and exhibiting high strength[13]. This method for obtaining highly oriented films and fibres from flexible-chain polymers is called "orientational crystallization" and is being developed by Frenkel and Baranov's group[14-16]. It allows to attain in a one-stage process without subsequent orientational drawing tensile strengths of up to 1400 MPa for fibres, up to 4000 MPa for single fibrils into which fibres are split and about 400 MPa for films at an elastic modulus of 30 000 MPa for fibres and 6000 MPa for films.

Independently, a number of other methods for the preparation from solutions and melts of high-strength and high-modulus materials with a relatively perfect structure exhibiting a high degree of orientation have been developed (cf. e.g. refs. 17–19). Although similar conclusions on the final structure of flexible-chain superfibres (i.e. fibres exhibiting a tenacity of the order of magnitude of 2000 MPa) have been reacted, theoretical approaches to the problem were very different. Crystallization from solutions and melts undergoing deformation will be considered in greater detail in Sect. 3. Here, some methods for the preparation of oriented systems containing ECC based on other principles will be mentioned.

Japanese researchers[20] have attained a high degree of uncoiling of molecules and a high orientation of the latters in fibres by extension below the glass transition temperature using the following technique: a melted polymer previously extended to high degrees of molecular orientation (not less than three times) was rapidly cooled to temperatures below the glass transition temperature. Subsequently, repeated drawing and heat treatment were carried out at the same temperature; as a result, a strength of up to 1000 MPa was reached for polyethylene.

A method for obtaining high-strength and high-modulus fibres upon application of high pressures (400 MPa) and deep cooling at high filament drawing velocities of about 12 000 m/min (usual take-up velocities are 500 to 1000 m/min) should be mentioned. For example, nylon 6 fibres with tenacities 1300–1500 MPa and Young's moduli 35 000–40 000 MPa at the elongation at break of 30% were obtained[21].

Fairly recently, another method for obtaining polymer materials with uniaxial orientation has been developed. It is the "directed polymerization" i.e. the synthesis of polymers under conditions at which the material attains instanteneously the oriented structure. The formation of crystals from the macromolecules in an extended conformation occurs in those polymerizing systems simultaneously with polymerization[22].

The most widely used method for preparing extended-chain crystals involves solid-phase polymerization when the monomer exists as a single crystal. The polymerization of single crystals of the monomer permits the preparation of a polymer material with a maximum orientation: a polymeric single crystal composed of fully extended macromolecules. Such polymer crystals are needle-shaped[22].

The preparation of oriented polymers by the method of the directed polymerization is of interest since it is possible to avoid the complex process of "disentangling" the macromolecules already packed randomly in the bulk of the unoriented polymer. However, methods involving conversion of these needle-shaped crystals into actual fibres have not yet been developed.

3 Main Features of Crystallization from Oriented Solutions and Melts

In contrast to crystallization at rest, crystallization of flowing melts gives rise to one or several preferred crystallization directions. The process of crystallization occurs by an essentially different mechanism involving intermolecular nucleation[23] (in contrast to folding nucleation which proceeds intramolecularly) leading to the formation of extended-chain crystals which consist either of completely extended molecules or single extended parts of different macromolecules packed in parallel[1]. In these systems, the connection between the crystallites is ensured by the large number of tie chains and, hence, the material exhibits excellent mechanical properties.

The formation of fibrillar structures during the crystallization of deformed solutions and melts under various conditions of mechanical treatment was observed by many authors[22, 24, 25] who studied the crystallization in stirred or flowing solutions. In all cases

---

1 It should be noted that the concept "extended-chain crystal" (ECC) does not mean a crystal, in which not a single molecule can form a fold. ECC can contain some molecules with folded conformation but the folds play a role of defects

of solution deformation (under stirring[24, 25]) or under the conditions of stationary flow between the cylinders of a rotational viscometer[22] fibrous species with a characteristic "shish-kebab" structure were observed. These findings were surprising at first[22], but later such structures were found to be a typical morphological feature of samples crystallized under the conditions of deformation. Depending on temperature and the degree and method of deformation, these structures differ in detail but retain their main features: they consist of a long central "strand", a fibrillar crystalline nucleus of the ECC type which forms a kind of support for the growth of lamellar folded crystals oriented normally to the longitudinal axis of the nucleus. Moreover, a high one-dimensional "c" orientation of molecular axes is observed both in the fibrillar strand and in lamellae (Fig. 3a).

Supramolecular structures formed during the crystallization of the melt under a tensile stress have already been described by Keller and Machin[25]. These authors have proposed a model for the formation of structures of the "shish-kebab" type according to which crystallization occurs in two stages: in the first stage, the application of tensile stress leads to the extension of the molecules and the formation of a nucleus from ECC and the second stage involves epitaxial growth of folded-chain lamellae.

Porter carried out crystallization with the formation of ECC by forcing the polymer melt under a high pressure (190 MPa) through a capillary of the Instron rheometer at a temperature close to the melting temperature[18]. A transparent polyethylene strand obtained in this way exhibited a preferred orientation of the c-axes of the crystallites, a higher melting temperature than the conventional polyethylene and a very high elastic modulus (70 000 MPa). The investigation of transparent polyethylene threads showed that they contain two different structures: ECC and folded-chain crystals (FCC) (Fig. 3b), the latter being "fillers" in a fibrillar matrix containing 17–25% ECC. The chain extension is particularly effective, since the capillary entrance is of a conical shape which gives rise to a longitudinal velocity gradient. The molecules are arranged in the flow direction in which almost instantaneous crystallization of the melt occurs which is greatly supercooled at these flow rates. Extrusion of the hardened polymer through capillaries occurring after crystallization stops as a result of the increasing friction.

A technological variation that permits a continuous process is the polymer extrusion through a spinneret of the desired design allowing crystallization inside the spinneret[26]. In this way, the cold extrusion stage used in Porter's method can be avoided. The degree

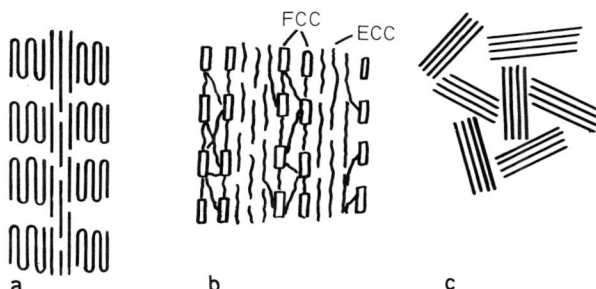

Fig. 3a–c. Supermolecular structures of polymers crystallized in various force fields: **a** structure of the "shish-kebab" type, **b** structure formed during crystallization in a capillary with a conical inlet and **c** structure of a polymer crystallized at hydrostatic compression at 4×10^8 Pa

of orientation of the macromolecules in a single fibre is even slightly higher than that obtained in the Instron rheometer and the required pressure decreases by a factor of 15.

The most important feature of polymers obtained by these methods is their high mechanical strength, primarily their elastic moduli and tenacities that, in some cases, approach the theoretical values. It has been recognized that these mechanical properties are uniquely related to the existence of ECC with very perfect chain packing in fibrillar crystals.

Many authors studying the formation of ECC from melts and solutions suggested that preliminary unfolding and extension of macromolecules occurs. Keller and Machin[25] have shown that in all known cases (including such extreme variants as the crystallization of natural rubber under extension and a polyethylene melt under flow) the same initial process of linear nucleation occurs and fibrillar structures is formed by the macromolecular chains oriented parallel to the fibrillar axes[27].

A characteristic feature of the structure of samples obtained under the conditions of molecular orientation is the presence of folded-chain crystals in addition to ECC. Kawai[22] has emphasized that "the process of crystallization from the melt under the conditions of molecular orientation can be regarded as a bicomponent crystallization in which, just as in the case of fibrous structures in the crystallization from solutions, the formation of crystals of the 'packet type' (ECC) occurs in the initial stage followed by the crystallization with folding".

In principle, it is possible to obtain ECC in the absence of molecular orientation if the crystallization is carried out very slowly at high temperatures close to the melting temperature. Thus, Mandelkern obtained polyethylene crystals similar to ECC in their thermodynamic characteristics by a 40 days' crystallization of an isotropic melt[28]. These experiments also characterize one of the possible paths of the generation of order in polymers: "order through fluctuations"[29] (see below).

The crystallization of polymer melts at high pressure (several thousand of atmospheres), just as the crystallization in flowing melts and stirred solutions, leads to the formation of ECC in which the molecules assume almost completely extended conformations. This suprising fact (the crystallization was carried out at hydrostatic compression) was first extablished experimentally by Wunderlich[30] and then confirmed by other authors[31-33]. However, diverse opinions exist about the mechanism of ECC formation.

Wunderlich[30] and Zubov[33] suppose that ECC under high pressures occur as a result of an isothermal thickening of folded-chain lamellae. However, this contradicts the later data of Wunderlich and of Japanese authors[31] who have shown that folded-chain crystals (FCC) are formed after ECC, when the melt is cooled. According to Kawai[22], crystallization under hydrostatic compression can he considered as a variant of the bicomponent crystallization.

Basset and Kalifa[32] and Takemura and co-workers[31] suppose that the formation of ECC is preceded by the appearance of a phase with properties intermediate between those of the melt and the orthorhombic solid phase. However, the opinions about the structure of this phase are very different. Basset and Kalifa treat it as a hexagonal solid phase which, on cooling,is transformed into an orthorhombic structure during the second crystallization stage undergoing a solid-solid transition. In their opinion, ECC are formed just during this transition.

Takemura and co-workers[31] have shown by optical measurements, X-ray diffraction and micro-DTA measurements that the high-pressure phase is liquid-crystalline, and that two transitions are observed during the melt cooling: melt → liquid-crystalline phase and liquid-crystalline phase → ECC followed by the formation of FCC. It should be noted that the mechanism of ECC formation and the phase transitions during this formation are analogous to the formation of ECC from solutions of rigid-chain polymers. The only difference is that in the latter case the transitions take place as a result of a concentration change whereas at high pressures it is the temperature which varies. Hence, for rigid-chain polymers the liquid-crystalline phase is lyotropic and exists over a definite concentration range, and the high-pressure phase is thermotropic and exists over a definite temperature and pressure range.

The thermodynamic analysis of conformational and structural transformations in the melt at high pressures[34] showed that the free volume and free energy minimum required for hydrostatic compression is attained as a result of the transition of the molecules in the melt into a more extended conformation (gauche → trans transitions) since the extended molecules ensure a more compact packing of the chains at compression. Chain uncoiling leads to a decrease in their flexibility parameter f with increasing pressure p:

$$f = \frac{(Z-2)\exp[(-\varepsilon - ap - 2bpf)/kT]}{1 + (Z-2)\exp[(-\varepsilon - ap - 2bpf)/kT]} \quad (5)$$

where a and b are constants characterizing the change in the specific volume during trans → gauche transitions. If the applied pressure is sufficient to reduce f to values lower than f_{cr} (see Eq. (3)), the melt is transformed into the liquid-crystalline state[31], and when the temperature decreases, formation of ECC from the liquid-crystalline phase occurs.

In spite of the presence of ECC, the sample exhibiting a domain structure remains unoriented on the macroscopic level. Figure 3c shows a great difference in the structures obtained, if molecular orientation exists and if hydrostatic compression is applied. Although the method of hydrostatic compression of the melt is of paramount importance from the scientific view point just for samples crystallized under pressure it was possible to prove unequivocally the existence of ECC), it does not allow a direct preparation of oriented samples of high strength (they are brittle and readily crumble to powder under minimum strain). However, the material obtained in this way can probably serve as a semi-finished product for further technological treatment that would improve its mechanical properties.

4 Thermodynamics of Crystallization of Flexible-Chain Polymers Under Conditions of Molecular Orientation

4.1 Effect of Molecular Orientation on Crystallization of Flexible-Chain Polymers

We carried out thermodynamic studies on the crystallization from melts of flexible-chain polymers uniaxially stretched at various degrees of molecular orientation in the melt and studied the effect of the stretching stress on thermodynamic parameters such as degree of crystallinity, melting temperature, etc., and on the morphological features of the system crystallized under conditions of molecular orientation[3, 13–16, 23, 35, 36].

Flexible-chain polymer molecules in the melt have the shape of a random coil and their flexibility is often estimated as the ratio of the mean-square end-to-end distance of the unperturbed real chain $\langle h^2 \rangle$ to the mean-square end-to-end distance of a freely jointed model chain $\langle h_0^2 \rangle$ of the same contour length the increase in the end-to-end distance being determined by the hindrance to rotation about valence bonds, i.e. the chemical nature of the chain[37]. However, this way of characterizing the flexibility takes into account only chain coiling and is in sufficient for the evaluation of the "induced" rigidity, when the increase in the end-to-end distance is due to external forces. In this case, one can describe the degree of coiling (uncoiling) of the molecule by another parameter β which is the ratio of the end-to-end distance h to the contour chain length L ($\beta = h/L$) and is at the same time a measure of molecular orientation in the melt. For macromolecules in the melt β is determined by Maxwell's distribution function $W(\beta)$ (see below) and in the absence of an external field usually ranges between $0 < \beta < 0.2$ (Fig. 4) for relatively long chains[37]. However, under conditions of molecular orientation, the distribution function is usually displaced towards higher β and the analysis of the crystallization process should be carried out over a wide range of β values.

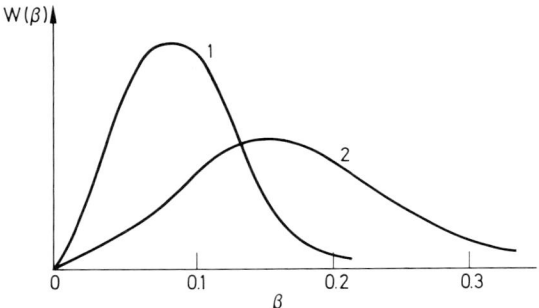

Fig. 4. Distribution function for the degrees of coiling β: *curve 1* for N = 100, *curve 2* for N = 30

Let us consider the process of isothermal crystallization under the conditions of molecular orientation taking a single-molecule model as an example and assuming that in the melt the ends of the molecule are at some distance, h, apart and are fixed in this position during crystallization. (The most real system corresponding to this model is a polymer sample with slight cross-linking or containing fairly long entangled molecules.) The change in the free energy of this system is attributed to the formation of the two main types of polymer crystals: folded-chain crystals (FCC) (Fig. 5 a) and extended-chain crystals (ECC) (Fig. 5 b) depending on the degree of molecular orientation of the melt, i.e. on β. FCC are characterized by the value of $\gamma = b/l$ (where l is the fold length and b the distance between the folds in the crystal). It should be noted that the crystallization of macromolecules with the formation of an extended-chain portion (an ECC element) does not mean a complete extension of the molecules up to $\beta = 1$. In fact, β always less than unity. In other words, complete chain extension in the melt is not attained and the extended part is always shorter than the whole macromolecule (Fig. 5).

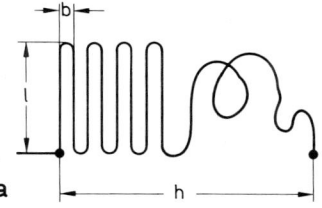

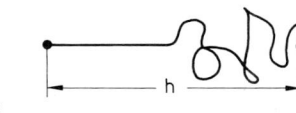

Fig. 5a, b. Models of the crystallization of flexible-chain polymers with the formation of **a** folded-chain crystals and **b** extended-chain crystals

The change in Gibbs free energy, ΔG, in the formation of FCC and ECC, depending on the drawing ratio of the melt β and the crystallization temperature, is given by[14]

$$\Delta G_{FCC} = -\alpha C\delta + \ln(1-\alpha) + \frac{\alpha N}{1-\alpha}(\beta^2 - 2\beta\gamma + \alpha\gamma^2) +$$
$$+ \alpha\gamma C'_\sigma(1+\delta) + (1+\delta)C''_\sigma/\gamma \quad (6)$$

$$\Delta G_{ECC} = -\alpha C\delta + \ln(1-\alpha) + \frac{\alpha N}{1-\alpha}(\beta^2 - 2\beta + \alpha) + C'''_\sigma\alpha(1+\delta) \quad (7)$$

where α is the degree of crystallinity, N the number of segments in the chain, δ the degree of the supercooling of the system equal to $\delta = (T_m^\circ - T)/T$ (where T_m° is the melting temperature of an ideal FCC, $\gamma \ll 1$, $\alpha = 1$, $\beta = 0$), C a constant depending on the nature of the polymer ($C = \frac{2A\Delta H_m N}{3kT_m^\circ N_A b_0}$, where A is the length of the statistical segment, b_0 the length of the monomer unit and N_A Avogadro's number) and C'_σ, C''_σ and C'''_σ are constants describing the contribution of the side and end surfaces to surface energies, respectively. The calculations were carried out for polyethylene using $A = 3 \times 10^{-9}$ m, $b_0 = 3 \times 10^{-10}$ m, $\Delta H_m = 4 \times 10^3$ J/mol and $T_m^\circ = 415$ K.

Figure 6 shows curves $\Delta G(\alpha)$, the changes in the free energy during folded-chain crystallization according to Eq. (6), for a number of β values (similar curves can be plotted using Eq. (7)). The plot shows that this function has a minimum at some value of α corresponding to the thermodynamic equilibrium value, $[\alpha]$, at a given temperature and chosen values of parameters β, γ and N. The equilibrium values, $[\alpha]$, determined from the condition $\partial \Delta G/\partial \alpha = 0$ by differentiation of Eqs. (6) and (7) are shown in Fig. 7. It can be seen that the equilibrium degree of crystallinity $[\alpha]$ for all β (except $\beta = 1$) is less than unity, i.e. the minimum of free energy is attained before the whole chain enters the crystal and there always remains an amorphous chain portion (not a part of the crystal) (Fig. 5). This implies that the entropy loss due to the strain on the amorphous chain part is compensated for by energy gain provided by the formation of the crystallite (as a result of increasing interchain interactions) at $[\alpha] < 1$. Further crys-

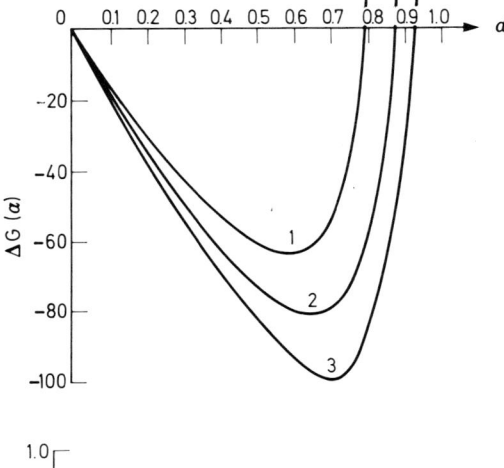

Fig. 6. Gibbs free energy vs. degree of crystallinity α, β: 1 0.4, 2 0.5 and 3 0.6. Crystallization temperature 380 K

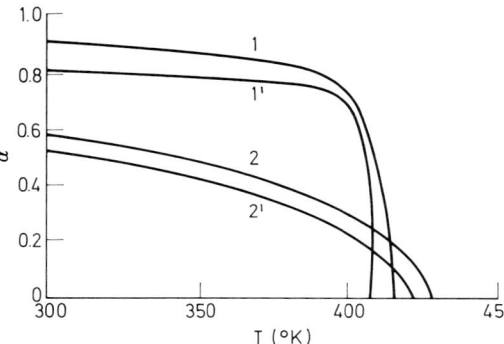

Fig. 7. Degree of crystallinity vs. temperature for FCC (*curves 1* and *2*) and ECC (*curves 1'* and *2'*) at different β values: $\beta_1 > \beta_2$

tallization is already related to the increasing free energy of the system (the right-hand-side of the curve in Fig. 6) and this increase causes termination of crystallization. Hence, the well-known experimental fact that flexible-chain crystallizing polymers contain amorphous regions is due to purely thermodynamic reasons, namely to the equilibrium coexistence of the crystallites and the amorphous chain parts rather than to kinetic hindrances (as has been assumed earlier).

We investigated the effect of molecular orientation on the process of crystallization for the formation of crystallites of both types, FCC and ECC. Fig. 8a shows the dependences of the equilibrium degree of crystallinity $[\alpha]$ on the molecular orientation in the melt calculated from the condition $\partial \Delta G/\partial \alpha = 0$ at constant γ, T and N. It can be seen from Fig. 8a that molecular orientation leads to a decrease in $[\alpha]$ during chain-folding crystallization whereas in "fibrillar" crystallization the degree of crystallinity increases with rising pre-extension of the melt approaching $[\alpha] = 1$ at $\beta = 1$. This finding is readily understandable, since the value of $\beta = 1$ means that the chains are completely extended, i.e. an element of a perfect crystalline structure is formed (or, when these elements are packed parallel, a perfect crystal is formed according to Flory-Mandelkern's model[38]).

We will compare the change in the free energy per macromolecule in the formation of folded-chain (curve 1) and "fibrillar" (curve 2) crystals as a function of β. Figure 8b shows these curves calculated for the corresponding equilibrium values $[\alpha]$. The increase

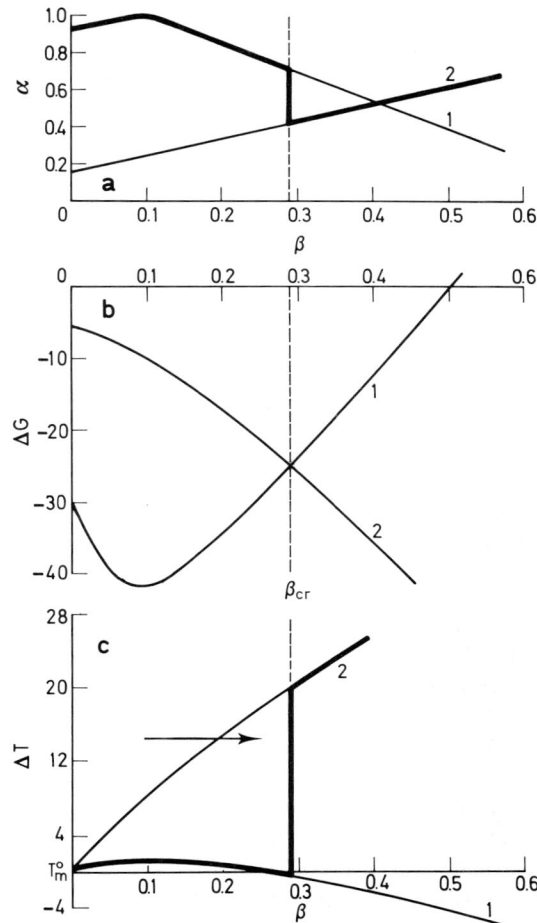

Fig. 8 a–c. Dependence of the degree of crystallinity (**a**), free energy (**b**) and melting temperature (**c**) on β. *1* folded-chain crystals, *2* fibrillar crystals; the *broken line* corresponds to $\beta = \beta_{cr}$

in the free energy of FCC with the degree of molecular orientation indicates that at high extensions the formation of these crystals becomes thermodynamically unfavourable: beginning from some value of β, these crystals become unstable even with respect to the melt the free energy of which was assumed to be 0 (curve 1 in Fig. 8b).

The decrease in the free energy of the molecule during the formation of a linear crystal with increasing β (Fig. 8b, curve 2) is due to the fact that molecular orientation favours chain unfolding. The free energy of this system tends to $-\infty$ at $\beta \to 1$, which shows that increasing extension causes crystallization at any temperature. If these degrees of extension in the melt can be attained, the flexibility of such completely extended molecules is certainly less than 0.63 and, as was shown in Sect. 3, the system in the extending field can spontaneously change to a state of parallel order without any decrease in temperature. In other cases, ordering ensures fixation of the system by intermolecular interactions even after the field has been removed which corresponds to the second crystallization stage (according to Flory[1]) without changing the entropy of the system.

Figure 8b shows that at low values of β, crystallization with the formation of FCC advantageously occurs thermodynamically (it corresponds to a lower free energy ΔG). At high degrees of molecular orientation, however, the free energy is of minor importance to extended-chain crystallization. Hence, although molecular orientation favours chain uncoiling, crystallization involving formation of ECC takes place only starting from some value of $\beta = \beta_{cr}$ (Fig. 8b) rather than at any degrees of molecular orientation. The critical values, β_{cr}, at which transition from one type of crystallization to the other type occurs depend on the characteristics of the polymer and the crystallization temperature. According to our calculations, for polyethylene $\beta_{cr} = 0.3$–0.4 which corresponds to relatively easily attainable degrees of deformation of macromolecular coils. At lower degrees of molecular orientation, predominantly FCC are formed. The fact that β_{cr} is much lower than unity is very important since it means that a complete chain extension is not indispensable for changing the character of crystallization.

To avoid misunderstanding, it should be emphasized that if the "transition" from one type of crystallization to the other one is considered, this does not imply a transformation of crystals of one type into the other one during stretching. In contrast, if the molecule enters a folded-chain crystal, it is virtually impossible to extend it. In this case, we raise the question, which of the two crystallization mechanisms controls the process at each given value of molecular orientation in the melt (this value being kept constant in the crystallization process during subsequent cooling of the system). At $\beta < \beta_{cr}$, only folded-chain crystals are formed whereas at $\beta > \beta_{cr}$ only fibrillar crystals result; at $\beta \approx \beta_{cr}$, crystals of both types can be formed.

The transition from one type of crystallization into the other during melt stretching has also been predicted theoretically[23] proceeding from the calculation of the free energy of nucleation in monomolecular nucleation (lamellar folded-chain nucleus) and multimolecular nucleation (fibrillar nucleus). It was shown (see p. 210) that transition from the first type of crystallization to the second type occurs if the degree of coiling of macromolecules decreases (which is analogous to an increase in the parameter β) regardless of whether the increase in β is due to rise in the inherent chain rigidity (as a result of strong hindrance to internal rotation) or to the extending stress upon application of external field (see above).

A comparison of Figs. 8a and 8b shows that the transition from one type of crystallization to the other is followed by a change in the slope of the dependence of the degree of crystallinity on β. The transition is revealed by an abrupt change in $[\alpha]$ at the transition point $\beta = \beta_{cr}$. This variation can result in either a decrease in the degree of crystallinity (Fig. 8a) or in its increase if the point $\beta = \beta_{cr}$ is located to the right of the intersection point of curves 1 and 2 which describe the dependence of $[\alpha]$ on β. The latter case is shown by the experimental dependences of the degrees of crystallinity (or the density related to it) on β (or the birefringence related to it) for polyethylene and polychloroprene crystallized under the conditions of molecular orientation (Fig. 9). The experimental dependences are in good agreement with the results of our calculations.

We calculated the melting temperature of a system characterized by predetermined values of parameters β, γ and N from the thermodynamic condition

$$\Delta G = \Delta H - T\Delta S \tag{8}$$

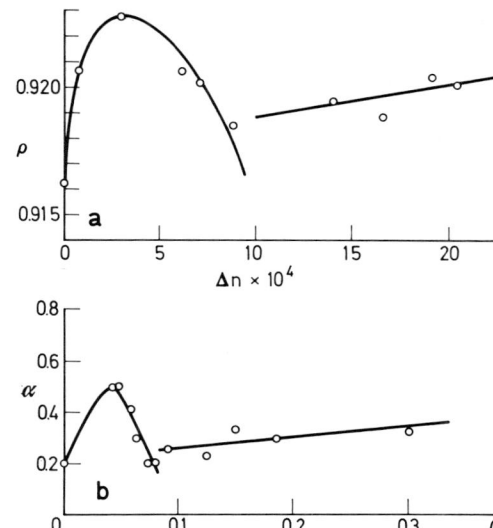

Fig. 9. a Density vs. birefringence (proportional to preorientation β) for a polyethylene melt. **b** Degree of crystallinity vs. β vor polychloroprene

at the equilibrium value of α, i.e. the conditions (8) and $\partial \Delta G/\partial \alpha = 0$ should be fulfilled simultaneously. Using Eqs. (6) and (7) for $\Delta G(\alpha)$, we obtain the melting temperature of crystals of both types as a function of molecular orientation β (Fig. 8c)

$$\frac{1}{T_m} - \frac{1}{T_m^\circ} = \frac{1}{CT_m^\circ}(N\beta^2 - 2N\beta\gamma - 1) \quad \text{(for FCC)} \tag{9}$$

$$\frac{1}{T_m} - \frac{1}{T_m^\circ} = \frac{1}{CT_m^\circ}(N\beta^2 - 2N\beta - 1) \quad \text{(for ECC)} \tag{10}$$

(for constant C see Eq. (6))

Figure 8c reveals that the melting temperature of folded-chain crystals is virtually independent of the degree of molecular orientation whereas the melting temperature of fibrillar crystals increases with it. Figure 8c also shows that at any value of β ECC are more stable than FCC: at any β, the melting temperature of fibrillar crystals is higher than that of folded-chain crystals. Hence, the transition from one type of crystallization to the other at $\beta = \beta_{cr}$ is accompanied by a sharp increase in the melting temperature (Fig. 8c): at $\beta = 0.3$, this temperature jump is about 20 K, according to our calculation. This result coincides with Clough's experimental data[39] that the melting temperature of polyethylene ECC was by 15–20 K higher than that of FCC. This abrupt rise in the melting temperature in the formation of ECC has also been observed for 1,4-polybutadiene crystallized under external stress[40] (Fig. 10).

Many authors (e.g. Pennings[41] and Wunderlich[42]) report that ECC are more stable thermodynamically than FCC reaching at this conclusion from the analysis of crystal formation at different temperatures: Pennings observed the formation of fibrils on stirring a polyethylene solution at 106 °C whereas at 90 °C in a non-agitated solution, only single folded-chain crystals were formed. Studying the melting process of the "shish-kebab" structures and fibrils, Wunderlich observed stable birefringence at much higher

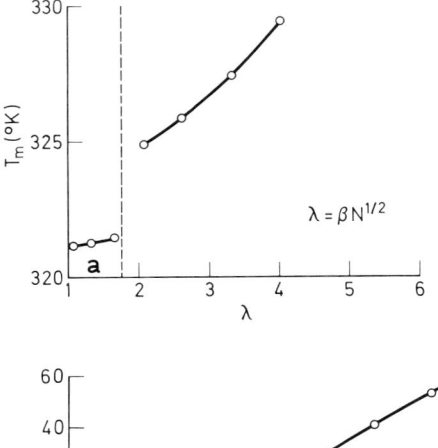
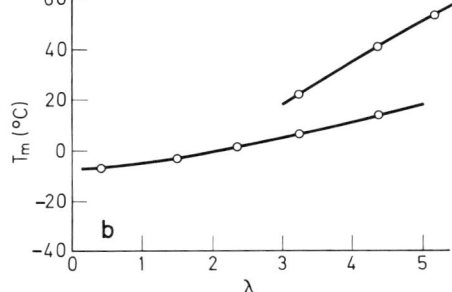

Fig. 10 a, b. Melting temperature vs. extension degree of the polymer for polychloroprene, (a) and cis-1,4-polybutadiene[40] (b)

temperatures than the melting temperature of samples crystallized in the absence of molecular orientation.

In Fig. 8c, FCC are stable in the region under curve 1, since this region is below their melting curve. The area above curve 2 corresponds to the melt and in the area between curves 1 and 2, extended-chain crystallites are stable (for FCC, this is already their melting range). If, therefore, the macromolecules are extended to $\beta > \beta_{cr}$ in the region between curves 1 and 2, it is possible to carry out extended-chain crystallization not by decreasing the temperature of the system (as usual), but as a result of a lowering of the free energy, if molecules are uncoiled by the action of a force field (no temperature decrease is required, since this region is below the melting temperature of ECC denoted by an arrow in Fig. 8c).

This method of crystallization leading to the formation of ECC was called *orientational* or directed orientational crystallization[16]. To carry out this crystallization it is necessary to achieve in the melt or in solution a molecular orientation that causes chain extension, i.e. an increase of parameter β.

Figure 4 (curve 1) shows that in the absence of extension the distribution function $W(\beta)$ lies in the range $0 < \beta < 0.2$ for relatively long chains. In other words, in the absence of external forces, crystallization of flexible-chain polymers always proceeds with the formation of FCC since in the unperturbed melt the values of β are lower than β_{cr}. For short chains, the function $W(\beta)$ is broader (at the same structural flexibility f) (Fig. 4, curve 2) and the chains are characterized by the values of $\beta > \beta_{cr}$, i.e. they can crystallize with the formation of ECC. Hence, at the same crystallization temperature, a molecular weight distribution gives rise to crystals of both types. Indeed, fractionation according to molecular weights during crystallization was performed experimentally[43].

The formation of ECC in the absence of external forces in low molecular weight polymers the length of which is insufficient for the appearance of the folding effect (the molecule in the melt is not yet a Gaussian coil) has been investigated using polyoxyethylene[44] and polydimethylsiloxane[45]. It was shown that there is a critical value of molecular weight below which the crystallization of molecules proceeds by the mechanism of ECC formation. This value can be estimated taking into account that the formation of ECC implies $\beta > \beta_{cr}$ ($\beta > 0.3$) for molecules in the melt. Since for flexible chains, $\langle \beta^2 \rangle^{1/2}$ is related to the number of segments N in the chain by the simple relation $\langle \beta^2 \rangle^{1/2} = 1/\sqrt{N}$, $\langle \beta^2 \rangle^{1/2} = 0.3$ corresponds to $N \sim 10$. Considering this value of N, the chain approaches a random Gaussian coil and, therefore, $N = 10$ may characterize the physical boundary between oligomers and polymers. The same values are critical with respect to the mechanism of crystallization involving chain-folding or uncoiling.

In the foregoing discussion, flexible-chain polymers with f close to unity were considered. The limitation of flexibility (decrease in f), just as the lowering of N, leads, according to Flory[1], to a decreas in β

$$\langle \beta^2 \rangle = \frac{1}{N} \cdot \frac{2-f}{f} \tag{11}$$

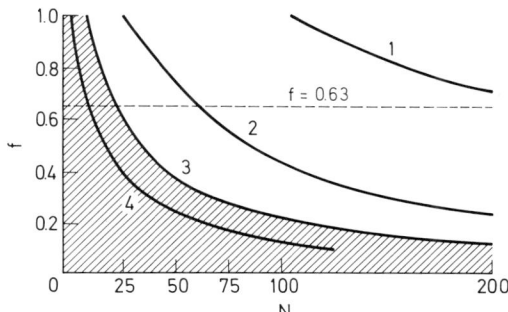

Fig. 11. Relationship between the chain flexibility f and chain length N. β: 1–0.1, 2–0.2, 3–0.3, 4–0.4

Figure 11 shows that it is possible to attain the critical values of $\beta = \langle \beta^2 \rangle^{1/2} = 0.3$ and to pass to the range of β values at which extended-chain crystallization occurs (shaded area in Fig. 11) in two ways: (1) by intersecting the curve $\beta = 0.3$ from right to left, i.e. by decreasing the chain length. In this way, we obtain the boundary of the transition from polymers to oligomers for each value of chain flexibility f and come to the natural conclusion that more flexible chains acquire the ability to fold during crystallization at lower lengths. (2) By intersecting the curve $\beta = 0.3$ from top to bottom, i.e. by decreasing chain flexibility. It is of no importance whether this diminution is related to a decrease in structural flexibility (on passing from flexible-chain to stiff-chain polymers) or to a lowering in chain flexibility in a stretching mechanical field (cf. Eq. (4) and Fig. 2).

4.2 Topomorphism of Polymer Crystals

Figure 11 shows that the molecular weight distribution in the melt (presence of short chains) can account for the coexistence of two types of crystals in the absence of molecular orientation or at a slight stretching of the melt. However, there is a purely thermodynamic reason for the appearance of this main structural feature of samples crystallized under conditions of molecular orientation, even at high degrees of orientation, when virtually the whole distribution function is displaced into the region of $\beta > \beta_{cr}$.

As has been shown above, crystallization under the conditions of molecular orientation cease for thermodynamic reasons and the degree of crystallinity reaches the equilibrium value $[\alpha] < 1$ so that an amorphous (noncrystallized) portion is always present. The evaluation of the stress of the remaining amorphous chains characterized by the value $\beta' = h'/L'$, the ratio of the end-to-end distance to the contour length of the amorphous chain part (Figs. 5 and 6), for various degrees of molecular orientation in the melt and crystallization temperatures have shown[14] that for the crystallization proceeding by the chain-unfolding mechanism the strain on the remaining chain portion is virtually independent of β and is entirely determined by the crystallization temperature. At high crystallization temperatures (above ~390 K for polyethylene), β' is always lower than β_{cr}. Hence, if crystallization is carried out at $\beta > \beta_{cr}$, the amorphous part of the chain with $\beta' < \beta_{cr}$ can form a folded-chain crystal after extended-chain crystallization at these degrees of melt stretching has occurred (according to Fig. 8b). At any value of $\beta > \beta_{cr}$ in the melt and the corresponding crystallization temperatures below the curve of FCC melting, curve 1 in Fig. 8c), a second stage of crystallization with the formation of FCC is possible.

Keller has reported the two-stage character of crystallization under the conditions of molecular orientation[46] and has shown by X-ray diffraction technique that at high temperatures intermolecular crystallization proceeds first and is followed by chain-folding crystallization. The presence of crystals of both types was observed by Clough[39] using DSC in films of cross-linked polyethylene isothermally crystallized at high extensions. The analysis of the crystallization kinetics showed that in the first and second stages of crystallization, ECC and lamellar crystals, respectively, are formed. At low extension, only the formation of FCC was observed which is in agreement with the foregoing prediction.

The coexistence of two types of structures in transparent polyethylene filaments obtained by Porter and co-workers[47] by forcing the melt through a capillary was proved by the existence of two peaks in the thermogram: a high-temperature peak for ECC and a low-temperature peak for FCC (cf. Fig. 8c).

The appearance of an additional melting temperature corresponding to the existence of ECC has also been observed[40]: at high extensions the polymer exhibits two melting temperatures corresponding to melting of crystals with different morphology. A low degrees of extension, only one low-temperature peak is observed corresponding to the melting of FCC (Fig. 8c).

The two-phase morphologic structure has also been observed in the electron micrographs of polyethylene films and fibers obtained by orientational crystallization[16] in which the amount of ECC was approximately 15 to 20% (the fraction of ECC in Porter's samples[47] was 17 to 25%).

Although the amount of ECC in polyethylene samples is not large, the mechanical properties of the material are markedly affected: the tenacity and elastic modulus are higher.

The existence of two types of crystals is an indication of conformational crystalline polymorphism: polymorphous states are associated with different ΔG values for different modes of packing of macromolecules in the crystals and different temperature ranges of thermodynamic preference. However, this polymorphism differs from the usual type in which the modifications exhibit different crystalline lattices and the corresponding transitions. ECC and FCC are characterized by a lattice with identical parameters for different modes of packing of molecules in the crystal, but their temperature ranges of growth and stability are different. However, there are no direct transitions from one modification into the other and only the regions of their coexistence at different ΔG are observed. We called this type of polymorphism "crystalline topomorphism" and the states were characterized as topomorphic.

To determine the regions of the thermodynamic stability of both types of crystals at various molecular orientations, one should consider the plot of Gibbs energys temperature at a certain value of β (Fig. 12) for the amorphous state (curve III), FCC (curve I) and ECC (curve II). The points of intersection of curves I and II with curve III are the melting temperatures of the respective crystal types. The point of intersection of curves I and II (point T*) located lower than both melting temperatures is the boundary of the ranges of thermodynamic preference of FCC and ECC. Although ECC are a type melting at a higher temperature (Fig. 8c), namely at $T < T^*$, FCC are thermodynamically more advantageous and in the temperature range $T^* < T < T_m^{II}$, ECC are thermodynamically more favourable. Hence, T* at a given value of β is the critical crystallization temperature (similar to β_{cr} at a given temperature, see Fig. 8b).

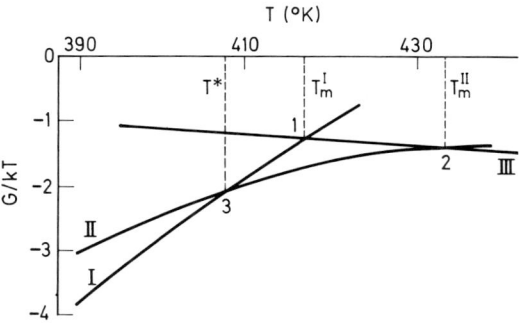

Fig. 12. Free energy vs. temperature for FCC-*curve I*, for ECC-*curve II* and for the amorphous state (melt)-*curve III*. Points 1 and 2 represent the melting temperatures of FCC and ECC, respectively, *point 3* separates the ranges of the predominant existence of FCC and ECC (see text)

Formally, T* could be regarded as a transition temperature of the crystal-crystal type. However, it was already mentioned that FCC are not converted into ECC on heating, i.e. there is no transition from curve I to curve II. If crystallization occurs at $T < T^*$, the FCC formed melt at T_m^I, i.e. the heating proceeds along curve 1 up to T_m^I. If crystallization is carried out at $T > T^*$, ECC are formed and the free energy increases on heating along curve II up to the melting temperature T_m^{II}.

Calculating the position of the points of intersection of curves $\Delta G(T)$ for various values of β, we plotted the phase diagram (to be precise, the topogram) with the coordi-

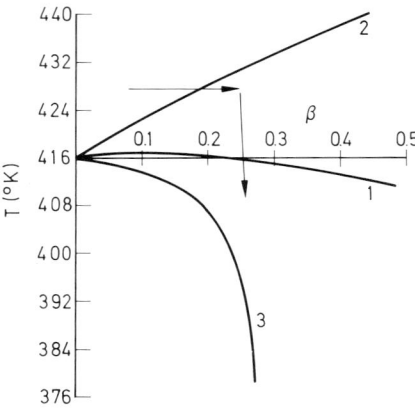

Fig. 13. Phase topogram. *Curves 1 and 2* – melting temperatures of FCC and ECC as a function of the degree of extension β; *curve 3* – dependence β_{cr} on crystallization temperature. *Arrows* show the way of orientational crystallization: (1) the formation of ECC-transition through *curve 2* in the process of melt extension (*horizontal arrow*) and (2) formation of FCC-transition through *curve 1* if the temperature decreases (*vertical arrow*)

nates temperature-β (Fig. 13). Curve 1 in the diagram is the melting curve of FCC (the melting temperature of FCC is point 1 in Fig. 12 vs. β), curve 2 describes the melting of ECC and curve 3 is the plot of T* vs. β or, which is the same, the dependence $\beta_{cr}(T)$. The range of the temperatures and values of β, in which the formation of FCC is thermodynamically preferred, is situated below curve 3 and curve 1 is the upper temperature limit of their stability, i.e. FCC formed in the region below curve 3 melt on passing through curve 1; the region between curves 3 and 2 is the range of the formation of ECC. The range between curves 3 and 1 represents the metastable states for FCC (part 1–3 in Fig. 12). This is the region of the phase diagram in which the above described two-stage crystallization proceeds (at high crystallization temperatures) and crystals of both types coexist.

Figure 12 and 13 show the fundamental instability (metastability) of FCC as a morphological form of crystallizing flexible-chain polymers. In fact, it is possible to pass from the melt into the region of thermodynamic preference of FCC only via the non-equilibrium crystallization: rapid supercooling below the melting temperatures of crystals of both types (in Fig. 12, T* is lower than T_m^I and T_m^{II} and in Fig. 13, the range of thermodynamic stability of FCC below curve 3 lies also below melting curves 1 and 2). The "equilibrium" (slow) method of crystallization should lead to the formation of ECC (see Sect 4.3). When the melt is cooled along curve 3 in Fig. 12, this curve first intersects curve 2, i.e. ECC are formed and in Fig. 13 this is the transition through curve 2. The fact that in the absence of stretching flexible-chain polymers always crystallize with the formation of FCC shows only that for these polymers, FCC are kinetically prefered over ECC (long flexible molecules are difficult to disentangle). Pre-orientation of the melt during extension decreases kinetic hindrances to a considerable extent. Moreover, the formation of FCC from the extended melt requires much more extensive supercooling than their formation from non-deformed melts (the higher β, the broader is the region between curves 2 and 3 in Fig. 13). This is the result of increasing thermodynamic preference for ECC as compared to FCC with rising β. Hence, under the conditions of molecular orientation, it is possible to carry out to some extent the transition of the system into the equilibrium crystalline modification (ECC) whereas the crystallization of an unpertubed melt yields only metastable FCC (under actual conditions). For this reason, Frenkel (following Prigogine et al.) called curve 1 a non-thermodynamic branch of crystallization in contrast to curve 2 – its thermodynamic branch[29, 48].

4.3 Formation of Extended-Chain Crystals (ECC) Under Equilibrium Conditions

Figure 13 shows that, in principle, ECC may be obtained in the absence of molecular orientation, if crystallization is carried out under equilibrium conditions, i.e. at long crystallization times and high temperatures close to T_m^o. In the region between curves 3 and 2, there is a small but non-zero probability of the formation of multimolecular nuclei from ECC at $\beta < \beta_{cr}$, since in this region ECC are thermodynamically stable at any β. Under these conditions, at low degrees of melt supercooling, i.e. at temperatures close to the melting temperature of ECC (Fig. 13, curve 2), the crystallization time is very long. At high temperatures, the viscosity of the system is relatively low, kinetic hindrances are weakened and transition into the equilibrium conformation (i.e. formation of ECC) becomes possible. In the region between curves 1 and 2, crystal nuclei of both types can be formed on the level of heterophase fluctuations, but the FCC nuclei are unstable and disappear because this temperature region is situated above the melting temperatures of FCC (curve 1). In contrast, the ECC nuclei in this range are thermodynamically stable: for ECC T_m (curve 2) is always higher than the temperature at which crystallization is carried out.

The parallel arrangements (of the nematic type) appearing as elements of the fluctuating structure at a given moment, initiate the growth of ECC according to Prigogine's mechanism[29] of "order through fluctuations". The tendency of these crystals to further growth in the transverse direction normal to the chain direction in the crystal is due to the fact that the probability of the location of a neighbouring extended portion in the same direction is higher than in other directions[49], if a region with an anisotropic distribution (i.e. an ordered region with a preferred molecular orientation, heterophase fluctuation) exists. The formation of extended portions means that β increases and the free energy of ECC decreases with rising β ($\Delta G_{ECC} \to -\infty$ at $\beta \to 1$, see Fig. 8b). Hence, the process of further uncoiling of the molecules is thermodynamically advantageous and allows the system to pass to a structure with a minimum of free volume and free energy, i.e. to ECC. This implies that the growth of these crystals can be promoted if the process is carried out slowly and the crystallization temperature is kept close to the effective melting temperature (according to Fig. 13, this temperature increases with chain extension, i.e. with β), ensuring at the same time the thermodynamic stability of the growing crystals (by keeping the temperature slightly below the melting temperature) and the kinetic "advantage" (by gradually raising the crystallization temperature along with the increase in the effective melting temperature as the crystals grow). If this hypothetical crystallization process is carried out (such process is in principle possible but difficult to control; hence Mandelkern and co-workers were not able to grow relatively perfect ECC[28]), the system can be transferred from the range of low β, corresponding to the unperturbed melt before crystallization, to the range of $\beta \approx 1$. In this way, crystals exhibiting the maximum melting temperature among all other possible types can be obtained (Fig. 13). Flory and Mandelkern called this temperature "equilibrium melting temperature".

4.4 Mechanism of the Formation of ECC Under the Conditions of Molecular Orientation

The growth of ECC under equilibrium conditions is too slow. Moreover, no macroscopic orientation appears and a structure of the type shown in Fig. 3c is formed. Therefore this procedure cannot be used in practice. Usually, under real conditions, macroscopically oriented ECC are obtained from the melt stretched to the values of $\beta > \beta_{cr}$ at relatively low crystallization temperatures. Under these conditions, the formation of ECC proceeds by another mechanism.

The application of an extending field to the melt simultaneously uncoils and orients the molecules. The extension of the molecules leads to a decrease in their effective flexibility (see Eq. (4)). The essential question is how high should the values of melt orientation and the changes in flexibility be to ensure formation of ECC in subsequent crystallization?

The formation of ECC is not only an extension of a portion of the macromolecule but also a mutual orientational ordering of these portions belonging to different molecules (intermolecular crystallization), as a result of which the structure of ECC is similar to that of a nematic liquid crystal. After the melt is supercooled below the melting temperature, the processes of mutual orientation related to the displacement of molecules virtually cannot occur because the viscosity of the system drastically increases and the chain mobility decreases. Hence, the state of one-dimensional orientational order should be already attained in the melt. During crystallization this ordering ensures the aggregation of extended portions to crystals of the ECC type fixed by intermolecular interactons on cooling.

It seems of importance to elucidate what the stresses are that should be applied to the melt in order to: (1) ensure formation of a nematic phase in the melt, and (2) attain values of $\beta > \beta_{cr}$ so that the crystallization caused by melt extension should proceed by the chain-extension mechanism. It is also desirable to answer the question whether the formation of the nematic phase is an indispensable intermediate stage preceding formation of ECC[51].

We studied the effect of the mechanical stretching field on the conformations of the macromolecules in the melt. It is known that for a freely jointed chain the Maxwell distribution of end-to-end distances holds[50].

$$W(h)dh = \left(\frac{3}{2\pi NA^2}\right)^{3/2} \exp\left(-\frac{3h^2}{2NA^2}\right) h^2 dh \qquad (12)$$

If the stretching force F is applied to a system of these molecules the distribution function is given by[51]

$$W(h)dh = 4\pi \left(\frac{3}{2\pi NA^2}\right)^{3/2} \exp\left(-\frac{F^2 NA^2}{6k^2T^2}\right) \exp\left(-\frac{3h^2}{2NA^2}\right) \times \\ \times h^2 \frac{Sh\,(Fh/kT)}{(Fh/kT)} dh \qquad (13)$$

and since $h = \beta L$, Eq. (13) becomes

$$W(\beta)d\beta = 4\pi L^3 \left(\frac{3}{2\pi NA^2}\right)^{3/2} \exp\left(-\frac{3}{2}N\beta^2\right)\beta^2 \frac{\text{Sh}(FL\beta/kT)}{FL\beta/kT}$$
$$\times \exp\left(-\frac{F^2L^2}{6Nk^2T^2}\right) \tag{14}$$

Introduction of the parameter $\varkappa = FL/kT$ characterizing the extending field gives

$$W(\beta)d\beta = 4\pi L^3 \left(\frac{3}{2\pi NA^2}\right)^{3/2} \exp\left(-\frac{3}{2}N\beta^2\right)\frac{\text{Sh}(\varkappa\beta)}{(\varkappa\beta)} \beta^2 \times \exp(-\varkappa^2/6N) \tag{15}$$

Figure 14 shows the displacement of the distribution function towards high β, i.e. the uncoiling of molecules under the influence of stretching for polyethylene ($A = 3 \times 10^{-9}$ m, $N = 100$ and $T = 420$ K). This displacement will be characterized by the position of the maximum of the distribution curve, the most probable value of β, i.e. β_m, as a function of $\varkappa$ (Fig. 15). Figure 15 also shows the values of stresses σ that should be applied to the melt to attain the corresponding values of $\varkappa$ ($\sigma = \varkappa kT/SL$, where S is the transverse cross-section of the molecule).

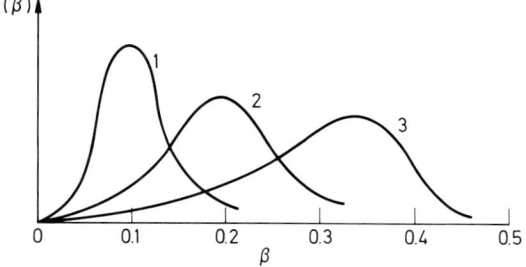

Fig. 14. Distribution functions for the degree of coiling at different values of parameter $\varkappa$ characterizing the extending field (*curve 1* – $\varkappa = 0$, *curve 2* – $\varkappa = 10$, *curve 3* – $\varkappa = 50$)

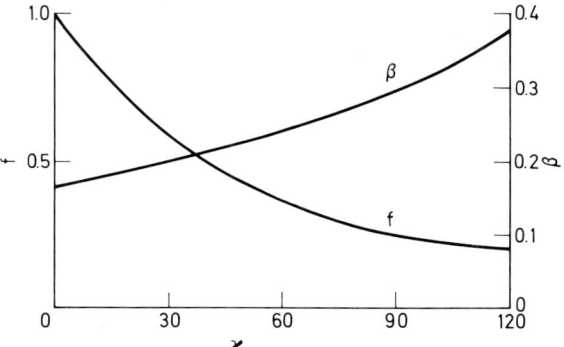

Fig. 15. Flexibility f and β as a function of $\varkappa$

When the chain undergoes extension, the stretching fields increases the number of trans isomers in the chain and, hence, decreases the effective chain flexibility which is related to β_m by the equation (see Eq. (11))

$$\beta_m^2 = \frac{2}{3N} \cdot \frac{2-f}{f} \tag{16}$$

Figure 15 describes the decrease in the flexibility f of the macromolecules during melt stretching (corresponding to an increase in β_m) with $\varkappa$. According to Flory's criterion, the diminution of the flexibility of molecules to the value of f < 0.63 leads to a spontaneous transition of the system into the state of parallel order. It can be seen in Fig. 15 that f = 0.63 is attained at $\varkappa_1 = 30$ or $\sigma_1 = 0.6 \times 10^7$ n/m²: at these stresses, the melt is organized into a nematic state.

The critical value of $\langle \beta^2 \rangle^{1/2} = 0.3$ corresponds to $\beta_m = 0.25$. Fig. 15 shows that these degrees of chain extension require stresses $\varkappa_2 = 70$ or $\sigma_2 = 1.5 \times 10^7$ n/m².

This analysis ($\sigma_2 > \sigma_1$) reveals that higher stresses are required for crystallization to occur by the extended-chain mechanism than for the transition of the melt into the liquid-crystalline state. Consequently, orientational crystallization is always preceded by the formation of the nematic phase as an intermediate stage preceding extended-chain crystallization. In other words, the formation of the nematic phase is a necessary but insufficient condition for carrying out orientational crystallization. This conclusion can also be demonstrated by using Figure 11: virtually, the whole range of values of $\beta > 0.3$ within which extended-chain crystallization occurs corresponds to f < 0.63 (except for very short chains for which Eq. (11) no longer holds and which are of no interest). In contrast, the values of f < 0.63 do not ensure extended-chain crystallization since they correspond to a considerable range of values of $\beta < 0.3$ (unshaded area in Fig. 11).

The existence of an intermediate liquid-crystalline phase in the melt of flexible-chain polymers has been demonstrated by many researchers[52-56]. Probably, the existence of this phase has been first assumed by Clough[52] on the basis of data on X-ray diffraction of orientationally crystallized cross-linked polyethylene. An additional reflexion was observed at temperatures above the melting temperature after the disappearance of the usual reflexion for crystalline polyethylene with orthorhombic packing and it was retained at temperatures 150–177 °C. This reflexion corresponding to hexagonal chain packing (and the melting peak) occurs only in the melting of strained samples and is not observed in the melting of isotropically crystallized samples. In other words, to retain this hexagonal phase, the stress within the sample should be maintained. The stress requirement indicates that chain packing above the melting temperature involves extended chains, and strain prevents their disordering during melting.

Investigating polypropylene melts, Smith has observed[53] that after the melting temperature has been reached, a whole series of mesomorphic transformations occurs rather than transition into the isotropic state and concluded that polypropylene melts are liquid-crystalline in nature. On the basis of IR spectra, Smith suggested that polypropylene molecules in the melt retain a helical conformation that exhibits high internal stability favouring the formation of the liquid-crystalline phase and the increase in its stability. He concluded that the liquid-crystalline state of polymers in the melt is much more important than has been supposed before. This conclusion seems to be particularly confirmed by the fact that, according to Flory, the minimum degree of asymmetry of molecules, i.e. their rigidity (1 – f) required for the transition into the liquid-crystalline state, is relatively low[2]. Thus, the formation of the liquid-crystalline phase in the melts of flexible-chain polymers is probably just as natural a phenomenon as in solutions of rigid-chain polymers. However, in solutions, this phase is usually nematic as is indicated by the increase in viscosity on the transition into the isotropic state (the viscosity of the nematic phase is always lower than that of the corresponding isotropic phase) whereas anisotropic polymer melts always exhibit higher viscosity than isotropic melts (as far as can be judged from the few data available so far). Generally speaking, this difference is an argument in favour of the assignment of the anisotropic phase in carbon-chain polymer melts to the smectic types (or one of the smectic types).

It should be emphasized that in solutions of rigid-chain polymers, the liquid-crystalline phase is formed in a system at rest when polymer concentration increases. Moreover, the lower the chain rigidity, the higher is the concentration of solution at which the transition into the ordered state occurs. Thus, semi-rigid polymers (e.g. polytetrafluoroethylene[2]) or flexible-chain polymers exhibiting additional internal stability (such as polypropylene[53] and polydiethylsiloxane[54]) can form an anisotropic phase only in the melt (i.e. at a polymer concentration equal to unity, "limiting case of solution") and at high temperatures ensuring chain mobility required for the transformation of the system into the ordered state. Finally, in flexible-chain polymer melts (polyethylene), the anisotropic phase can be observed only under the conditions of extending stress as has been reported by several authors[52, 55, 56]. Undeformed polyethylene melts are isotropic although the stresses required for the transition into the mesophase are low.

It should be noted that the relative "accessibility" of the transition into the oriented state observed for polymers of various rigidity under appropriate conditions is due to the internal anisotropy of macromolecules caused by their chain structure (see Sect. 1 of this paper and monographs[2, 3]).

The recent paper by Krüger and coworkers[57] deserves a comment. They investigated the melts of a number of crystallizable (including polyethylene) and non-crystallizable polymers by the method of Brillouene spectroscopy and found that the temperature gradient of the velocity of sound for crystallizable polymers exhibits a break at temperatures exceeding their melting temperature by 60 to 110 °C. These authors interpreted the liquid phase between the melting temperature and the temperature of the additional transition at which a break in the dependences of density, refractive index and viscosity on temperature is also observed, as a phase with a local nematic structure. The determination of the Gruneisen constant indicates that a sharp change in the intermolecular interactions is observed at the temperature of the "additional" transition. The authors suggested that at a temperature much higher than T_m, cooperative changes occur in the structure of various polymer melts, similar to the changes observed for low molecular weight liquid crystals. This important conclusion should be checked by other methods. Nevertheless, this paper seems to be very interesting because the behaviour of different polymer melts at these high temperatures was investigated for the first time.

Pennings carried out detailed investigations of the melting of fibrillar polyethylene crystals in fibers grown from dilute solutions in xylene[55]. He found three melting peaks for these samples and identified them as follows: the lowest temperature peak was assigned to the melting of FCC, the second to the transition of ECC into the intermediate hexagonal phase and the third at the highest temperature to the melting of this phase in which the heat evolution was closely related to chain randomization in the melt and to the formation of high-energy gauche conformers.

The generation of an intermediate phase during melting under isometric conditions of orientationally crystallized polyethylene has also been observed[56] at temperatures exceeding the melting temperature of ECC. The authors suppose that the mesophase obtained is similar to the high-baric phase formed during crystallization of polyethylene under hydrostatic compression[31] and precedes the formation of ECC.

Hence, it can be concluded that the formed intermediate oriented phase is an indispensable stage preceding extended-chain crystallization so that this type of crystallization occurs in two stages: the first stage involves formation of the oriented phase during melt deformation and the second formation of ECC from this phase on cooling.

4.5 Melting of ECC

The processes of the formation and melting of ECC will be demonstrated by using the Gibbs free energy-temperature diagram (Fig. 16). In the absence of molecular orientation, the isotropic melt (line A_1) crystallizes (at point 1), when the temperature decreases and FCC are formed (FCC line) (ECC at 4 cannot be formed for kinetic reasons), point 1 being their melting temperature. The application of a stretching field (or the introduction of molecular orientation by another method) leads to a forced transition of the isotropic

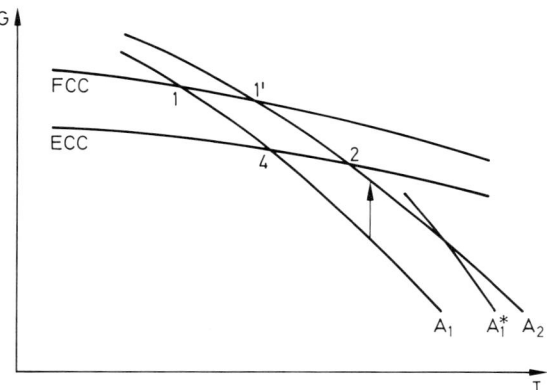

Fig. 16. Gibbs energy-temperature diagram if FCC and ECC are present in the system. A_1-isotropic (undeformed) melt, A_2-deformed melt (nematic phase); *points 1* and *4* – melting temperatures of FCC and ECC under unconstrained conditions (transition into isotropic melt); *points 1'* and *2* – melting temperatures of FCC and ECC under isometric conditions (transition into nematic phase), *point 3* – melting temperature of nematic phase (transition into isotropic melt but not completely randomized)

melt into the oriented state (to line A_2) the free energy of which is higher than that of the undeformed melt and which is the thermotropic liquid-crystalline (intermediate) phase. Under these conditions (when molecular orientation exists), the formation of ECC is favoured thermodynamically since their free energy is lower than that of FCC for obvious reasons: they have no defective ab faces (Fig. 17), their energy of the interfaces with the amorphous regions is lower, internal strains induced by the chains are absent, etc. (cf. Figs. 2 and 17). Hence, after the first stage (A_1 – A_2 transition), crystallization occurs (second stage): i.e. formation of ECC from the intermediate phase as the temperature decreases (transition along the line A_2 to the ECC line with crystallization at point 2).

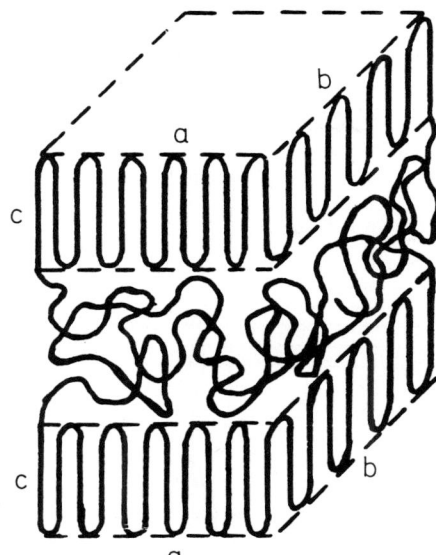

Fig. 17. Schematic representation of the supramolecular structure of polymers containing lamellar crystals

During heating of the orientationally crystallized sample under isometric conditions (dimensions of the sample are fixed during melting, its shrinkage is prevented and strains are maintained), melting of ECC occurs at point 2 with the formation of the liquid-crystalline intermediate phase that will be stable up to point 3 where it melts, i.e. it undergoes transition into the high-energy amorphous phase (the melt) A_1^*. This transition is defined by Pennings[55] as chain randomization with the generation of high-energy gauche conformers (phase A_1^* is isotropic but this melt is not completely randomized). By removing stresses, it is possible to pass to the isotropic equilibrium melt (to line A_1) at temperatures above point 3. If the stresses are kept constant, heating above point 3 leads to chemical degradation of the sample.

If this sample contains also folded-chain crystals (reasons for their appearance during orientational crystallization were stated before), under isometric conditions they undergo melting at a higher temperature (at point 1' with respect to the oriented melt with transition to line A_2) than under the conditions of "free" heating (point 1 with transition in the isotropic melt to line A_1).

During heating under unconstrained conditions (when the sample can undergo shrinkage), ECC will also melt at a lower temperature at point 4 with transition directly into the isotropic melt (on line A_1) and the existence of the intermediate phase has not been reported[52, 55]. It is quite clear that two melting peaks are observed (point 1 and 4), when heating occurs under unconstrained conditions whereas melting under isometric conditions yields three peaks (points 1', 2 and 3), the first two peaks corresponding to a superheating compared to points 1 and 4 (the scheme of both types of melting is shown in Fig. 18).

We investigated the character of the phase transition in the melting of both crystalline modifications by considering the dependences $\Delta G (\alpha, T)$ according to Eqs. (6) and (7) in the range of melting temperatures and solving simultaneously the equations $\Delta G = 0$ and $\partial \Delta G / \partial \alpha = 0$, and for FCC the equation $\partial \Delta G / \partial \gamma = 0$, too. For FCC, at the melting point,

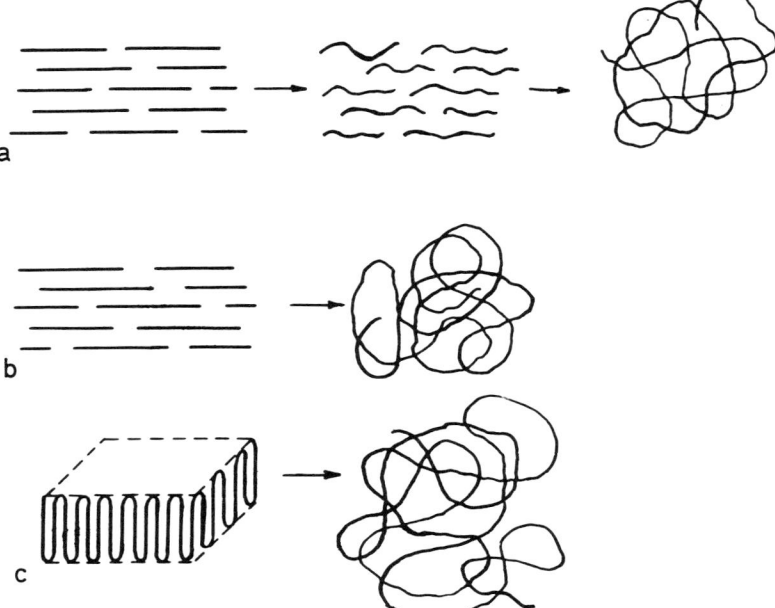

Fig. 18 a–c. Schematic representation of ECC melting under isometric conditions (**a**), of ECC melting if samples are allowed to shrink during melting (**b**), of melting of a lamellar crystal (**c**)

we obtained the equilibrium of two phases with $\alpha = 0$ and $\alpha \neq 0$ (Fig. 7) and, correspondingly, a break in the first derivative $\partial \Delta G/\partial T$, i.e. melting of FCC is a first-order phase transition following the Ehrenfest's criterion. In fact, the transition shown schematically in Fig. 18c is related to a change in the symmetry elements of the system and may be included in the order-disorder transitions even if it is the transition into a not entirely randomized melt (which is a high energy isotropic phase).

Melting of ECC involving transition into the isotropic melt was shown by Flory to be a first-order process. It can be seen in Fig. 18b that there occurs a transition from a "complete" order to a fully random chain arrangement in the isotropic melt (Fig. 16, point 4).

The transition from the isotropic melt to the liquid-crystalline state that occurs when the melt is extended represents an isotropic state → anisotropic state transition which should be considered a first-order transition[2], since it is related to the generation of the orientational order (according to J. Frenkel, the melting of the anisotropic phase is a variant of the "orientational melting"[59]. The investigation of the ECC → liquid-crystalline phase transition (Fig. 16, point 2) by analysing the $\partial \Delta G_{ECC}/\partial T$ derivative revealed that the formation of ECC is a transition of an order higher than one. Figure 7 shows that the degree of crystallinity increases gradually after transition through the melting temperature and phase equilibrium has not been observed at any point. Since this transition is not related to a change in the symmetry elements of the system (Fig. 18a), it should be at least a second-order process as is indicated by the break in the second derivative $\partial^2 \Delta G_{ECC}/\partial T^2$. Since very little is known at present about the transitions crystal

→ thermotropic liquid-crystalline phase in flexible-chain polymers and, in general, about the transitions between mesophases in polymers, the order of transitions in these cases is still open to question. The transitions from the anisotropic into the isotropic state are however, first-order phase transitions.

Although crystals obtained from the oriented melt contain extended chains, they are less perfect than ideal ECC which, in principle, can be produced under equilibrium conditions. As was already shown, the degree of crystallinity of ECC obtained by extension is less than unity (Fig. 8a). Hence, remaining amorphous chains always exist and can form folds or have not crystallized at all. Since at strong supercooling of the polymer, the viscosity of the system is very high, further chain uncoiling and the achievement of minimum free volume (as it occurs under equilibrium conditions) are impossible; thus, the material contains defects in the form of folded-chain crystals and non-crystallized chain portions. The melting temperature of these crystals is lower than that of perfect extended-chain crystals obtained under equilibrium conditions.

As compared to ECC produced under equilibrium conditions, ECC formed at a considerable supercooling are at thermodynamic equilibrium only from the standpoint of thermokinetics[60]. Indeed, under chosen conditions (β and crystallization temperatures), these crystals exhibit some equilibrium degree of crystallinity at which a minimum free energy of the system is attained compared to all other possible states. In this sense, the system is in a state of thermodynamic equilibrium and is stable, i.e. it will maintain this state for any period of time after the field is removed. However, with respect to crystals with completely extended chains obtained under equilibrium conditions, this system corresponds only to a relative minimum of free energy, i.e. its state is metastable from the standpoint of equilibrium thermodynamics[60, 61].

5 Structure and Properties of Systems Obtained Under Conditions of Molecular Orientation

Many papers deal with the crystallization of polymer melts and solutions under the conditions of molecular orientation achieved by the methods described above. Various physical methods have been used in these investigations: electron microscopy, X-ray diffraction, birefringence, differential scanning calorimetry, etc. As a result, the properties of these systems have been described in detail and definite conclusions concerning their structure have been drawn (e.g.[4, 13–19, 39, 52]).

As already mentioned, the crystallization of flexible-chain polymers in the absence of molecular orientation leads to the formation of semicrystalline systems containing spherulites. At low degrees of molecular orientation, the polymer also consists of FCC but in this case spherulites are flattened with respect to the orientation direction. The degree of spherulite flattening increases with growing deformation, and the orientation of the crystallites gradually changes from the isotropic **a**-orientation to the **c**-orientation. Electron micrographs show lamellar folded structures extended in the direction normal to that of the molecular orientation[16, 60]. Hence, the structure of the samples differs greatly from the spherulite structure: relatively long lamellae are arranged on each other and the orientations of the molecular axes in the crystallites coincides with the direction of melt extension.

If the degrees of molecular orientation are sufficiently high for the condition $\beta < \beta_{cr}$ to be valid, interchain crystallization with the formation of one-dimensionally oriented structures, ECC, becomes possible. The presence of structures of this type is indicated first of all by the appearance in the thermograms of a high temperature melting peak corresponding to ECC and located above the "conditional" melting peak (i.e. FCC) since at any degree of molecular orientation, the melting temperature of ECC is higher than of FCC (Figs. 8c and 13). When ECC appear, the morphology of the sample as a whole changes and in all cases the samples obtained under the conditions of a considerable molecular orientation exhibit a pronounced fibrillar structure on very different levels from a few millimeters to several nanometers. If the molecular orientation increases, crystallized samples display morphological transitions from the spherulitic structure with isotropic orientation of the crystallites to flattened spherulites with the preferred chain orientation in the stretching direction and then to structures consisting of lamellae forming parallel layers. The latter are extended normally to the stretching direction of the melt, i.e. completely oriented in this direction. At $\beta > \beta_{cr}$, a transition to fibrillar structure occurs with fundamentally different mode of crystallization i.e. a transition from chain folding to their unfolding.

The existence in the samples of crystalline fibrillar structures with extended chains oriented in the direction of molecular orientation leads to peculiar features of X-ray scattering. As already mentioned, during crystallization under the conditions of molecular orientation flexible-chain polymers form a complex structure consisting of crystalline modes of two types. This is shown in the thermograms of these samples; they always contain two melting peaks, a low-temperature peak (FCC) and a high-temperature peak (ECC)[16, 39]. The two-phase structure of these samples is also indicated by their X-ray diffraction patterns: for polyethylene[16] at room temperature, the wide-angle X-ray scattering shows a typical c-texture and the small-angle scattering exhibits reflexions in the form of layer lines, due to long periods and characteristic of ordinary oriented samples containing FCC. These small-angle reflexions disappear at $T_1 = 136$ °C which is in good agreement with the appearance of the low temperature peak in the thermogram and implies melting of FCC. On further heating above T_1, the wide-angle reflexions remain and equatorial reflexions of the thombic shape appear in the small-angle scattering pattern. They disappear at $T_2 = 142$ °C simultaneously with the disppearance of wide-angle reflexions. At the same temperature, the second endothermic DTA peak is observed; it is assigned to the melting of second-type structures, ECC. The electron micrographs of samples crystallized from the oriented melt exhibit continuous and slightly tortuous fibrillar structures located in the direction of orientation and connected with each other by ties so that the system of fibrils as a whole forms a peculiar spatial network[4, 16].

Similar results were also obtained in the study of polypropylene for which the high-temperature peak is located much further from the low-temperature peak than for polyethylene and all the above effects are more pronounced. Meridional reflexion indicating, just as for polyethylene, the periodicity along the direction of the orientation disappear at 173–175 °C and an intense equational scattering appears simultaneously. A further increase in temperature from 175 to 195 °C weakens the intensity of this scattering and at T = 195 °C both small- and wide-angle reflexions disappear completely. Moreover, a distinct c-texture is retained over the whole range of temperatures, but the intensity of wide-angle scattering gradually decreases. The temperature of 175 °C corresponds to the conditional polypropylene melting temperature. Hence, it is natural to suggest that ECC melting occurs at this temperature. However, even beyond this temperature polypropylene samples exhibit a partially crystalline structure and are in a strained state. Some strength and rigidity of samples is retained up to the temperature of the high-temperature endothermic peak in DTA[63] (Fig. 19). This behaviour in thermomechanical experiments could be expected for rigid-chain polymers (if they would melt). Common oriented samples of flexible-chain polymers are characterized by a contraction (according to thermomechanical curves, samples are contracted 20–30 times) and by a drastic decrease in the elastic modulus at the melting temperature of FCC. According to Keller[64], the foregoing deformation mechanism similar to that for low molecular weight solids is a proof of the presence of the ECC type of structures in the sample.

The most important conclusion of the foregoing investigations is evidence of the presence of two types of structures in samples crystallized from the oriented melt: FCC that amount to 85–90% of the total mass of the sample and ECC with the melting temperatures exceeding by 5–6 K for polyethylene and 15–20 K for polypropylene the

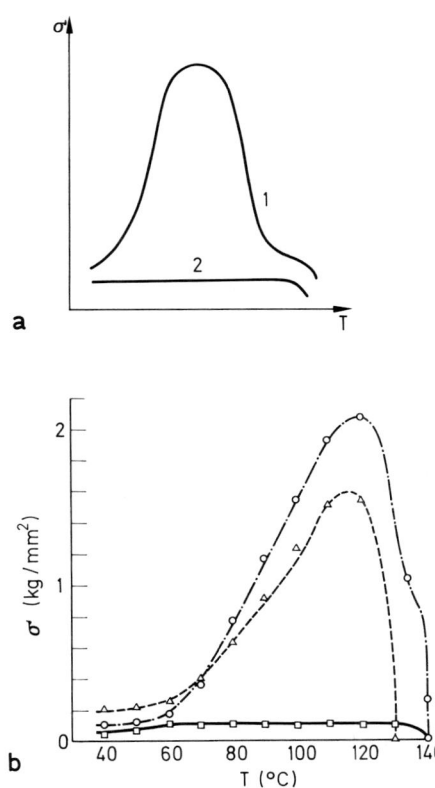

Fig. 19a, b. Curves of isometric heating: **a** reference sample (*curve 1*) and sample obtained if molecular orientation exists and containing a spatial framework (*curve 2*) (schematic representation); **b** data for high density polyethylene (--△-- - ordinary fiber, —○— - fiber obtained by orientational crystallization, —□— - fibrils torn out of a broken fiber

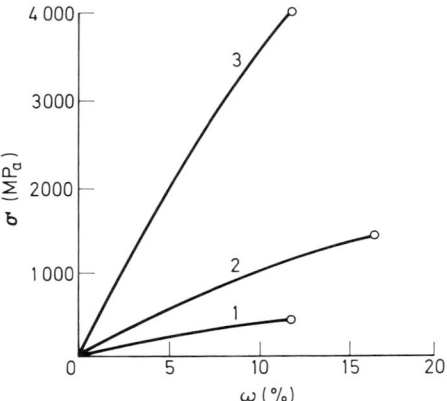

Fig. 20. Stress(σ)-strain (ω) dependence for high density polyethylene samples. *1* reference sample, *2* sample obtained if molecular orientation exists, *3* super high tenacity fiberfibril. Asterisks denote the points of fiber failure

usual melting temperature of these polymers. Although ECC constitute only a small fraction of the sample (10–15 wt.-%), even this amount is sufficient to change markedly the mechanical characteristics of the samples, namely to obtain fibers and films of high tenacity and elastic modulus such as polyethylene fibers with a tensile strength of about 10 MPa and a modulus of 600 to 700 MPa (the tenacity of single thin fibrils attains 40 MPa) (Fig. 20) and polypropylene fibers with a tensile strength of 12 MPa and a modulus of 200 to 300 MPa.

All these facts and also relatively low values of tenacity and modulus of films and fibers obtained by orientational crystallization as compared to the theoretical values suggest the following model of the structure of these systems: approximately 80–90% of the sample consists of the usual semicrystalline matrix with FCC and a relatively low tie chain concentration. However, this matrix virtually does not react to external load. This load is "taken up" by the ECC formed during orientational crystallization. They form a continuous spatial framework taking up external loads at low deformations and preventing to a considerable extent the development of deformations or internal strains on heating. It is precisely this self-reinforcement of the polymer by an ECC framework in which virtually all chains are tie chains and which ensures a high tenacity and elastic modulus and also low elongation at break in the direction of molecular orientation (Fig. 20) and a virtual absence of shrinkage up to T_m.

What is the reason for the limitation of a further ECC growth and what is the explanation of the fact that ECC form only the framework of the system? Let us consider Fig. 13 and keep in mind that the real orientational crystallization is simply a transition through curve 2 (horizontal arrow in Fig. 13): the transition from the melt range to the range of the thermodynamic stability of ECC (below curve $\overline{2}$) caused by the deformation of the system. After crystallization at the intersection point with curve 2 has occurred, it is difficult to attain a further increase in β, since the effective degree of supercooling of the system (at a constant temperature) drastically increases when the melting temperature of ECC rises with β (along curve 2), and hardening begins. Hence, the values of β attained during orientational crystallization (without a decrease in temperature) are determined by the intersection point with curve 2, i.e. by the extension temperature of the melt. The higher this temperature, the higher is β at the intersection point. However, the rise in the temperature of the melt is limited by a decrease in its strength and, therefore, it is impossible to raise essentially the stretching temperature.

As our evaluations show[51], in Baranov's experiments[16, 56, 63] it was possible to carry out orientational crystallization with the formation of ECC at β of about 0.2–0.25. When the temperature decreases (vertical arrow in Fig. 13), the growth of ECC starting at the intersection point with curve 2 continues but only up to the intersection with curve 1. Below this curve, the local decrease in strain near the ECC being formed and the corresponding diminution in β to the values of $\beta < \beta_{cr}$ (i.e. the decrease in molecular orientation in the remaining bulk of the uncrystallized sample) leads to an intense folding rather than to further chain extension (see Sect. 4, secondary crystallization) and, hence, the volume fraction of ECC remains small. Consequently, at temperatures below T_m of FCC, the total crystallinity of the sample increases as a result of the formation of FCC rather than ECC. Fig. 7 shows that the values of $\beta \approx 0.2$–0.25 attained in practice correspond to the degree of crystallinity of ECC (i.e. its fraction in the sample) at $T = T_m$ of FCC of about 0.2. This is in agreement with the estimate of the fraction of the framework in the crystallized sample made by us.

This model of the structure of orientationally crystallized samples based on experimental data is in good agreement with the results of the foregoing thermodynamic analysis which resulted in relationships describing the formation of two structures, FCC and ECC, during the crystallization of strongly oriented melts of flexible-chain polymers.

It should be emphasized that the presence of ECC in polymers crystallized from the melts undergoing deformation is one of the most fundamental problems of orientational crystallization. Although direct observations of ECC in these polymers have not been made, the presence of a high-temperature endothermic peak in the thermograms is a fairly convincing argument for the existence of ECC. However, the most convincing evidence of the existence of the ECC "framework" is the invariability of the elastic modulus and the sample size on heating close to T_m of ECC (Fig. 19). An important argument is also the formation of an intermediate liquid-crystalline phase during the melting of orientationally crystallized polyethylene and polypropylene[56] under isometric conditions. It is known that this phase is not formed in the melting of FCC. However, our calculations reported above show that the formation of the intermediate oriented phase is a necessary stage preceding extended-chain crystallization.

6 Conclusions: Prospects of Processing of High-Tenacity Polymer Materials

The problem of obtaining high tenacity polymer films and fibers is directly related to the feasibility of crystallization by the mechanism of ECC formation. For rigid-chain polymers, their inherently low f values, certainly lower than f_{cr}, predetermine the spontaneous formation of nematic domains in solution (where the molecules are kinetically able to exhibit their rigidity) which serve as stocks for the subsequently formed oriented structure. The orientation process during the solution spinning of fibers from these polymers involves "assembling" of a macroscopically one-dimensionally oriented system from these "stocks". The only technological problem is to orient them in one direction and to remove the solvent, i.e. to obtain the structure of the type shown in Fig. 21 a (rather than

Fig. 21 a–c. Schematic representation supramolecular structure of a crystalline rigid-chain polymer (**a**), an idealized ECC of a flexible-chain polymer (**b**) and an orientationally crystallized sample with a spatial ECC framework (**c**)

type 3c). For flexible-chain polymers with f higher than f_{cr}, it is necessary to extend the molecules up to $\beta > \beta_{cr}$ (it was shown above that in this case the value of $f < f_{cr}$ is certainly attained) in order to ensure extended-chain crystallization and to obtain an idealized structure of the type shown in Fig. 21b. The evident similarity between Figs. 21a and 21b demonstrates the equivalence principle formulated by S. Frenkel[60, 61, 65]. According to this principle, systems of rigid-chain polymer at rest and of extended flexible-chain polymers can be equivalent, if the same degrees of extension of molecules are attained (in rigid-chain polymers as a result of their chemical nature and in flexible-chain polymers as a result of decreasing f in an extending field).

At present, it is known that the structures of the ECC type (Figs 3 and 21) can be obtained in principle for all linear crystallizable polymers. However, in practice, ECC does not occur although, as follows from the preceding considerations, the formation of linear single crystals of macroscopic size (100% ECC) is not forbidden for any fundamental thermodynamic or thermokinetic reasons[60, 65]. It should be noted that the attained tenacities of rigid- and flexible-chain polymer fibers are almost identical. The reasons for a relatively "low" tenacity of fibers from rigid-chain polymers and for the adequacy of the model in Fig. 21a have been analyzed in detail in Ref. 65.

In conclusion, the fundamental features of various methods for obtaining high strength systems from flexible-chain polymers should also be mentioned. Since the presence of ECC leads to an increase in the fraction of tie chains in crystallized samples (their number can be increased by other methods not related to a direct formation of ECC, e.g. by orientational drawing investigated by Marikhin and Myasnikova[4]), the main technological problem of increasing the strength of flexible-chain polymers is the development of methods for raising the volume fraction of ECC and in orientational crystallization for increasing the absolute amount of the fraction occupied by the ECC framework (Fig. 21 c). This framework, as already mentioned, amounts to 10 to 20 vol-% only. However, even such small portion of structures of this type (as compared to samples obtained by the conventional technology) ensures a marked improvement in the mechanical properties. For some polymer articles this improvement may be quite sufficient. By analysing the orientational crystallization in the preceding section, some reasons for the limitation of the increase in the fraction of the ECC framework were discussed by using the phase diagram in Fig. 13. Probably, it is possible to find a solution to this problem by studying the relationship between the rate of molecular orientation (i.e. the extension rate of the melt prior to crystallization) and the crystallization rate, i.e. by increasing the degree of melt extension as a result of a "deeper" penetration below curve 2 along the β axis (Fig. 13), if possible prior to crystallization.

It should be noted that the fraction of ECC in samples obtained by other methods described in Sect. 2 is approximately as small as that of the framework in the orientationally crystallized samples. These methods differ in details but depend on the mechanical treatment of the crystallizing system and are therefore given the common name "stress-induced crystallization". Although the structure of the samples obtained by these methods has some features in common with that of orientationally crystallized samples, the thermodynamics and kinetics of orientational crystallization are fundamentally different from the mechanism of stress-induced crystallization.

First of all the term "stress-induced crystallization" includes crystallization occuring at any extensions or deformations both large and small (in the latter case, ECC are not formed and an ordinary oriented sample is obtained). In contrast, orientational crystallizattion is a crystallization that occurs at melt extensions corresponding to $\beta > \beta_{cr}$, when chains are considerably extended prior to crystallization and the formation of an intermediate oriented phase is followed by crystallization from the preoriented state. Hence, orientational crystallization proceeds in two steps: the first step is the transition of the isotropic melt into the nematic phase (first-order transition of the order-disorder type) and the second involves crystallization with the formation of ECC from the nematic phase (second- or higher-order transition not related to the change in the symmetry elements of the system).

Various methods of stress-induced crystallization cannot be modelled by a single general mechanism because their details are different. However, in all cases the nucleation rate greatly increases as a result of mechanical treatment, and a considerable portion of tie chains exists owing to the presence of an ECC "strand". It is connected with a high local orientation due to the geometrical features of the mechanical field in contrast to the orientational crystallization coupled with the macroscopic orientation in the sample. Application of the methods of stress-induced crystallization according to Pennings[17], in which slow epitaxial linear growth occurs at a low velocity gradient and a low polymer concentration in solution, and to Porter[18], which is the limiting case of cold extrusion, lead to the formation of "shish-kebab" structures containing a central ECC nucleus on which FCC "grow" by the mechanism of epitaxial growth. In orientational crystallization a two-phase structure is also produced: the sample contains both ECC and FCC, but FCC form the matrix for the framework and do not yield epitaxially grown lamellae, i.e. the sample structure is not of the shish-kebab type.

Both stress-induced crystallization and orientational crystallization can be used for the preparation of polymer materials with mechanical property values (e.g. tenacities and elastic moduli) much higher than those for polymer films and fibers obtained by conventional processing. We believe that the advantage of orientational crystallization over more complex methods consists in the possibility of obtaining samples of elastic moduli and tenacities in a one-step continuous process.

The problem of orientational crystallization cannot yet be considered to be completely solved both in the technological and physical sense. Its further development and an improvement of the mechanical properties will require a considerable effort and a more profound investigation of the mechanism of this process.

7 References

1. Flory, P. J.: Proc. Roy. Soc. (London) A. *234*, 60 (1956)
2. Papkov, S. P., Kulichikhin, V. G.: Liquid-crystalline state of polymers (in Russian), Moscow: Khimiya 1977
3. Elyashevich, G. K., Frenkel, S. Ya., in: Orientational phenomena in polymer solutions and melts (in Russian), Moscow: "Khimiya, p. 9–90, 1980. See also: Elyashevich, G. K., Frenkel, S. Ya.: Vysokomol. Soedin. *21 B*, 920 (1979)
4. Marikhin, V. A., Myasnikova, L. P.: Supermolecular structure of polymers (in Russian), Leningrad: Khimiya 1977
5. Sakurada, J., Ito, T., Nakamae, K.: J. Polym. Sci. *C15*, 75 (1966)
6. Novak, I. I., Vettegren, V. I.: Vysokomol. Soedin. *6*, 706 (1964)
7. Glenz, W., Peterlin, A.: J. Macromol. Sci. *4*, 473 (1970)
8. Bonart, R., Hosemann, R.: Makromol. Chem. *39*, 105 (1960)
9. Zhurkov, S. N., Slutsker, A. I., Yastrebinsky, A. A.: Dokl. Akad. Nauk SSSR, *153*, 305 (1963)
10. Flory, P. J.: J. Am. Chem. Soc. *84*, 2857 (1962)
11. Peterlin, A.: Polym. Eng. Sci. *14*, 627 (1974)
12. Regel, V. R., Slutsker, A. I., Tomashevsky, E. E.: Kinetic nature of strength of solids (in Russian), Moscow: Nauka 1974
13. Frenkel, S.: J. Polym. Sci.: Polym. Symposia *58*, 195 (1977)
14. Elyashevich, G. K., Baranov, V. G., Frenkel, S. Ya.: J. Macromol. Sci.-Physics *B 13*, 255 (1977)
15. Elyashevich, G. K. et al.: Fizika Tv. Tela *18*, 2475 (1976)
16. Baranov, V. G.: Khim. Volokna *N 3*, 14 (1977)
17. Pennings, A. J. et al.: J. Polym. Sci. *C 38*, 167 (1972)
18. Southern, J. H. et al.: Makromol. Chem. *162*, 19 (1972)
19. Keller, A.: J. Polym. Sci. Polym. Symp. *58*, 395 (1977)
20. US patent 3,946,094
21. Text. Ind. *3*, 28 (1973)
22. Tudze, T., Kawai, T.: Physical chemistry of polymers (Russian transl.), Moscow: Khimiya 1977
23. Zurabian, R. S. et al.: J. Polym. Sci.: Polym. Symp. *C 44*, 163 (1974)
24. Pennings, A. J., Kiel, A. M.: Kolloid Z., Z. Polym. *222*, N 1, 1 (1968)
25. Keller, A., Machin, M. J.: J. Macromol. Sci. *B 1*, 41 (1967)
26. Collier, J. R.: Polym. Eng. Sci. *16*, 204 (1976)
27. Peterlin, A.: Polymer Eng. Sci.: *16*, 126 (1976)
28. Mandelkern, L.: J. Polym. Sci. *B 5*, 1 (1967)
29. Prigogine, I., Lefevere, R.: Synergetics, Coop. Phenomena Multi-Comp. Syst., 1st Proc. Symp., 1972, Maken H. B. (ed.), p. 125–135
30. Wunderlich, A., Arakawa, T.: J. Polym. Sci. *A 2*, 3697 (1964)

31. a) Yasuniwa, M. et al.: Japan J. Appl. Phys. *15*, 1421 (1976)
    b) Tanaka, T., Takemura, T.: Polym. J. *12*, 255 (1980)
32. Bassett, D. C., Kalifa, B. A.: Polymer *17*, 275 (1976)
33. Zubov, Yu. A., Ozerin, N. A., Bakeev, N. F.: Dokl. Akad. Nauk. SSSR *221*, 121 (1975)
34. Elyashevich, G. K., Poddubny, V. I., Baranov, V. G.: Dokl. Akad. Nauk. SSSR *236*, 1373 (1978)
35. Elyashevich, G. K., Baranov, V. G., Frenkel, S. Ya.: Fizika Tv. Tela *16*, 2071 (1974)
36. Poddubny, V. I. et al.: Polym. Eng. Sci. *20*, 206 (1980)
37. Flory, P. J.: Principles of polymer chemistry, New York: Cornell Univ. Press 1953
38. Mandelkern, L.: Crystallization of polymers, New York: McGraw-Hill 1964
39. Clough, S. B.: J. Macromol. Sci. *B 4*, 199 (1970)
40. Cesari, M. et al.: J. Polym. Sci., Polym. Lett. *14*, 107 (1976)
41. Pennings, A. J.: J. Phys. Chem. Solids *1967*, 389
42. Wunderlich, B.: Polymer *5*, 611 (1964)
43. Kardos, J. L. et al.: J. Polym. Sci. *A 2*, 2061 (1971)
44. Godovsky, Yu. K. et al.: Vysokomol. Soedin. *B 13*, (1971); *A 15*, 813 (1973)
45. Levin, V. Yu. et al.: Vysokomol. Soedin. *B 17*, 244 (1975)
46. Hill, M. J., Keller, A.: J. Macromol. Sci. *B 3*, 153 (1969)
47. Perkins, W. G. et al.: Polym. Eng. Sci. *16*, 200 (1976)
48. Frenkel, S., Baranov, V. G.: Brit. Polym. J. *16*, 228 (1977)
49. Wulf, A., Rocco, A. G. de: J. Chem. Phys. *55*, 12 (1971)
50. Volkenstein, M. V.: Configurational statistics of polymer chains, New York: Interscience 1963
51. Litvina, T. G., Elyashevich, G. K., Baranov, V. G.: Vysokomol. Soedin. *A 24* (1982)
52. Clough, S. B.: Polym. Lett. *8*, 519 (1970)
53. Smit, P. P. A.: Kolloid-Z., Z. Polym. *250*, 8 (1972)
54. Beatty, C. L. et al.: Macromolecules *8*, 547 (1975)
55. Pennings, A. J. and Zwijnenburg, A.: J. Polym. Sci.: Polym. Phys. Ed. *7*, 1011 (1979)
56. Poddubny, V. I. et al.: Vysokomol. Soedin. *B 21*, 818 (1979)
57. Krüger, J. K. et al.: Polymer *21*, 620 (1980)
58. Flory, P. J.: J. Chem. Phys. *15*, 397 (1947)
59. Frenkel, J.: Kinetic theory of liquids, New York: Dover Publ. 1955
60. Frenkel, S. Ya., Elyashevich, G. K.: Vysokomol. Soedin. *A 13*, 493 (1971)
61. Frenkel, S. Ya., Elyashevich, G. K., in: Relaxation phenomena in polymers, pp. 229, 234, 240, Leningrad; Khimiya 1972
62. Elyashevich, G. K., Frenkel, S. Ya.: Vysokomol. Soedin. *A 15*, 2752 (1973)
63. Gerassimov, V. S. et al.: Vysokomol. Soedin. *B 18*, 316 (1976)
64. Keller, A.: J. Polym. Sci., Polym. Symp. *58*, 395 (1977)
65. Frenkel, S.: Pure Appl. Chem. *38*, 117 (1974)

Received January 5, 1981
K. Dušek (editor)

Author Index Volumes 1–43

Allegra, G. and *Bassi, I. W.:* Isomorphism in Synthetic Macromolecular Systems. Vol. 6, pp. 549–574.
Andrews, E. H.: Molecular Fracture in Polymers. Vol. 27, pp. 1–66.
Anufrieva, E. V. and *Gotlib, Yu. Ya.:* Investigation of Polymers in Solution by Polarized Luminescence. Vol. 40, pp.1–68.
Ayrey, G.: The Use of Isotopes in Polymer Analysis. Vol. 6, pp. 128–148.
Baldwin, R. L.: Sedimentation of High Polymers. Vol. 1, pp. 451–511.
Basedow, A, M. and *Ebert, K.:* Ultrasonic Degradation of Polymers in Solution. Vol. 22, pp. 83–148.
Batz, H.-G.: Polymeric Drugs. Vol. 23, pp. 25–53.
Bekturov, E. A. and *Bimendina, L. A.:* Interpolymer Complexes. Vol. 41, pp. 99–147.
Bergsma, F. and *Kruissink, Ch. A.:* Ion-Exchange Membranes. Vol 2, pp. 307–362.
Berlin, Al. Al., Volfson, S. A., and *Enikolopian, N. S.:* Kinetics of Polymerization Processes. Vol. 38, pp. 89–140.
Berry, G. C. and *Fox, T. G.:* The Viscosity of Polymers and Their Concentrated Solutions. Vol. 5, pp. 261–357.
Bevington, J. C.: Isotopic Methods in Polymer Chemistry. Vol. 2, pp. 1–17.
Bird, R. B., Warner, Jr., H. R., and *Evans, D. C.:* Kinetik Theory and Rheology of Dumbbell Suspensions with Brownian Motion. Vol. 8, pp. 1–90.
Biswas, M. and *Maity, C.:* Molecular Sieves as Polymerization Catalysts. Vol. 31, pp. 47–88.
Block, H.: The Nature and Application of Electrical Phenomena in Polymers. Vol. 33, pp. 93–167.
Böhm, L. L., Chmeliř, M., Löhr, G., Schmitt, B. J. und *Schulz, G. V.:* Zustände und Reaktionen des Carbanions bei der anionischen Polymerisation des Styrols. Vol. 9, pp. 1–45.
Bovey, F. A. and *Tiers, G. V. D.:* The High Resolution Nuclear Magnetic Resonance Spectroscopy of Polymers. Vol. 3, pp. 139–195.
Braun, J.-M. and *Guillet, J. E.:* Study of Polymers by Inverse Gas Chromatography. Vol. 21, pp. 107–145.
Breitenbach, J. W., Olaj, O. F. und *Sommer, F.:* Polymerisationsanregung durch Elektrolyse. Vol. 9, pp. 47–227.
Bresler, S. E. and *Kazbekov, E. N.:* Macroradical Reactivity Studied by Electron Spin Resonance. Vol. 3, pp. 688–711.
Bucknall, C. B.: Fracture and Failure of Multiphase Polymers and Polymer Composites. Vol. 27, pp. 121–148.
Bywater, S.: Polymerization Initiated by Lithium and Its Compounds. Vol. 4, pp. 66–110.
Bywater, S.: Preparation and Properties of Star-branched Polymers. Vol. 30, pp. 89–116.
Carrick, W. L.: The Mechanism of Olefin Polymerization by Ziegler-Natta Catalysts. Vol. 12, pp. 65–86.
Casale, A. and *Porter, R. S.:* Mechanical Synthesis of Block and Graft Copolymers. Vol. 17, pp. 1–71.
Cerf, R.: La dynamique des solutions de macromolecules dans un champ de vitesses. Vol. 1, pp. 382–450.
Cesca, S., Priola, A. and *Bruzzone, M.:* Synthesis and Modification of Polymers Containing a System of Conjugated Double Bonds. Vol. 32, pp. 1–67.
Cicchetti, O.: Mechanisms of Oxidative Photodegradation and of UV Stabilization of Polyolefins. Vol. 7, pp. 70–112.

Clark, D. T.: ESCA Applied to Polymers. Vol. 24, pp. 125–188.
Coleman, Jr., L. E. and *Meinhardt, N. A.:* Polymerization Reactions of Vinyl Ketones. Vol. 1, pp. 159–179.
Crescenzi, V.: Some Recent Studies of Polyelectrolyte Solutions. Vol. 5, pp. 358–386.
Davydov, B. E. and *Krentsel, B. A.:* Progress in the Chemistry of Polyconjugated Systems. Vol. 25, pp. 1–46.
Dole, M.: Calorimetric Studies of States and Transitions in Solid High Polymers. Vol. 2, pp. 221–274.
Dreyfuss, P. and *Dreyfuss, M. P.:* Polytetrahydrofuran. Vol. 4, pp. 528–590.
Dušek, K. and *Prins, W.:* Structure and Elasticity of Non-Crystalline Polymer Networks. Vol. 6, pp. 1–102.
Eastham, A. M.: Some Aspects of the Polymerization of Cyclic Ethers. Vol. 2, pp. 18–50.
Ehrlich, P. and *Mortimer, G. A.:* Fundamentals of the Free-Radical Polymerization of Ethylene. Vol. 7, pp. 386–448.
Eisenberg, A.: Ionic Forces in Polymers. Vol. 5, pp. 59–112.
Elias, H.-G., Bareiss, R. und *Watterson, J. G.:* Mittelwerte des Molekulargewichts und anderer Eigenschaften. Vol. 11, pp. 111–204.
Elyashevich, G. K.: Thermodynamics and Kinetics of Orientational Crystallization of Flexible-Chain Polymers. Vol. 43, pp. 205–245.
Fischer, H.: Freie Radikale während der Polymerisation, nachgewiesen und identifiziert durch Elektronenspinresonanz. Vol. 5, pp. 463–530.
Fradet, A. and *Maréchal, E.:* Kinetics and Mechanisms of Polyesterifications. I. Reactions of Diols with Diacids. Vol. 43, pp. 51–142.
Fujita, H.: Diffusion in Polymer-Diluent Systems. Vol. 3, pp. 1–47.
Funke, W.: Über die Strukturaufklärung vernetzter Makromoleküle, insbesondere vernetzter Polyesterharze, mit chemischen Methoden. Vol. 4, pp. 157–235.
Gal'braikh, L. S. and *Rogovin, Z. A.:* Chemical Transformations of Cellulose. Vol. 14, pp. 87–130.
Gallot, B. R. M.: Preparation and Study of Block Copolymers with Ordered Structures, Vol. 29, pp. 85–156.
Gandini, A.: The Behaviour of Furan Derivatives in Polymerization Reactions. Vol. 25, pp. 47–96.
Gandini, A. and *Cheradame, H.:* Cationic Polymerization. Initiation with Alkenyl Monomers. Vol. 34/35, pp. 1–289.
Geckeler, K., Pillai, V. N. R., and *Mutter, M.:* Applications of Soluble Polymeric Supports. Vol. 39, pp. 65–94.
Gerrens, H.: Kinetik der Emulsionspolymerisation. Vol. 1, pp. 234–328.
Ghiggino, K. P., Roberts, A. J. and *Phillips, D.:* Time-Resolved Fluorescence Techniques in Polymer and Biopolymer Studies. Vol. 40, pp. 69–167.
Goethals, E. J.: The Formation of Cyclic Oligomers in the Cationic Polymerization of Heterocycles. Vol. 23, pp. 103–130.
Graessley, W. W.: The Etanglement Concept in Polymer Rheology. Vol. 16, pp. 1–179.
Hagihara, N., Sonogashira, K. and *Takahashi, S.:* Linear Polymers Containing Transition Metals in the Main Chain. Vol. 41, pp. 149–179.
Hasegawa, M.: Four-Center Photopolymerization in the Crystalline State. Vol. 42, pp. 1–49.
Hay, A. S.: Aromatic Polyethers. Vol. 4, pp. 496–527.
Hayakawa, R. and *Wada, Y.:* Piezoelectricity and Related Properties of Polymer Films. Vol. 11, pp. 1–55.
Heidemann, E. and *Roth, W.:* Synthesis and Investigation of Collagen Model Peptides. Vol. 43, pp. 143–203.
Heitz, W.: Polymeric Reagents. Polymer Design, Scope, and Limitations. Vol. 23, pp. 1–23.
Helfferich, F.: Ionenaustausch. Vol. 1, pp. 329–381.
Hendra, P. J.: Laser-Raman Spectra of Polymers. Vol. 6, pp. 151–169.
Henrici-Olivé, G. und *Olivé, S.:* Kettenübertragung bei der radikalischen Polymerisation. Vol. 2, pp. 496–577.
Henrici-Olivé, G. und *Olivé, S.:* Koordinative Polymerisation an löslichen Übergangsmetall-Katalysatoren. Vol. 6, pp. 421–472.
Henrici-Olivé, G. and *Olivé, S.:* Oligomerization of Ethylene with Soluble Transition-Metal Catalysts. Vol. 15, pp. 1–30.

Henrici-Olivé, G. and *Olivé, S.:* Molecular Interactions and Macroscopic Properties of Polyacrylonitrile and Model Substances. Vol. 32, pp. 123–152.
Hermans, Jr., J., Lohr, D. and *Ferro, D.:* Treatment of the Folding and Unfolding of Protein Molecules in Solution According to a Lattic Model. Vol. 9, pp. 229–283.
Holzmüller, W.: Molecular Mobility, Deformation and Relaxation Processes in Polymers. Vol. 26, pp. 1–62.
Hutchison, J. and *Ledwith, A.:* Photoinitiation of Vinyl Polymerization by Aromatic Carbonyl Compounds. Vol. 14, pp. 49–86.
Iizuka, E.: Properties of Liquid Crystals of Polypeptides: with Stress on the Electromagnetic Orientation. Vol. 20, pp. 79–107.
Ikada, Y.: Characterization of Graft Copolymers. Vol. 29, pp. 47–84.
Imanishi, Y.: Syntheses, Conformation, and Reactions of Cyclic Peptides. Vol. 20, pp. 1–77.
Inagaki, H.: Polymer Separation and Characterization by Thin-Layer Chromatography. Vol. 24, pp. 189–237.
Inoue, S.: Asymmetric Reactions of Synthetic Polypeptides. Vol. 21, pp. 77–106.
Ise, N.: Polymerizations under an Electric Field. Vol. 6, pp. 347–376.
Ise, N.: The Mean Activity Coefficient of Polyelectrolytes in Aqueous Solutions and Its Related Properties. Vol. 7, pp. 536–593.
Isihara, A.: Intramolecular Statistics of a Flexible Chain Molecule. Vol. 7, pp. 449–476.
Isihara, A.: Irreversible Processes in Solutions of Chain Polymers. Vol. 5, pp. 531–567.
Isihara, A. and *Guth, E.:* Theory of Dilute Macromolecular Solutions. Vol. 5, pp. 233–260.
Janeschitz-Kriegl, H.: Flow Birefringence of Elastico-Viscous Polymer Systems. Vol. 6, pp. 170–318.
Jenkins, R. and *Porter, R. S.:* Unpertubed Dimensions of Stereoregular Polymers. Vol. 36, pp. 1–20.
Jenngins, B. R.: Electro-Optic Methods for Characterizing Macromolecules in Dilute Solution. Vol. 22, pp. 61–81.
Johnston, D. S.: Macrozwitterion Polymerization. Vol. 42, pp. 51–106.
Kamachi, M.: Influence of Solvent on Free Radical Polymerization of Vinyl Compounds. Vol. 38, pp. 55–87.
Kawabata, S. and *Kawai, H.:* Strain Energy Density Functions of Rubber Vulcanizates from Biaxial Extension. Vol. 24, pp. 89–124.
Kennedy, J. P. and *Chou, T.:* Poly(isobutylene-*co*-β-Pinene): A New Sulfur Vulcanizable, Ozone Resistant Elastomer by Cationic Isomerization Copolymerization. Vol. 21, pp. 1–39.
Kennedy, J. P. and *Delvaux, J. M.:* Synthesis, Characterization and Morphology of Poly(butadiene-*g*-Styrene). Vol. 38, pp. 141–163.
Kennedy, J. P. and *Gillham, J. K.:* Cationic Polymerization of Olefins with Alkylaluminium Initiators. Vol. 10, pp. 1–33.
Kennedy, J. P. and *Johnston, J. E.:* The Cationic Isomerization Polymerization of 3-Methyl-1-butene and 4-Methyl-1-pentene. Vol. 19, pp. 57–95.
Kennedy, J. P. and *Langer, Jr., A. W.:* Recent Advances in Cationic Polymerization. Vol. 3, pp. 508–580.
Kennedy, J. P. and *Otsu, T.:* Polymerization with Isomerization of Monomer Preceding Propagation. Vol. 7, pp. 369–385.
Kennedy, J. P. and *Rengachary, S.:* Correlation Between Cationic Model and Polymerization Reactions of Olefins. Vol. 14, pp. 1–48.
Kennedy, J. P. and *Trivedi, P. D.:* Cationic Olefin Polymerization Using Alkyl Halide – Alkylaluminum Initiator Systems. I. Reactivity Studies. II. Molecular Weight Studies. Vol. 28, pp. 83–151.
Kennedy, J. P., Chang, V. S. C. and *Guyot, A.:* Carbocationic Synthesis and Characterization of Polyolefins with Si–H and Si–Cl Head Groups. Vol. 43, pp. 1–50.
Khoklov, A. R. and *Grosberg, A. Yu.:* Statistical Theory of Polymeric Lyotropic Liquid Crystals. Vol. 41, pp. 53–97.
Kissin, Yu. V.: Structures of Copolymers of High Olefins. Vol. 15, pp. 91–155.
Kitagawa, T. and *Miyazawa, T.:* Neutron Scattering and Normal Vibrations of Polymers. Vol. 9, pp. 335–414.

Kitamaru, R. and *Horii, F.:* NMR Approach to the Phase Structure of Linear Polyethylene. Vol. 26., pp. 139–180.

Knappe, W.: Wärmeleitung in Polymeren. Vol. 7, pp. 477–535.

Koningsveld, R.: Preparative and Analytical Aspects of Polymer Fractionation. Vol. 7.

Kovacs, A. J.: Transition vitreuse dans les polymers amorphes. Etude phénoménologique. Vol. 3, pp. 394–507.

Krässig, H. A.: Graft Co-Polymerization of Cellulose and Its Derivatives. Vol. 4, pp. 111–156.

Kraus, G.: Reinforcement of Elastomers by Carbon Black. Vol. 8, pp. 155–237.

Kreutz, W. and *Welte, W.:* A General Theory for the Evaluation of X-Ray Diagrams of Biomembranes and Other Lamellar Systems. Vol. 30, pp. 161–225.

Krimm, S.: Infrared Spectra of High Polymers. Vol. 2, pp. 51–72.

Kuhn, W., Ramel, A., Walters, D. H., Ebner, G. and *Kuhn, H. J.:* The Production of Mechanical Energy from Different Forms of Chemical Energy with Homogeneous and Cross-Striated High Polymer Systems. Vol. 1, pp. 540–592.

Kunitake, T. and *Okahata, Y.:* Catalytic Hydrolysis by Synthetic Polymers. Vol. 20, pp. 159–221.

Kurata, M. and *Stockmayer, W. H.:* Intrinsic Viscosities and Unperturbed Dimensions of Long Chain Molecules. Vol. 3, pp. 196–312.

Ledwith, A. and *Sherrington, D. C.:* Stable Organic Cation Salts: Ion Pair Equilibria and Use in Cationic Polymerization. Vol. 19, pp. 1–56.

Lee, C.-D. S. and *Daly, W. H.:* Mercaptan-Containing Polymers. Vol. 15, pp. 61–90.

Lipatov, Y. S.: Relaxation and Viscoelastic Properties of Heterogeneous Polymeric Compositions. Vol. 22, pp. 1–59.

Lipatov, Y. S.: The Iso-Free-Volume State and Glass Transitions in Amorphous Polymers: New Development of the Theory. Vol. 26, pp. 63–104.

Mano, E. B. and *Coutinho, F. M. B.:* Grafting on Polyamides. Vol. 19, pp. 97–116.

Mengoli, G.: Feasibility of Polymer Film Coating Through Electroinitiated Polymerization in Aqueous Medium. Vol. 33, pp. 1–31.

Meyerhoff, G.: Die viscosimetrische Molekulargewichtsbestimmung von Polymeren. Vol. 3, pp. 59–105.

Millich, F.: Rigid Rods and the Characterization of Polyisocyanides. Vol. 19, pp. 117–141.

Morawetz, H.: Specific Ion Binding by Polyelectrolytes. Vol. 1, pp. 1–34.

Morin, B. P., Breusova, I. P. and *Rogovin, Z. A.:* Structural and Chemical Modifications of Cellulose by Graft Copolymerization. Vol. 42, pp. 139–166.

Mulvaney, J. E., Oversberger, C. C. and *Schiller, A. M.:* Anionic Polymerization. Vol. 3, pp. 106–138.

Okubo, T. and *Ise, N.:* Synthetic Polyelectrolytes as Models of Nucleic Acids and Esterases. Vol. 25, pp. 135–181.

Osaki, K.: Viscoelastic Properties of Dilute Polymer Solutions. Vol. 12, pp. 1–64.

Oster, G. and *Nishijima, Y.:* Fluorescence Methods in Polymer Science. Vol. 3, pp. 313–331.

Overberger, C. G. and *Moore, J. A.:* Ladder Polymers. Vol. 7, pp. 113–150.

Patat, F., Killmann, E. und *Schiebener, C.:* Die Absorption von Makromolekülen aus Lösung. Vol. 3, pp. 332–393.

Penczek, S., Kubisa, P. and *Matyjaszewski, K.:* Cationic Ring-Opening Polymerization of Heterocyclic Monomers. Vol. 37, pp. 1–149.

Peticolas, W. L.: Inelastic Laser Light Scattering from Biological and Synthetic Polymers. Vol. 9, pp. 285–333.

Pino, P.: Optically Active Addition Polymers. Vol. 4, pp. 393–456.

Plate, N. A. and *Noah, O. V.:* A Theoretical Consideration of the Kinetics and Statistics of Reactions of Functional Groups of Macromolecules. Vol. 31, pp. 133–173.

Plesch, P. H.: The Propagation Rate-Constants in Cationic Polymerisations. Vol. 8, pp. 137–154.

Porod, G.: Anwendung und Ergebnisse der Röntgenkleinwinkelstreuung in festen Hochpolymeren. Vol. 2, pp. 363–400.

Pospíšil, J.: Transformations of Phenolic Antioxidants and the Role of Their Products in the Long-Term Properties of Polyolefins. Vol. 36, pp. 69–133.

Postelnek, W., Coleman, L. E., and *Lovelace, A. M.:* Fluorine-Containing Polymers. I. Fluorinated Vinyl Polymers with Functional Groups, Condensation Polymers, and Styrene Polymers. Vol. 1, pp. 75–113.

Rempp, P., Herz, J., and *Borchard, W.:* Model Networks. Vol. 26, pp. 107–137.
Rigbi, Z.: Reinforcement of Rubber by Carbon Black. Vol. 36, pp. 21–68.
Rogovin, Z. A. and *Gabrielyan, G. A.:* Chemical Modifications of Fibre Forming Polymers and Copolymers of Acrylonitrile. Vol. 25, pp. 97–134.
Roha, M.: Ionic Factors in Steric Control. Vol. 4, pp. 353–392.
Roha, M.: The Chemistry of Coordinate Polymerization of Dienes. Vol. 1, pp. 512–539.
Safford, G. J. and *Naumann, A. W.:* Low Frequency Motions in Polymers as Measured by Neutron Inelastic Scattering. Vol. 5, pp. 1–27.
Schuerch, C.: The Chemical Synthesis and Properties of Polysaccharides of Biomedical Interest. Vol. 10, pp. 173–194.
Schulz, R. C. und *Kaiser, E.:* Synthese und Eigenschaften von optisch aktiven Polymeren. Vol. 4, pp. 236–315.
Seanor, D. A.: Charge Transfer in Polymers. Vol. 4, pp. 317–352.
Seidl, J., Malinský, J., Dušek, K. und *Heitz, W.:* Makroporöse Styrol-Divinylbenzol-Copolymere und ihre Verwendung in der Chromatographie und zur Darstellung von Ionenaustauschern. Vol. 5, pp. 113–213.
Semjonow, V.: Schmelzviskositäten hochpolymerer Stoffe. Vol. 5, pp. 387–450.
Semlyen, J. A.: Ring-Chain Equilibria and the Conformations of Polymer Chains. Vol. 21, pp. 41–75.
Sharkey, W. H.: Polymerizations Through the Carbon-Sulphur Double Bond. Vol. 17, pp. 73–103.
Shimidzu, T.: Cooperative Actions in the Nucleophile-Containing Polymers. Vol. 23, pp. 55–102.
Shutov, F. A.: Foamed Polymers Based on Reactive Oligomers, Vol. 39, pp. 1–64.
Silvestri, G., Gambino, S., and *Filardo, G.:* Electrochemical Production of Initiators for Polymerization Processes. Vol. 38, pp. 27–54.
Slichter, W. P.: The Study of High Polymers by Nuclear Magnetic Resonance. Vol. 1, pp. 35–74.
Small, P. A.: Long-Chain Branching in Polymers. Vol. 18.
Smets, G.: Block and Graft Copolymers. Vol. 2, pp. 173–220.
Sohma, J. and *Sakaguchi, M.:* ESR Studies on Polymer Radicals Produced by Mechanical Destruction and Their Reactivity. Vol. 20, pp. 109–158.
Sotobayashi, H. und *Springer, J.:* Oligomere in verdünnten Lösungen. Vol. 6, pp. 473–548.
Sperati, C. A. and *Starkweather, Jr., H. W.:* Fluorine-Containing Polymers. II. Polytetrafluoroethylene. Vol. 2, pp. 465–495.
Sprung, M. M.: Recent Progress in Silicone Chemistry. I. Hydrolysis of Reactive Silane Intermediates. Vol. 2, pp. 442–464.
Stahl, E. and *Brüderle, V.:* Polymer Analysis by Thermofractography. Vol. 30, pp. 1–88.
Stannett, V. T., Koros, W. J., Paul, D. R., Lonsdale, H. K., and *Baker, R. W.:* Recent Advances in Membrane Science and Technology. Vol. 32, pp. 69–121.
Stille, J. K.: Diels-Alder Polymerization. Vol. 3, pp. 48–58.
Stolka, M. and *Pai, D.:* Polymers with Photoconductive Properties. Vol. 29, pp. 1–45.
Subramanian, R. V.: Electroinitiated Polymerization on Electrodes. Vol. 33, pp. 33–58.
Sumitomo, H. and *Okada, M.:* Ring-Opening Polymerization of Bicyclic Acetals, Oxalactone, and Oxalactam. Vol. 28, pp. 47–82.
Szegö, L.: Modified Polyethylene Terephthalate Fibers. Vol. 31, pp. 89–131.
Szwarc, M.: Termination of Anionic Polymerization. Vol. 2, pp. 275–306.
Szwarc, M.: The Kinetics and Mechanism of N-carboxy-α-amino-acid Anhydride (NCA) Polymerization to Poly-amino Acids. Vol. 4, pp. 1–65.
Szwarc, M.: Thermodynamics of Polymerization with Special Emphasis on Living Polymers. Vol. 4, pp. 457–495.
Takemoto, K. and *Inaki, Y.:* Synthetic Nucleic Acid Analogs. Preparation and Interactions. Vol. 41, pp. 1–51.
Tani, H.: Stereospecific Polymerization of Aldehydes and Epoxides. Vol. 11, pp. 57–110.
Tate, B. E.: Polymerization of Itaconic Acid and Derivatives. Vol. 5, pp. 214–232.
Tazuke, S.: Photosensitized Charge Transfer Polymerization. Vol. 6, pp. 321–346.
Teramoto, A. and *Fujita, H.:* Conformation-dependent Properties of Synthetic Polypeptides in the Helix-Coil Transition Region. Vol. 18, pp. 65–149.
Thomas, W. M.: Mechanism of Acrylonitrile Polymerization. Vol. 2, pp. 401–441.

Tobolsky, A. V. and *DuPré, D. B.:* Macromolecular Relaxation in the Damped Torsional Oscillator and Statistical Segment Models. Vol. 6, pp. 103–127.
Tosi, C. and *Ciampelli, F.:* Applications of Infrared Spectroscopy to Ethylene-Propylene Copolymers. Vol. 12, pp. 87–130.
Tosi, C.: Sequence Distribution in Copolymers: Numerical Tables. Vol. 5, pp. 451–462.
Tsuchida, E. and *Nishide, H.:* Polymer-Metal Complexes and Their Catalytic Activity. Vol. 24, pp. 1–87.
Tsuji, K.: ESR Study of Photodegradation of Polymers. Vol. 12, pp. 131–190.
Tsvetkov, V. and *Andreeva, L.:* Flow and Electric Birefringence in Rigid-Chain Polymer Solutions. Vol. 39, pp. 95–207.
Tuzar, Z., Kratochvíl, P., and *Bohdanecký, M.:* Dilute Solution Properties of Aliphatic Polyamides. Vol. 30, pp. 117–159.
Valvassori, A. and *Sartori, G.:* Present Status of the Multicomponent Copolymerization Theory. Vol. 5, pp. 28–58.
Voorn, M. J.: Phase Separation in Polymer Solutions. Vol. 1, pp. 192–233.
Werber, F. X.: Polymerization of Olefins on Supported Catalysts. Vol. 1, pp. 180–191.
Wichterle, O., Šebenda, J., and *Králíček, J.:* The Anionic Polymerization of Caprolactam. Vol. 2, pp. 578–595.
Wilkes, G. L.: The Measurement of Molecular Orientation in Polymeric Solids. Vol. 8, pp. 91–136.
Williams, G.: Molecular Aspects of Multiple Dielectric Relaxation Processes in Solid Polymers. Vol. 33, pp. 59–92.
Williams, J. G.: Applications of Linear Fracture Mechanics. Vol. 27, pp. 67–120.
Wöhrle, D.: Polymere aus Nitrilen. Vol. 10, pp. 35–107.
Wolf, B. A.: Zur Thermodynamik der enthalpisch und der entropisch bedingten Entmischung von Polymerlösungen. Vol. 10, pp. 109–171.
Woodward, A. E. and *Sauer, J. A.:* The Dynamic Mechanical Properties of High Polymers at Low Temperatures. Vol. 1, pp. 114–158.
Wunderlich, B. and *Baur, H.:* Heat Capacities of Linear High Polymers. Vol. 7, pp. 151–368.
Wunderlich, B.: Crystallization During Polymerization. Vol. 5, pp. 568–619.
Wrasidlo, W.: Thermal Analysis of Polymers. Vol. 13, pp. 1–99.
Yamashita, Y.: Random and Black Copolymers by Ring-Opening Polymerization. Vol. 28, pp. 1–46.
Yamazaki, N.: Electrolytically Initiated Polymerization. Vol. 6, pp. 377–400.
Yamazaki, N. and *Higashi, F.:* New Condensation Polymerizations by Means of Phosphorus Compounds. Vol. 38, pp. 1–25.
Yokoyama, Y. and *Hall H. K.:* Ring-Opening Polymerization of Atom-Bridged and Bond-Bridged Bicyclic Ethers, Acetals and Orthoesters. Vol. 42, pp. 107–138.
Yoshida, H. and *Hayashi, K.:* Initiation Process of Radiation-induced Ionic Polymerization as Studied by Electron Spin Resonance. Vol. 6, pp. 401–420.
Yuki, H. and *Hatada, K.:* Stereospecific Polymerization of Alpha-Substituted Acrylic Acid Esters. Vol. 31, pp. 1–45.
Zachmann, H. G.: Das Kristallisations- und Schmelzverhalten hochpolymerer Stoffe. Vol. 3, pp. 581–687.
Zambelli, A. and *Tosi, C.:* Stereochemistry of Propylene Polymerization. Vol. 15, pp. 31–60.

Catalysis · Science and Technology

Editors: J. R. Anderson, M. Boudart

Volume 1

1981. 107 figures, approx. 58 tables. X, 309 pages
ISBN 3-540-10353-8
Distribution rights for all socialist countries:
Akademie-Verlag, Berlin

Contents:
H. Heinemann: **History of Industrial Catalysis**
The first chapter reviews industrial catalytic developments, which have been commercialized during the last forty years. Emphasis is put on heterogeneous catalytic processes, largely in the petroleum, petrochemical and automotive industries, where the largest scale applications have occurred. Homogeneous catalytic processes are briefly treated and polymerization catalysis is mentioned. The author concentrates on major inventions and novel process chemistry and engineering (79 references).

J. C. R. Turner: **An Introduction to the Theory of Catalytic Reactors**
The second chapter introduces to the catalytic chemist those aspects of chemical reaction engineering involved in any industrial application of a catalytic chemical reaction (19 references).

A. Ozaki, K. Aika: **Catalytic Activation of Dinitrogen**
The third chapter is a comprehensive and critical review of studies on the catalytic activation of dinitrogen, including chemisorption and coordination of dinitrogen, kinetics and mechanism of ammonia synthesis, chemical and instrumental characterization of active catalysts, and homogeneous activation of dinitrogen including metal complexes (353 references).

M. E. Dry: **The Fischer-Tropsch Synthesis**
The fourth chapter concentrates mainly on the development of the Fischer-Tropsch process from the late 1950's to 1979. During this period the Sasol plant was the only Fischer-Tropsch process in operation and hence a large part of this review deals with the information generated at Sasol. The various types of reactors are compared and discussed (198 references).

J. J. Sinfelt: **Catalytic Reforming of Hydrocarbons**
The fifth chapter discusses the catalytic reforming of hydrocarbons from the point of view of the individual types of chemical reactions involved in the process and the nature of the catalysts employed. Some consideration is also given to technological aspects of catalytic reforming (103 references).

Springer-Verlag
Berlin Heidelberg New York

Volume 2

1981. 145 figures, approx. 30 tables.
Approx. 280 pages
ISBN 3-540-10593-X
Distribution rights for all socialist countries:
Akademie-Verlag, Berlin

Contents:
G.-M. Schwab: **History of Concepts in Catalysis**
The concept of catalysis can be attributed to J. Berzelius (1838), whose formulation was based on the manifold observations made in the 17th and 18th centuries. This article traces the development of this and related theories along with the scientific research and empirical material from which they are drawn.

J. Haber: **Crystallography of Catalyst Types**
Structural properties of metals and their substitutional and interstitial alloys, transition metal oxides as well as alumina, silica, aluminosilicates and phosphates are discussed. Implications of point and extended defects for catalysis are emphasized and the problem of the structure and composition of the surface as compared to the bulk is considered.

G. Froment, L. Hosten: **Catalytic Kinetics: Modelling**
The text reviews the methodology of kinetic analysis for simple as well as complex reactions. Attention is focused on the differential and integral methods of kinetic modelling. The statistical testing of the model and the parameter estimates required by the stochastic character of experimental data is described in detail and illustrated by several practical examples. Sequential experimental design procedures for discrimination between rival models and for obtaining parameter estimates with the greatest attainable precision are developed and applied to real cases.

A. J. Lecloux: **Texture of Catalysts**
Useful guidelines and methods for a systematic investigation and a coherent description of catalyst texture are proposed in this contribution. Such a description requires the specification of a very large number of parameters and implies the use of "models" involving assumptions and simplifications. The general approach for determining the porous texture of solids is based on techniques, whose results are cross analyzed in such a way that a self-consistent picture of the porous texture of solids is obtained.

K. Tanabe: **Solid Acid and Base Catalysts**
This chapter deals with the types of solid acids and bases, the acidic and basic properties, and the structure of acidic and basic sites. The chemical principles of the determination of acid-base properties and the mechanism for the generation of acidity and basicity are also described. How acidid and basic properties are controlled chemically is discussed in connection with the preparation method of solid acids and bases.

Reactivity and Structure

Concepts in Organic Chemistry

Editors: K. Hafner, J.-M. Lehn, C. W. Rees, P. v. R. Schleyer, B. M. Trost, R. Zahradník

This series will not only deal with problems of the reactivity and structure of organic compounds but also consider synthetical-preparative aspects.
Suggestions as to topics will always be welcome.

Volume 1: J. Tsuji
Organic Synthesis
by Means of Transition Metal Complexes
A Systematic Approach
1975. 4 tables. IX, 199 pages
ISBN 3-540-07227-6

Volume 2: K. Fukui
Theory of Orientation and Stereoselection
1975. 72 figures, 2 tables. VII, 134 pages
ISBN 3-540-07426-0

Volume 3: H. Kwart, K. King
d-Orbitals in the Chemistry of Silicon, Phosphorus and Sulfur
1977. 4 figures, 10 tables. VIII, 220 pages
ISBN 3-540-07953-X

Volume 4: W. P. Weber, G. W. Gokel
Phase Transfer Catalysis in Organic Synthesis
1977. 100 tables. XV, 280 pages
ISBN 3-540-08377-4

Volume 5: N. D. Epiotis
Theory of Organic Reactions
1978. 69 figures, 47 tables. XIV, 290 pages
ISBN 3-540-08551-3

Volume 6: M. L. Bender, M. Komiyama
Cyclodextrin Chemistry
1978. 14 figures, 37 tables. X, 96 pages
ISBN 3-540-08577-7

Volume 7: D. I. Davies, M. J. Parrott
Free Radicals in Organic Synthesis
1978. 1 figure. XII, 169 pages
ISBN 3-540-08723-0

Volume 8: C. Birr
Aspects of the Merrifield Peptide Synthesis
1978. 62 figures, 6 tables. VIII, 102 pages
ISBN 3-540-08872-5

Volume 9: J. R. Blackborow, D. Young
Metal Vapour Synthesis in Organometallic Chemistry
1979. 36 figures, 32 tables. XIII, 202 pages
ISBN 3-540-09330-3

Volume 10: J. Tsuji
Organic Synthesis with Palladium Compounds
1980. 9 tables. XII, 207 pages
ISBN 3-540-09767-8

Volume 11:
New Syntheses with Carbon Monoxide
Editor: J. Falbe
With contributions by H. Bahrmann, B. Cornils, C. D. Frohning, A. Mullen
1980. 118 figures, 127 tables. XIV, 465 pages
ISBN 3-540-09674-4

Volume 12: J. Fabian, H. Hartmann
Light Absorption of Organic Colorants
Theoretical Treatment and Empirical Rules
1980. 76 figures, 48 tables. VIII, 245 pages
ISBN 3-540-09914-X

Springer-Verlag
Berlin
Heidelberg
New York